AGING
Volume 8

Physiology and Cell Biology of Aging

Aging Series

Volume 9: Sensory Systems and Communication in the Elderly
J. M. Ordy, and K. R. Brizzee, editors, due 1979

Volume 8: Physiology and Cell Biology of Aging
A. Cherkin, C. E. Finch, N. Kharasch, T. Makinodan, F. L. Scott, and B. S. Strehler, editors, 248 pp., 1979

Volume 7: Alzheimer's Disease: Senile Dementia and Related Disorders
R. Katzman, R. D. Terry, and K. L. Bick, editors, 452 pp., 1978

Volume 6: Aging in Muscle
G. Kaldor and W. J. DiBattista, editors, 240 pp., 1978

Volume 5: Geriatric Endocrinology
R. B. Greenblatt, editor, 256 pp., 1978

Volume 4: The Aging Reproductive System
E. L. Schneider, editor, 291 pp., 1978

Volume 3: Neurobiology of Aging
R. D. Terry and S. Gershon, editors, 452 pp., 1976

Volume 2: Genesis and Treatment of Psychologic Disorders in the Elderly
S. Gershon and A. Raskin, editors, 288 pp., 1975

Volume 1: Clinical, Morphologic, and Neurochemical Aspects in the Aging Central Nervous System
H. Brody, D. Harman, and J. M. Ordy, editors, 240 pp., 1975

Aging
Volume 8

Physiology and Cell Biology of Aging

Editors:

Arthur Cherkin, Ph.D.
Director, Geriatric Research, Education, and Clinical Center
V. A. Medical Center
Los Angeles, California

Norman Kharasch, Ph.D.
Professor of Biomedical Chemistry
University of Southern California

Francis L. Scott, Ph.D.
Pennwalt Corporation
Pharmaceutical Division
Rochester, New York

Caleb E. Finch, Ph.D.
Director, Neurobiology Laboratory
Associate Professor, Gerontology and Biological Science
Andrus Gerontology Center
University of Southern California
Los Angeles, California

Takashi Makinodan, Ph.D.
Geriatric Research, Education, and Clinical Center
V. A. Wadsworth Hospital Center and Department of Medicine
University of California at Los Angeles
Los Angeles, California

Bernard Strehler, Ph.D.
Professor of Biology
University of Southern California
Los Angeles, California

Raven Press ▪ New York

Raven Press, 1140 Avenue of the Americas, New York, New York 10036

Made in the United States of America

International Standard Book Number 0–89004–2837
Library of Congress Catalog Card Number 77–94148

Preface

This volume provides a comprehensive review of the effects of aging on a variety of biological functions. Beginning with a discussion of the current status of aging research, topics discussed include the decline and restoration of the immune response, hormonal changes with aging, the basis of memory changes with age, and pharmacologic treatment of organic brain syndrome and the effects of altered physiology in the elderly upon pharmacologic action.

This volume will be of interest to gerontologists, geriatricians, psychiatrists, clinical and basic pharmacologists, and workers in the fields of immunology and endocrinology interested in physiological changes with age in these systems.

The Editors

Acknowledgment

This volume is based on the Intra-Science Symposium on the Biochemistry of Aging.

The annual Intra-Science Symposia seek to explore new directions of research in a given field and to honor the leading researchers instrumental in its development.

We wish to acknowledge the participating scientists and corporate friends of the Intra-Science Research Foundation who have generously contributed their enthusiastic support in the preparation of this volume.

In addition, the support of the members of the Board of Directors of the Intra-Science Research Foundation and those of the Science Council of the Foundation is deeply acknowledged.

Contents

Contributors

Richard C. Adelman
Fels Research Institute
and
Department of Biochemistry
Temple University School of Medicine
and
Division of Biomedical Research
Philadelphia Geriatric Center
and
Temple University Institute on Aging
Philadelphia, Pennsylvania 19141

Lynn S. Baker
Geriatric Research
Education and Clinical Center
V. A. Wadsworth Hospital Center
Los Angeles, California 90073
and
The Department of Medicine
University of California at Los Angeles
Los Angeles, California 90024

Edward L. Bennett
Laboratory of Chemical Biodynamics
Lawrence Berkeley Laboratory
Berkeley, California 94720

Carl W. Cotman
Department of Psychobiology
University of California
Irvine, California

H. Emmenegger
Institute of Basic Medical Research
Sandoz, Ltd.
Basel, Switzerland

Caleb E. Finch
Andrus Gerontology Center
University of Southern California
Los Angeles, California 90007

D. Gershon
Department of Biology
Technion-Israel Institute of Technology
Haifa, Israel

Allan L. Goldstein
Division of Biochemistry
University of Texas Medical Branch
Galveston, Texas 77550

P. Gygax
Institute of Basic Medical Research
Sandoz, Ltd.
Basel, Switzerland

Leonard Hayflick
Children's Hospital Medical Center
Bruce Lyon Memorial Research Laboratory
Oakland, California 94609

Carel F. Hollander
Institute for Experimental Gerontology TNO
Rijswijk, The Netherlands

H. H. Huang
Department of Physiology
Neuroendocrine Research Laboratory
Michigan State University
East Lansing, Michigan 48824

P. Iwangoff
Institute of Basic Medical Research
Sandoz, Ltd.
Basel, Switzerland

Marguerite M. B. Kay
Geriatric Research
Education and Clinical Center
V. A. Wadsworth Hospital Center
Los Angeles, California 90073
and
The Department of Medicine
University of California at Los Angeles
Los Angeles, California 90024

Rachmiel Levine
Department of Metabolism and Endocrinology
City of Hope National Medical Center
Duarte, California 91010

Teresa L. K. Low
Division of Biochemistry
University of Texas Medical Branch
Galveston, Texas 77550

Alice M. Magner
Departments of Neurology and Neuroscience
Harvard Medical School
Children's Hospital Medical Center
Boston, Massachusetts 02115

Takashi Makinodan
Geriatric Research
Education and Clinical Center
V. A. Wadsworth Hospital Center
Los Angeles, California 90073
and
The Department of Medicine
University of California at Los Angeles
Los Angeles, California 90024

W. Meier-Ruge
Institute of Basic Medical Research
Sandoz, Ltd.
Basel, Switzerland

Joseph Meites
Department of Physiology
Neuroendocrine Research Laboratory
Michigan State University
East Lansing, Michigan 48824

U. Reiss
Department of Biology
Technion-Israel Institute of Technology
Haifa, Israel

A. Reznick
Department of Biology
Technion-Israel Institute of Technology
Haifa, Israel

Mark R. Rosenzweig
Department of Psychology
University of California
Berkeley, California 94720

Jeffrey L. Rossio
Division of Biochemistry
University of Texas Medical Branch
Galveston, Texas 77550

Stephen W. Scheff
Department of Psychobiology
University of California
Irvine, California 92717

Francis L. Scott
Pennwalt Corporation
Pharmaceutical Division
Rochester, New York 14603

Dennis J. Selkoe
Departments of Neurology and Neuroscience
Harvard Medical School
Children's Hospital Medical Center
Boston, Massachusetts 02115

Michael L. Shelanski
Departments of Neurology and Neuroscience
Harvard Medical School
Children's Hospital Medical Center
Boston, Massachusetts 02115

J. W. Simpkins
Department of Physiology
Neuroendocrine Research Laboratory
Michigan State University
East Lansing, Michigan 48824

Bernard L. Strehler
University of Southern California
Los Angeles, California 90007

Gary B. Thurman
Division of Biochemistry
University of Texas Medical Branch
Galveston, Texas 77550

Glenn E. Trivers
Division of Biochemistry
University of Texas Medical Branch
Galveston, Texas 77550

Robert E. Vestal
Department of Medicine
University of Washington School of Medicine
Seattle, Washington 98195
and
The Medical Service
V. A. Hospital
Boise, Idaho 83702

Physiology and Cell Biology of Aging (Aging, Volume 8), edited by A. Cherkin, et al.
Raven Press, New York © 1979.

The Future and Aging Research

Bernard L. Strehler

University of Southern California, Los Angeles, California, 90007

Perhaps no area of scientific inquiry will have a greater impact on the quality of life of present and future generations than will research into the molecular, cellular, and systemic changes that gradually lead to the mortality of the individual. The almost certain extension of the healthy human life-span by 10 to 20 years will unquestionably increase the options of individuals for more fulfilling lives and inevitably lead to a restructuring of the means by which resources are distributed, such as social security and medical care. These options should be recognized, and their attendant problems anticipated before they are upon us.

In this introduction, I shall briefly summarize the highlights of our present understanding of the human aging process and engage in speculations about the future. Given the rapid advances that have occurred in biomedical science in the last third of a century, it seems virtually certain that before or shortly after the year 2000 A.D. (barring an apocalyptic interruption of human history), our species will understand both the essential principles and the majority of details of the chemical events that have heretofore led to individual aging and death. The prediction of the consequences of this understanding is at best uncertain prophecy, at worst pseudoscientific chicanery: one cannot predict the consequences of knowledge that one does not possess.

There are three key variables that will determine when and to what degree the human condition will be improved through the acquisition of the prerequisite knowledge.

1. The investment we now make in research at the most basic level. This will be influenced in turn by (a) the support for such research by private and governmental agencies, particularly the infant National Institute of Aging; (b) the degree to which Congress and the executive branch respond to the patent need for adequate funding of imaginative research; and (c) the involvement of a greatly increased number of dedicated and talented researchers in the key unresolved problems in the biology (including the psychobiology) of aging.
2. The nature of the findings.
3. The degree to which understanding will permit modification, reversal or, less probably, the complete arrest of the process, i.e., nonfinite life-spans for our species.

One can construct a matrix relating these interdependent variables and assign probabilities to each variable and to their combinatorial probabilities. If one accepts 2,000 A.D. as an arbitrary projection point, the first variable has a probability lying between 0.2 and 0.8, the second between 0.5 and 0.9, and the third between 0.5 and 0.05.

For a 20% extension of healthy life-span, the combined probabilities are about 0.8; for a 50 to 100 year extension, the value is about 0.4. For an indefinite extension of life-span, the probability is less than 0.05, unless both successful cloning of an individual and the transplantation of the essence of selfness and key memories are defined as indefinite extension, in which case the probability of the last is increased to between 0.2 and 0.6. The key unknown that man can modify is the first variable, the rate at which knowledge is acquired. Although it is still too early to assign a definite probability to this variable (in terms of percentage of some hypothetical optimum), a value between 0.2 and 0.4 seems likely, in view of the administrative impediments that currently exist within the structure of the Department of Health, Education and Welfare.

The most promising areas include the following.

1. Understanding the mechanism of failure of the immune system, its consequences, and possible modifications. [Makinodan *(this volume)* will address this in greater detail.]
2. The discovery that certain kinds of genetic materials are evidently lost (up to 50% lost by age 90) in nondividing human tissues.
3. The clarification of the means by which the selective expression of genes is controlled during development and aging.
4. The posttranslational modification of proteinaceous gene products. (This will be discussed in detail by Gershon et al.)
5. The role of core body temperature on aging rates.
6. Membrane modifications, particularly those involved in endocrine coordination of the body. (These topics are treated elsewhere in this volume by Adelman and Finch, among others.)
7. An understanding of the means by which memory is stored, retrieved, consolidated, and lost during aging.

In summary, many promising areas of cell biochemistry research have been uncovered in recent years. The expansion and exploitation of these findings auger well for the future of the field and of the species. With luck, our grandchildren will be the clear beneficiaries of our work; but foot dragging, lack of competence, and lack of resources will perhaps allow only our great-grandchildren or their children to reap the benefits of the research that has already been done.

BIBLIOGRAPHY

1. Finch, C., and Hayflick, L. (1977): *Handbook of Aging: Biological Aspects,*. Van Nostrand, New York.
2. Rosenfeld, A. (1976): *Pro-Longevity.* Knopf, New York.
3. Strehler, B. (1977): *Time, Cells and Aging.* 2nd ed. Academic Press, New York and London.
4. Viidik, A., ed. (1978): *Lectures in Gerontology.* Academic Press, New York and London.

Physiology and Cell Biology of Aging
(Aging, Volume 8), edited by A. Cherkin, et al.
Raven Press, New York © 1979.

Cell Aging

Leonard Hayflick

Children's Hospital Medical Center, Bruce Lyon Memorial Research Laboratory, Oakland, California 94609

Is the inevitability of aging and death of individual cells in an organism predetermined? It is generally believed that whatever causes physiologic decrements with time in the whole organism undoubtedly does not produce similar changes, and at the same rate, in each cell composing that organism. If the rates of aging vary between organs, tissues and their constitutive cells, the root causes of aging may occur as a consequence of decrements in some few cell types in which the rate is fastest and the effects greatest. Are normal somatic cells predestined to undergo irreversible functional decrements that presage aging in the whole organism?

There are at least two ways in which this question has been put to the test: first, eukaryotic cells have been grown and studied in cell culture; and second, vertebrate cells containing specific markers that allow them to be distinguished from host cells have been serially transplanted in isogenic animals. The goals of such studies, as they pertain to the science of gerontology, have been directed toward answering this fundamental question: Can vertebrate cells, functioning and replicating under ideal conditions, escape from the inevitability of aging and death that is universally characteristic of the whole animals from which they were derived?

BACKGROUND

Among the studies undertaken in cell culture, one investigation has stood out as the classic response to this intriguing question. In the early part of this century, Alexis Carrel, a noted cell culturist, described experiments purporting to show that fibroblast cells derived from chick heart tissue could be cultured seriatim indefinitely. The culture was voluntarily terminated after 34 years. The importance of this experiment to gerontologists was the implication that if cells, released from *in vivo* control, could divide and function normally for periods in excess of the life-span of a species, then either the type of cells cultured play no part in the aging phenomenon, or aging is the result of changes occurring at the supracellular level (i.e., aging would be the result of decrements that occur only in organized tissue or whole organs as a result of the physiologic interactions between those organized cell hierarchies). The inference would be that aging is not, per se, the result of events occurring at the cellular level. In the years following Carrel's observations, support for his experimental results seemed to be forthcoming from the many laboratories in which that cultured cell populations, derived from many tissues of a variety of animal species and from man, seemed to have the striking ability to replicate indefinitely. These cell populations, which arise spontaneously from cultures, number in the hundreds and are best known by the prototype cell lines HeLa (derived from a human cervical carcinoma in 1952) and L cells (derived from mouse mesenchyme in 1943). They continue to flourish even to this day in cell culture laboratories throughout the world. Nevertheless, what seemed to be incontrovertible evidence for the potential immortality of vertebrate cells soon fell to new insights and a preponderance of opposing evidence.

AGING UNDER GLASS

Of central importance to the question is whether the cell populations studied *in vitro* are composed of normal or abnormal cells. Clearly, the aging of animals occurs in normal cell populations, and if one is to equate the behavior of normal cells *in vivo* to similar cells *in vitro,* the latter must be shown to be normal as well. It is for this reason that the "immortal" cell lines described above (of which the HeLa cell and L cell populations are prototypes) must be excluded from consideration since they are composed of cells that are abnormal in one or more important properties. For example, the cells composing immortal cell lines vary in their chromosomal constitution in such a way that they do not reveal either the exact number or the precise morphology of chromosomes characteristic of the cells composing the tissue from which they originally descended. Human hematopoietic cell lines, often shown to be very nearly normal karyologically, harbor a cancer virus (the Epstein-Barr virus) not found in normal cells. Other so-called immortal diploid cell lines were all examined karyologically early in their passage history when they would be expected to be diploid. When examined after approximately one year of continuous cultivation, all are found to be karyologically abnormal.

When inoculated into laboratory animals, many of these cell lines also give rise to tumors, and some reveal biochemical properties uncharacteristic of the cells composing the tissue of origin. The widespread use of these cell lines for a variety of research purposes in laboratories throughout the world is subject to the criticism that, in most cases, they are not characteristic of any cell type found in human or animal tissue. Many experimental data generated from the use of such cell populations cannot be extrapolated to apply to cells that characterize the animal species from which they were originally descendant. Consequently, the use of "immortal" cell lines for many purposes is questionable since such cells undoubtedly represent laboratory artifacts whose behavior may be unrelated to cells *in vivo.*

This fundamental flaw in interpreting normal cell behavior *in vitro* can in fact be circumvented since cell populations entirely typical of normal cells found *in vivo* can be cultured and, with respect to gerontologic inquiry, the findings are profoundly different from the behavior of abnormal cell lines.

Several years ago, Moorhead and I found that cultured normal human embryonic fibroblasts underwent a finite number of serial subcultivations or population doublings, and then died (12). We demonstrated that when such cells were grown under the most favorable conditions, death was inevitable after about 50 population doublings (the phase III phenomenon). We also showed that the death of these normal cells was not due to some trivial cause involving medium components or cultural conditions, but that the death of cultured normal cells was an inherent property of the cells themselves (12,13). That observation has now been confirmed in hundreds of laboratories in which variations in medium components and cultural conditions have been as numerous as the laboratories in which the studies were done.

Since normal diploid cell strains have a limited doubling potential *in vitro*, studies of any single strain would be severely curtailed if it were not possible to preserve these cells at subzero temperatures for apparently indefinite periods. The reconstitution of frozen human fetal diploid cell strains has revealed that regardless of the doubling level reached by the population at the time it is preserved, a total of about 50 doublings can be expected when the doublings made before and after preservation are combined (12,13). Storage of human diploid cell strains merely arrests the cells at a particular population doubling level, but does not influence the total number of expected doublings.

In the last 16 years we have reconstituted a total of 130 ampules of our human diploid cell strain WI-38, which was placed in liquid-nitrogen storage at the eighth doubling level. Since 1962, one ampule has been reconstituted approximately each month, and all have yielded cell populations that have undergone 50 ± 10 cumulative population doublings. This represents the longest period that viable normal human cells have been arrested at subzero temperatures.

Since normal human embryo fibroblasts are able to undergo only a fixed number of reproductive cycles *in vitro*, we postulated that this observation might be interpreted as a manifestation of aging at the cellular level. Subsequent experimental data have tended to support the validity of this notion.

In order to distinguish this area of gerontological research, which concerns itself with studies on cultured cells, I have suggested the name "cytogerontology" (14,15). This would serve to distinguish these studies from those done at levels of greater complexity *in vivo* at the supracellular level.

DONOR AGE VERSUS CELL-DOUBLING POTENTIAL

Since cultured normal human cells derived from embryonic tissue have a finite proliferative capacity of about 50 population doublings, and if this represents cellular aging, it would be important to determine the proliferative capacity of normal cells derived from human adults of varying ages. My first report of such studies did indeed show a diminished proliferative capacity for cultured normal human adult fibroblasts in which 14 to 29 doublings occurred in cells derived from eight adult donors (13). This figure was comparable to a range of 35 to 63 doublings found in cells cultured from 13 human embryos.

Since these studies, three other reports have appeared that not only confirmed the principle observed but significantly extended it (23,26,36). Thus, it is now generally believed that there is an inverse relation between the age of a human donor and the *in vitro* proliferative capacity of at least two cell types—fibroblasts derived from skin and lung, and cells derived from liver.

DO CELLS IN VIVO DIVIDE MORE OFTEN THAN CELLS IN VITRO?

If normal human embryonic fibroblasts grown *in vitro* have a finite lifetime of 50 ± 10 doublings, how does one account for the lifelong multiplication of bone-marrow cells and of intestinal and skin epithelium *in vivo?* Presumably, these cell populations undergo far more than 50 population doublings during an individual's lifetime. There are several answers to this question, all of which are based on the number of cells produced by a primary cell population that undergoes the maximum 50 doublings. The answer, as we reported some years ago, is 20 million metric tons of cells (12), certainly a quantity sufficient to account for all cells generated during the lifetime of an individual. In a more elegant consideration of this problem, Kay (20) offered the following explanation: Cells can divide in several ways, the two extremes of which are (a) tangential division, in which a stem cell multiplies to produce another stem cell and one differentiated cell; and (b) logarithmic, in which, if all mitoses are synchronous, a single cell divides, yielding a doubled number of cells at each division, similar to our hypothetical model for the ultimate yield of 20 million tons of WI-38 cells if all cells remain after 50 population doublings. Kay pointed out that a maintained cell output in such rapidly dividing tissue as bone marrow could be maintained by an asynchronous division of cells within the logarithmic model. The essential factor is a variation in the rate of primitive stem-cell division so as to produce a continuous release of mature differentiated cells. The advantage of this system (called "clonal succession") over the tangential model is a reduction in the number of cell generations required to yield a given population of mature cells, thus allowing a much closer adherence to the original genetic message (20).

Only 54 population doublings would be required to produce, from a single cell, by the logarithmic pattern, the total output of human erythrocytes and leukocytes during 60 years of life. This figure is surprisingly close to the 50 ± 10 doubling limit that we found for normal human fibroblasts grown *in vitro.* The tangential system would require about 12,000 doublings. Thus, the asynchronous logarithmic division model can account for all cells produced during an individual's lifetime since it would contain several generations of dormant ancestral cells lingering, for example, at the 25th population doubling that would be successively promoted to form clones of maturing stem cells.

If all animal cell types were continually renewed without loss of function or capacity for self-renewal, one would expect the organs composed of such cells to function normally indefinitely, and their host to live forever. Unhappily, however, renewal cell populations do not occur in most tissues, and when they do, a proliferative finitude is often manifest. The important question, then, is whether by transferring marked cells to younger animals seriatim it is possible to circumvent experimentally the aging and death of normal animal cells that result from the aging and death of the "host." If such experiments could be conducted, an *in vivo* counterpart of *in vitro* experiments would be available, and normal cells transplanted serially to proper inbred hosts would, like their *in vitro* counterparts, be expected to age and die. Such experiments would largely rule out the objections to *in vitro* findings that are based on the artificiality of cell replication *in vitro*. The question could be answered by serial orthotopic transplantation of normal somatic tissue to new, young, inbred hosts each time the recipient approached old age. Under these conditions, would transplanted normal cells of age-chimeras proliferate indefinitely?

THE FINITE LIFE-SPANS OF NORMAL CELLS *IN VIVO*

Data from seven different laboratories in which mammary tissues, skin, hematopoietic cells, and leukocytes were employed indicate that normal cells, serially transplanted to isogenic hosts, do not survive indefinitely (17). Furthermore, the trauma of transplantation does not appear to influence the results, and finally, in heterochronic transplants, survival time is related to the age of the grafted tissue. Under similar conditions of tissue transplantation, cancer-cell populations, like immortal cell lines, can be serially passed indefinitely. The implications of this observation may be that acquiring the potential for unlimited cell division, or escaping from senescent changes by mammalian cells *in vitro* or *in vivo,* can only be achieved by cells that have acquired some or all properties of cancer cells. Paradoxically, this hypothesis leads to the conclusion that, for mammalian somatic cells to become biologically "immortal," they must first be induced to an abnormal or neoplastic state either *in vivo* or *in vitro,* whereupon they can then be subcultivated or transplanted indefinitely.

The study of single antibody-forming cell clones *in vivo* has shown that these cells are also capable of only a finite ability to replicate after serial transfer *in vivo* (44,45). Harrison (9,10), however, reported that when marrow cell transplants from young and old normal donors are made to a genetically anemic recipient mouse strain, the old as well as the young transplants populate the recipients, curing the anemia. He further reported that such transplants to anemic mice can be made to function normally over a period of 73 months (10). After five successive transplantations, most of the mice expired, once again exhibiting the finitude of normal cell proliferation *in vivo* (11). The fact that this normal hematopoietic cell population produced red blood cells far beyond the normal life-span of the mouse is in keeping with speculations that aging is not necessarily timed at the same rate in all tissues. Furthermore, aging may not be due to the loss of cell division capacity, but is probably due to other functional decrements that are known to occur prior to the loss of ability to replicate. The loss of capacity to divide is presumably an extreme limit, capable only of demonstration *in vitro* and by serial transplantation *in vivo.*

ORGAN CLOCKS?

Is it possible that a limit on cell proliferation or function in some strategic organ could orchestrate the entire phenomenon of senescence? Burnet (2,3,4) speculated that, if this is so, the most likely organ is the thymus and its dependent tissues. Burnet reasoned that aging is largely the result of somatic mutations that are mediated by autoimmune processes and that are influenced by progressive weakening of the function of immunological surveillance. He further argued that weakening of immunological surveillance may be related to weakness of the thymus-dependent immune system. He concluded that the thymus and its dependent tissues are subjective to a proliferative limit similar to the phase III phenomenon or senescence *in vitro* described by us for human cells. Whether the role played by the thymus and its dependent tissues as the pacemaker in senescence is important still remains to be established.

LATENT PERIOD OF EXPLANTED CELLS VS DONOR AGE

Since 1925, it has been known that the time lapse between introduction of embryo tissue in culture and cell migration from the explant increases with embryo age (6,42). More recently, investigators have shown again that the time necessary for the first cells to emigrate from rat tissue explants grown *in vitro* (the latent period) correlates inversely with the age of the donor (32,38,39,40). Similar observations have been reported for chicken cells (5,22), and a linear increase in the latent period has been found to occur in explants cultured from human donors ranging in age from newborn to 80 years (41,43).

FUNCTIONAL AND BIOCHEMICAL DECREMENTS IN CULTURED NORMAL HUMAN CELLS

It is unlikely that animals age because one or more important cell populations lose their proliferative capacity. It is more probable that, as we have shown, normal cells have a finite capacity for replication, and that this finite limit is rarely if ever reached by cells *in vivo* but is demonstrable *in vitro.* I have therefore suggested that other functional losses that occur in cells before the cessation of division capacity produce physiologic decrements in animals long before their normal cells have reached their maximum proliferative capacity. Indeed, it is now becoming recognized that many functional changes that take place in normal human cells grown *in vitro* are expressed well before the cells lose their capacity to replicate (Table 1). These changes, which herald the approaching loss of division capacity, are more likely to have the central role in the expression of aging and to result in the death of the individual animal well before its cells fail to divide.

To be sure, the several classes of cells that are incapable of division in mature animals (e.g., neurons and muscle cells) may have a greater effect on the expression of age changes than the cell classes in which division commonly occurs. It is important, therefore, to indicate that the cessation of mitotic activity is only one functional decrement whose genetic basis may be similar to functional decrements known to occur in nondividing cells. Therefore, the same kind of gene action resulting in physiologic decrements in aging nondividing cells is believed to occur in aging cells that can divide. Thus, it is not my contention that age changes result necessarily from loss of cell-division capacity, but that they result simply in loss of function in any class of cells. That function might be measured as reduced division capacity, or as any number of the myriad functional decrements characteristic of aging cells. The genetic changes leading to these decrements is postulated to be the common denominator so that the measurement of loss of population doubling potential *in vitro* may have the same basis as the loss of other cell functions characteristic of nondividing cells. Therefore, an understanding of the mechanism by which cultured normal cells lose their capacity to replicate could provide insights into the causes of decrements in other functional properties that are characteristic of nondividing cells and that may be even more direct causes of biologic aging.

LOCATION OF THE "CLOCK"

In an effort to locate and understand the mechanism controlling the finite replicative capacity of cultured normal cells, Wright and I have recently completed a series of experiments that bear on this important question. When cultured cells are treated with cytochalasin B and are then centrifuged, it is possible to obtain millions of anucleate cells (cytoplasts) (46,47). Cytoplasts remain viable for several days, allowing sufficient time to fuse them with inactivated Sendai virus to whole cells producing heteroplasmons (as distinguished from heterokaryons, which result from fusions made between whole cells only). With these techniques it is possible to determine whether the "clock" that dictates a cell's replicative capacity is located in the nucleus or in the cytoplasm. This question was approached by fusing cytoplasts derived from young cells to whole old cells, and vice versa. By determining the remaining number of population doublings traversed by these heteroplasmons, it was possible to determine the influence of young or old human cytoplasm on opposite aged whole human cells. Data derived from such studies have been interpreted to mean that the "clock"

TABLE 1. *Metabolic and cell characteristics that increase, decrease, or do not change as normal human fibroblasts age* in vitro[a]

Increase	Decrease	No change
Glycogen content	Glycolytic enzymes	Glycolysis
Lipid content	Pentose phosphate shunt	Permeability to glucose
Lipid synthesis	Mucopolysaccharide synthesis	Respiration
Protein content	Transaminases	Respiratory enzymes
RNA content	Collagen synthesis	Permeability to amino acids
RNA turnover	DNA content	Glutamic dehydrogenase
Lysosomes and lysosomal enzymes	Nucleic acid synthesis	Nucleohistone content
Heat lability of G-6-PD and 6-phosphogluconate dehydrogenase	Collagen synthesis and collagenolytic activity	Alkaline phosphatase
Proportion of RNA and histone in chromatin	Lactic dehydrogenase isoenzyme pattern	Soluble RNAse, soluble DNAse, soluble seryl transfer-RNA synthetase, soluble and chromatin-associated DNA polymerase
Activity of "chromatin-associated enzymes" (RNAse, DNAse, protease, nucleoside triphosphatase, and DPN pyrophosphorylase)	Ribosomal RNA content	Mean temperature of denaturation of DNA and chromatin
5′-MNase activity	Incorporation of tritiated thymidine	No. of mitochondria
Esterase activity	RNA-synthesizing activity of chromatin	HLA specificities (mass cultures)
Acid phosphatase band 3—glucuronidase activity	Alkaline phosphatase	Virus susceptibility
Membrane-associated ATPase activity	Specific activity of lactic dehydrogenase	Poliovirus and herpesvirus titer, mutation rate, and protein chemistry
Cell size and volume	Rate of histone acetylation	Cell viability at subzero temperatures
No. and size of lysosomes	No. of cells in proliferating pool	Diploidy (only in phase III)
No. of residual bodies	Population doubling potential as function of donor age	Histone/DNA ratio
Cytoplasmic microfibrils, constricted and "empty"	Proportion of mitochondria with completely transverse cristae	
Endoplasmic reticulum	HLA specificities (cloned cells)	
Cyclic-AMP level/mg protein	Adherence to polymerizing fibrin and influence on fibrin retraction	
	Cyclic-AMP levels (molar values)	

[a] References to each characteristic can be found in (17).

is located in the nucleus (46,47). With the more recent development of techniques by which viable eukaryotic nuclei can be isolated and themselves reinserted into cytoplasts, an even more direct answer to this question is now possible, since nuclei from old cells can be inserted into young cytoplasm, and vice versa. Muggleton-Harris and I recently performed these experiments, in which viable normal human cells have been reconstructed from isolated nuclei and cytoplasts (33). Nuclei from young and old WI-38 cells have been inserted into opposite aged cytoplasts, yielding viable whole cells with the capacity to replicate. Results from these studies suggest that the "clock" that determines proliferative capacity may be located in the nucleus, although further experimentation is still necessary to confirm this.

EVENTS OCCURRING IN LATE PHASE III

A period of decreasing rate of cell proliferation (phase III) after a period of rapid cell proliferation (phase II) is characteristic of cultured normal human diploid cell strains (12,13). This finite lifetime has also been observed in other normal cell cultures, such as those from chickens, mice, tortoises, and mink (17). Although a variety of changes is known to occur as cultured normal human cells traverse phase II to phase III (Table 1), the molecular mechanisms underlying these changes are incompletely understood.

In an effort to provide information about the cause of these decrements, Matsumura et al. (30) have investigated the properties of normal human cell cultures after the cessation of replicative capacity (phase III) or late phase III. WI-38, a normal human embryonic diploid cell strain, can be maintained in phase III for months (13). The survival, ability to incorporate tritiated thymidine, and morphology of WI-38 in late phase III have not been systematically described. The purpose of our study was to examine these variables in late phase III WI-38. Some cultures were kept for more than one year, during which time we did not find any sign of spontaneous transformation.

Various morphological changes were noticed in late phase III cultures held for several months. These changes consisted of enlargement of the cytoplasmic region, alteration in cell shape from spindle-like to flat and polygonal, vacuolation of the cytoplasmic region, accumulation of debris in the culture, and alteration in nuclear morphology. Some of these characteristics were not always found in all cultures. For example, whereas vacuolation of the cytoplasm was significant in some cells that had been kept without transfer for months, this characteristic was not present in other cells of the same culture series. Vacuolation of the cytoplasm and the accumulation of debris in a culture often disappeared after transfer of the culture in which trypsin was used.

However, one characteristic, the alteration of nuclear morphology, was reproducible. It was common to find that more than 10% of cells contained a lobed nucleus and/or more than one nuclear region, which is uncommon in phase II cultures. Among the irregularities in nuclear morphology, the most common was the appearance of two nuclear regions. (All cells that contained two or more nuclear regions will be referred to as *multinucleated cells.*) Since the DNA content of each cell was not determined, the possibility of simple fission of nuclei without accompanying DNA synthesis cannot be excluded. In all four experiments that were conducted, the proportion of multinucleated cells increased progressively. Although there were fluctuations in the proportion, 40 to 50 percent of the cells were found to be multinucleated by 15 weeks incubation.

Where the number of grains far exceeded the number expected from repair synthesis, late phase III WI-38 cultures were examined for incorporation of ^{3}H-TdR. We regarded those labeled cells as being in, having been in, or having passed through S-phase during exposure to ^{3}H-TdR. Both single nucleated cells and multinucleated cells were labeled. Cells that had one part of their nuclear regions labeled and the other part unlabeled were rarely seen.

About 70% of cells in a senescent culture remained unlabeled during a 9-day exposure to ^{3}H-TdR. If labeled cells clustered on the substrate surface, it was assumed that a few cells with proliferative potential were present. However, we did not find any noticeable clustering of labeled cells, but found them scattered all over the substrate surface among nonlabeled cells.

The incorporation of ^{3}H-TdR was found to continue in a proportion of cells until the end of the observation period. About 60 to 90% of the cells did not incorporate ^{3}H-TdR during the 6-day exposure. Within the range of a large fluctuation, no apparent correlation was observed between the percentage of unlabeled cells corrected for cell proliferation and the time spent in late phase III.

Late phase III may be characterized as a time in which there is a slow increase or decrease of cell numbers that continues for more than 6 months. Individual cells in a population may proliferate slowly, survive without any cell division, or die. A considerable percentage of cells continued to synthesize DNA.

Many cells in a late phase III population and many cells in the proliferating (phase II) population appear not only to differ quantitatively in their rate of proliferation, rate of DNA synthesis, and cell size, but also to differ in such qualitative properties as nuclear morphology.

The presence in early phase III cultures of cells suspected to be similar to those senescent cells described by us here has been reported. In early phase III, (a) binucleate cells have been observed (24); (b) aneuploidy was found to increase; (c) cells with a DNA content 2 to 4 times that of diploid cells have been detected; (d) cells that have little or no ability to synthesize DNA have been reported, and were found to increase in number toward the end of early phase III (7,29); and (e) nondividing or slowly dividing cells were found to increase in number (1,31,37). Present evidence seems to support the hypothesis that subpopulation(s) similar to the major population of cells in a late phase III culture may be present in a proliferating phase II population.

Based on these considerations, it seems that late phase III populations consist of a variety of cells, each of which has its own history, and that a mass culture of such cells reflects the average history of the population.

The biological significance of cells in a late phase III culture may be interpreted in several ways that are not mutually exclusive. The cells may be at a particular stage of differentiation (27,28); they may have accumulated errors in regulatory mechanisms leading to ultimate cell death (34); or they may have reached the end of a programmed series of genetic events (13,16,17). Our results do not exclude any of these possibilities.

A period of "crisis," which appears after infection of human diploid cells with SV_{40}, precedes the period of infinite proliferation of transformed cells (8). There appears to be a similarity in morphology between the cells in a late phase III culture and transformed cells in "crisis" (8). Abnormalities in nuclear morphology during crisis, such as the presence of lobed nuclei and multinucleation, are also seen in the present results where late phase III cells have been maintained for several months without introduction of SV_{40}. Additional events necessary to bring late phase III cells into conditions of immortality may be supplied by part of the SV_{40} genome.

DNA SYNTHESIS IN WI-38 DURING *IN VITRO* AGING

Variations in the rate of proliferation and metabolic activity could occur in at least two incompatible ways: (a) A population at one moment in its history may consist of cells, or subpopulations of cells, each of which is at a different stage in its life history [based on this interpretation, a model of limited proliferation potential was developed by Holliday et al. (19)]; or (b) the variation in some properties of individual cells, such as the capacity to enter S-phase (25), may not be directly related to the cell's limited proliferation potential.

In support of (a), there is some evidence that cells plated at low density contain nondividing cells, the proportion of which increases with the population doubling level (PDL) of a culture (1,28,31,37). By autoradiography studies, it was observed that the percent of labeled cells (labeling index) reached a plateau after a 24- to 30-hr exposure to tritiated thymidine (^{3}H-TdR), and that the labeling index decreased as a function of PDL (7). It has also been observed that individual cells that had been cloned from a cell population varied in their doubling potential (31,37).

However, the theoretical basis for (a) does not seem to be well established. It has been reported that by reducing the cell density and by extending the time of exposure to ^{3}H-TdR, a labeling index close to 100% was obtained even with cultures that had a very high PDL (7). This may suggest that an increased sensitivity to cell density is the primary consideration, instead of an increase in the proportion of cells incapable of DNA synthesis. It has also been reported that when mitotic cells are collected from random cultures, the fraction of those mitotic cells that enter S-phase thereafter is kept constant regardless of PDL.

In an effort to resolve the difference in these results, we assessed those factors described below (in addition to PDL) that influence cell proliferation (29). Cell density, depletion in some medium components during incubation, and concentration of thymidine (by autoradiography), among other variables, may influence the cell cycle through some cell regulatory mechanism. The dilution of nonincorporating cells by proliferating cells and the effect of ^{3}H-TdR (24) on cell proliferation are factors that must be considered in an autoradiography study. Because of their more limited capacity to adjust to culture conditions, cells inoculated at a low density may behave in a different way from those in mass culture.

We measured the ability of a WI-38 population to synthesize DNA by means of autoradiography, and studied the effect on labeling index of thymidine concentration, concentration of ^{3}H-TdR, and cell proliferation (29). We also determined the percent of cells in a population which did not incorporate ^{3}H-TdR during the period of exposure.

The transportation through the cell membrane and the uptake into cytoplasm of thymidine depend on thymidine concentration, and thus may affect the labeling index. However, it seems improbable from the results presented by Cristofalo (7), and from our results (29), that the thymidine concentration (approximately $2.5–5 \times 10^{-8}$M) is the major factor affecting the determination of the labeling index as a function of PDL. It would appear reasonable to regard the labeling index as a parameter that directly reflects the percentage of cells that are, or have been, in S-phase during continuous exposure to ^{3}H-TdR.

As to the technique of autoradiography, our results (29) demonstrate that the labeling index is a stable parameter within a range of several modifications of fixation methods, time of exposure in the dark, and concentration of ^{3}H-TdR. Another factor possibly affecting the labeling index is the selective death of nonlabeled cells. If nonlabeled cells are those committed to die at the very end of their life history, then the labeling index also depends on the survival time of nonlabeled cells. Evidence so far obtained suggests that this is not the case, at least under some culture conditions (30). ^{3}H-TdR might also cause the death of nonlabeled cells. However, we regard this effect as only secondary in importance compared to the effect of ^{3}H-TdR on the labeled cells, and we did not investigate it.

It seems probable from our results that the following six factors influence the labeling index through their effect on those cells capable of entering S-phase during continuous exposure to ^{3}H-TdR: (a) cell proliferation rate, (b) time of incubation with ^{3}H-TdR, (c) tritium concentration, (d) cell density, (e) medium renewal, and (f) possible death of labeled cells during incubation.

The variable, percentage of nonlabeled cells, corrected for cell proliferation, provides an estimation of the proportion of cells (in a population of inoculated cells) that do not enter S-phase during exposure to ^{3}H-TdR (29). When the cells are exposed to ^{3}H-TdR, no matter how any of the above factors influences the labeled cells, that parameter remains constant so long as it does not influence nonlabeled cells. The percentage of nonlabeled cells (corrected for cell proliferation) for middle PDL cells was found to decrease rapidly during the first day of incubation, and from then on extremely slowly (29). Hence, at any moment in its history, a population may consist of cells, or subpopulations of cells, each of which is at a different stage in its own life history. One of the simplest conclusions that follows from this is that WI-38 at middle PDLs consists of at least two subpopulations, one with a high rate of entrance into S-phase, and the other with a very low rate of entrance into S-phase. Whether each of the two subpopulations consists of smaller subpopulations is not certain, but it is highly probable (37).

This may resolve the apparent discrepancies in previous reports. Some studies suggest the presence of subpopulations in a proliferating population with respect to the capacity of cells to synthesize DNA, whereas other studies do not (7,25). A high labeling index obtained from late passage cells after a long exposure to ^{3}H-TdR (0.01 Ci/ml) (25) is probably due both to the slow rate of cell entrance into S-phase in one subpopulation, and to dilution by the other subpopulation. Contrariwise, the appearance of a plateau in the labeling index (7), observed for late passage cells, may be due to 0.1 Ci/ml ^{3}H-TdR, which inhibits the proliferation of labeled cells. These findings are in accordance with what was observed in sparse cultures by Merz and Ross (31) and by Smith and Hayflick (37), considering the possibility that the nondividing cells observed might, in part, have the capacity to synthesize DNA.

COMPARISON BY AUTORADIOGRAPHY OF MACROMOLECULAR BIOSYNTHESIS IN PHASE II AND PHASE III WI-38

The incorporation of radioactive thymidine and DNA biosynthesis was shown to decline dramatically in aging WI-38 fibroblasts (7,24). Likewise, the autoradiographic observations of Macieira-Coelho, et al. (24) suggested decreased RNA synthesis in aging fibroblasts. Protein synthesis was also found to be affected in aging cultured cells. The rate of incorporation into proteins by the cells of radioactive amino acids was found to decline, and conformational changes were shown to occur in proteins synthesized in aging cultures (18). Little is known about the effect of aging on lipid synthesis by human fibroblasts. The total lipid content of WI-38 cells was found to increase with age, and some qualitative differences in lipid species were also detected on comparison of "young" and "old" cultures (21).

We compared the biosynthetic abilities of WI-38 fibroblasts in "young" and "old" cultures by autoradiography of cells grown with labelled precursors of DNA, RNA, protein, and lipids (35). The data indicated that aging of the cultures affects lipid synthesis much less than it does DNA, RNA, and protein synthesis.

Although the number of grains per area of cells grown with the same labeled precursor differed from one experiment to the other, the comparison of the "young" and "old" cultures in each of three experiments consistently showed a much lower incorporation of radioactivity from protein hydrolysate, uridine, and thymidine by cells in old cultures. On the other hand, the incorporation of radioactivity from acetate and oleic acid by cells in old cultures was only 10 to 20% lower than in cells of young cultures. A rough estimate of the number of grains did not reveal apparent differences between old and young cultures with respect to the incorporation of labelled cholesterol.

Comparison of the results of the three experiments indicated a gradual decrease in the ability of the cells to synthesize proteins and nucleic acids. However, lipid synthesis (as reflected by the incorporation of radioactivity from acetate and oleic acid), and the incorporation of exogenous cholesterol did not appear to be affected by aging to the same degree as were protein and nucleic acid synthesis (35). The smaller decrease in the rate of lipid synthesis compared to that of protein synthesis in aging of WI-38 cultures would be expected to result in cells richer in lipid, but in fact, the lipid content of WI-38 cells from old cultures is known to be higher than that of cells from young cultures (21).

THEORIES OF AGING

Most gerontologists agree that there is probably no single cause of aging. The explanation that probably comes closest to a unifying theory consists of concepts based on genetic instability as a cause of aging (15,16). The genetic contribution to the aging process also appears to be foremost in the determination of a life-span that is characteristic of each species. The genetic basis for aging is partly predicated on the observation that the range of variation in the maximum life-span among different species is much greater than the range of individual life-spans within the same species. One fundamental problem in relating genetic processes to aging lies in separating the genetic basis of differentiation from a possible genetic basis for aging, or the concept of "first we ripen, then we rot."

Genetic instability as a cause of age changes might include the progressive accumulation of faulty copying in dividing or otherwise functioning cells, or the accumulation of errors in information-containing molecules.

Either the progressive accumulation of errors in the function of fixed postmitotics or actively dividing cells could act as a clock. This process would initiate secondary effects which would ultimately be manifest as biologic aging. Thus, aging could be a special case of morphogenesis: cells may be programmed simply to run out of program.

Probably no other area of biologic inquiry is susceptible to as many theories as is the science of biogerontology, not only because of a lack of sufficient fundamental data but also because manifestations of biologic changes in time affect almost all biologic systems from the molecular level to the level of the whole organism. It is therefore easy to construct a theory of aging based on a biologic decrement that may be observed to occur in time in any system at the level of the cell, tissue, organ, or whole animal. The important question will always be whether the change observed is a direct cause of aging, or if it is the result of changes that may be occurring at a more fundamental level.

If modern concepts of biologic development are rooted in signals originating from information-containing molecules, it seems reasonable to attribute postdevelopmental changes to a similar system of signals occurring at the molecular level. This assumes that the switching on and switching off of genes during developmental processes also determines age changes (i.e., age changes, like developmental changes, are "programmed" into the original pool of genetic information and are "played out" in an orderly sequence just as developmental changes are). The graying of hair is not generally considered a disease associated with the passage of time, but is regarded as a highly predictable event that occurs late in life after the genetic expression of many other programmed developmental events that occur in orderly sequences.

This example might be analogous to the attribution of aging to a similar series of orderly programmed genetic events that shut down or slow down essential physiologic phenomena after the onset of postreproductive age. This programming may be the result of specific gene determinants that, like the end of a tape recording, simply trigger a sequence of events to shut the machine down. Alternatively, the universality of aging might be attributed to functional failures arising from the random accumulation of "noise" in some vulnerable parts of the system that ultimately interfere with optimum function and produce all the well known physiologic decrements.

If the noise is randomly accumulated, why do members of each species appear to age at specific, highly predictable times? The time period during which noise accumulates and becomes manifest in some functional decrement could be called "the mean time to failure." The mean time to failure of the average automobile may be five to six years, which may vary as a function of the competence of repair processes. Barring total replacement of all vital elements, deterioration is inevitable. Similarly, failure of cell function may occur at predictable times, depending on the fidelity of the synthesizing machinery and on the degree of perfection of cellular repair systems. Since biologic systems do not appear to function perfectly and indefinitely, one must conclude that the ultimate death of a cell, or its loss of function, is genetically programmed and has a mean time to failure. The mean time to failure may apply to single cells, tissues, organs, and to the intact animal itself. The genetic apparatus may simply run out of accurate programmed information that might result in different mean times to failure for all the dependent biologic systems. The existence of different species life spans may be the reflection of more perfect repair systems in animals of greater longevity.

Finally, one must consider the two cell lineages that seem to have escaped from the inevitability of aging or death—germ cells (precursors of egg and sperm cells) and continuously reproducing cancer-cell populations. The immortality of cancer-cell populations may be explained by the suggestion that genetic information is exchanged between somatic cells or viruses and somatic cells in the same way that the genetic cards are reshuffled when egg and sperm fuse. Thus, exchange of genetic information may serve somehow to reprogram or reset the biologic clock. By this mechanism, species survival is guaranteed, but the individual members are ultimately programmed to fail.

ACKNOWLEDGMENTS

This work was supported in part by Contract NAS 2–9658 from the National Aeronautics and Space Administration, Ames Research Center, Moffett Field, California, and the Glenn Foundation for Medical Research, Manhasset, New York.

REFERENCES

1. Absher, P. M., and Absher, R. G. (1976): Clonal variation and aging of diploid fibroblasts. *Exp. Cell Res.,* 103:247–255.
2. Burnet, F. M. (1970): An immunological approach to aging. *Lancet,* 2:358–360.
3. Burnet, F. M. (1971): *Genes, Dreams and Realities,* p. 232. Medical and Technical Publishing Co., Aylesbury, England.
4. Burnet, F. M. (1974): *Intrinsic Mutagenesis: A Genetic Approach to Ageing,* p. 244. John Wiley and Sons, New York.
5. Chaytor, D. E. B. (1962): Mitotic index in vitro of embryonic heart fibroblasts of different donor ages. *Exp. Cell Res.,* 28:212–213.
6. Cohn, A. E., and Murray, H. A. (1925): Physiological ontogeny. A. Chicken embryos. IV. The negative acceleration of growth with age as demonstrated by tissue cultures. *J. Exp. Med.,* 42:275–290.
7. Cristofalo, V. J., and Sharf, B. B. (1973): Cellular senescence and DNA synthesis. *Exp. Cell Res.,* 76:419–427.
8. Girardi, A. J., Jensen, F. C., and Koprowski, H. (1965): SV_{40}-Induced transformation of human diploid cells. Crisis and recovery. *J. Cell. Comp. Physiol.,* 65:69–78.
9. Harrison, D. E. (1972): Normal function of transplanted mouse erythrocyte precursors for 21 months beyond donor life spans. *Nat. New Biol.,* 237:220–222.
10. Harrison, D. E. (1973): Normal production of erythrocytes by mouse marrow continuous for 73 months. *Proc. Natl. Acad. Sci. USA,* 70:3184–3188.
11. Harrison, D. E. (1975): Normal function of transplanted marrow cell lines from aged mice. *J. Gerontol.,* 30:279–285.
12. Hayflick, L., and Moorhead, P. S. (1961): The serial cultivation of human diploid cell strains. *Exp. Cell Res.,* 25:585–621.
13. Hayflick, L. (1965): The limited in vitro lifetime of human diploid cell strains. *Exp. Cell Res.,* 37:614–636.
14. Hayflick, L. (1974): Cytogerontology. In: *Theoretical Aspects of Aging,* edited by M. Rockstein, p. 83. Academic Press, New York.
15. Hayflick, L. (1974): Biomedical gerontology, current theories of biological aging. *The Gerontologist,* 14:No. 5.
16. Hayflick, L. (1975): Current theories of biological aging. *Fed. Proc.,* 34:9–13.
17. Hayflick, L. (1977): The cellular basis for biological aging. In: *Handbook of the Biology of Aging,* edited by C. Finch and L. Hayflick, pp. 159–186. Van Nostrand-Reinhold, New York.
18. Holliday, R., and Tarrant, G. M. (1972): Altered enzymes in ageing human fibroblasts. *Nature,* 238:26–30.
19. Holliday, R., Huschtscha, L. I., Tarrant, G. M., and Kirkwood, T. B. L. (1977): Testing the commitment theory of cellular aging. *Nature,* 198:366–372.
20. Kay, H. E. M. (1965): How many cell generations? *Lancet,* 2:418–419.

21. Kritchevsky, D., and Howard, B. V. (1966): The lipids of human diploid cell strain WI-38. *Ann. Med. Exp. Biol. Fenniae (Helsinki),* 44:343–347.
22. Lefford, F. (1964): The effect of donor age on the emigration of cells from chick embryo explants in vitro. *Exp. Cell Res.,* 35:557–571.
23. Le Guilly, Y., Simon, M., Lenoir, P., and Bourel, M. (1973): Long-term culture of human adult liver cells: morphological changes related to in vitro senescence and effect of donor's age on growth potential. *Gerontologia,* 19:303–313.
24. Macieira-Coelho, A., Ponten, J., and Philipson, L. (1966): The division cycle and RNA-synthesis in diploid human cells at different passage levels in vitro. *Exp. Cell Res.,* 42:673–684.
25. Macieira-Coelho, A. (1974): Are non-dividing cells present in ageing cell cultures? *Nature,* 248:421–422.
26. Martin, G. M., Sprague, C. A., and Epstein, C. J. (1970): Replicative life-span of cultivated human cells. Effects of donor's age, tissue, and genotype. *Lab. Invest.,* 23:86–92.
27. Martin, G. M., Sprague, C. A., Norwood, T. H., Prendergrass, W. R., Bornstein, P., Hoehn, H., and Arend, W. P. (1975): Do hyperplastoid cell lines "differentiate themselves to death?" *Adv. Exp. Med. Biol.,* 53:67–90.
28. Martin, G. (1977): Cellular aging—clonal senescence. *Am. J. Pathol.,* 89:484–511.
29. Matsumura, T., Pfendt, E. A., and Hayflick, L. (1979): DNA synthesis in the human diploid cell strain WI-38 during in vitro aging. *J. Gerontol. (in press).*
30. Matsumura, T., Zerrudo, Z., and Hayflick, L. (1978): Senescent human fibroblasts in culture. *J. Gerontol. (in press).*
31. Merz, G., and Ross, J. D. (1969): Viability of human diploid cells as a function of in vitro age. *J. Cellular Physiol.,* 74:219.
32. Michl, J., Soukupova, M., and Holečková, E. (1968): Ageing of cells in cell and tissue culture. *Exp. Gerontol.,* 3:129–134.
33. Muggleton-Harris, A. L., and Hayflick, L. (1976): Cellular aging studied by the reconstruction of replicating cells from nuclei and cytoplasms isolated from normal human diploid cells. *Exp. Cell Res.,* 103:321–330.
34. Orgel, L. E. (1973): Aging of clones of mammalian cells. *Nature,* 243:441–445.
35. Razin, S., Pfendt, E. A., Matsumura, T., and Hayflick, L. (1977): Comparison by macromolecular biosynthesis in "young" and "old" human diploid fibroblast cultures. *Mech. Ageing Dev.,* 6:379–384.
36. Schneider, E. L., and Mitsui, Y. (1976): The relationship between in vitro cellular aging and in vivo human age. *Proc. Natl. Acad. Sci. USA,* 73:3584–3588.
37. Smith, J. R., and Hayflick, L. (1974): Variation in the life-span of clones derived from human diploid cell strains. *J. Cell Biol.,* 62:48–53.
38. Soukupova, M., and Holečková, E. (1964): The latent period of explanted organs of newborn, adult and senile rats. *Exp. Cell Res.,* 33:361–367.
39. Soukupova, M., Holečková, E., and Cinnerova, O. (1965): Behaviour of explanted kidney cells from young, adult and old rats. *Gerontologia,* 11:141–152.
40. Soukupova, M., Hněvkovský, P., Chvapil, M., and Hruza, Z. (1968): Effect of collagenase on the behaviour of cells from young and old donors in culture. *Exp. Gerontol.,* 3:135–139.
41. Soukupova, M., and Hněvkovský, P. (1972): The influence of donor age on the behaviour of human embryonic tissue in vitro. *Physiol. Bohemoslov.,* 21:485–488.
42. Suzuki, Y. (1926): *Mitt. Allg. Path. Sendai,* 2:191.
43. Waters, H., and Walford, R. L. (1970): Latent period for outgrowth of human skin explants as a function of age. *J. Gerontol.,* 25:381–383.
44. Williamson, A. R. (1972): Extent and control of antibody diversity. *Biochem. J.,* 130:325–333.
45. Williamson, A. R., and Askonas, B. A. (1972): Senescence of an antibody-forming cell clone. *Nature,* 238:337–339.
46. Wright, W. E., and Hayflick, L. (1975): Nuclear control of cellular aging demonstrated by hybridization of anucleate and whole cultured normal human fibroblasts. *Exp. Cell Res.,* 96:113–121.
47. Wright, W. E., and Hayflick, L. (1975): Contributions of cytoplasmic factors to in vitro cellular senescence. *Fed. Proc.,* 34, 76–79.

Physiology and Cell Biology of Aging
(Aging, Volume 8), edited by A. Cherkin, et al.
Raven Press, New York © 1979.

Characterization and Possible Effects of Age-Associated Alterations in Enzymes and Proteins

D. Gershon,* A. Reznick, and U. Reiss

Department of Biology, Technion-Israel Institute of Technology, Haifa, Israel

THE OCCURRENCE AND DISTRIBUTION OF ALTERED ENZYME MOLECULES IN AGING ORGANISMS

Most enzymes studied so far have shown an accumulation of modified inactive or partially active molecules as a function of age (1,2). This has been found by determining catalytic activity per unit of enzyme antigen in animals of various ages. The loss of activity per antigenic unit for the various enzymes ranges from 30 to 70%. The enzymes so far investigated that show this phenomenon are aldolase A (3), aldolase B (4), cytosol superoxide dismutase (5,6), catalase (7), tyrosine aminotransferase (8), and lactic dehydrogenase (9), all from various tissues of the mouse and the rat. In addition, similar observations have been made on the nematode enzymes isocitrate lyase (10,11), aldolase (12), enolase (13), 3-phosphoglycerate kinase (14) and elongation factor 1 (15). Well-established exceptions that do not exhibit a decline in specific activity per antigenic unit are triosephosphate isomerase (TPI) of nematodes (16) and ornithine decarboxylase of rat liver (17,18). In the former, however, a decline in the total amount of the enzyme was observed, and in the latter case the oldest rats tested were only 24 months old. The lack of detection of inactive forms of TPI could be due to their preferential degradation, as will be discussed later.

Several parameters of the enzymes, derived from animals of various ages, were studied: affinity for substrate (Km), affinity for specific inhibitors (Ki), electrophoretic mobility, molecular weight, antigenic identity, and heat stability (1,2). Age-related differences could be discerned only in heat stability and specific catalytic activity per unit antigen or per mg purified enzyme. Despite concerted efforts, no changes in other properties could be observed between purified enzymes from young and old animals. Thus, the chemical alterations that lead to complete or partial inactivation of enzyme molecules are of a very subtle nature.

---

* On sabbatical leave at the Roche Institute of Molecular Biology, Nutley, New Jersey

TYPES OF MODIFICATIONS IN AGE-RELATED ALTERATIONS OF ENZYME MOLECULES

Chemical modifications leading to total or partial loss of activity can be divided into two broad categories according to the level at which they take place: a) transcriptional and/or translational modification leading to amino acid substitutions during synthesis; and b) posttranslational modifications, which involve alterations of amino acid residues already incorporated in peptides.

Predictably, random amino acid substitutions in proteins should in most cases cause net charge differences due to changes of basic to neutral or acidic amino acids and all the alternative combinations. Such charge differences were neither observed by acrylamide gel electrophoresis nor more recently by a very sensitive system of isoelectric focusing in acrylamide gels. Gorin et al. (19) have shown that under conditions in which the five isozymic forms of rabbit muscle aldolase that differ from each other by only one charge could be easily separated, no separation could be observed between "young" and "old" forms of purified rat superoxide dismutase (SOD) or nematode aldolase. Also, Sharma et al. (13) could not detect any charge changes in nematode enolase from animals of various ages. It is therefore unlikely that amino acid substitutions constitute a significant cause of the observed age-dependent enzyme alterations. Thus, it is suggested that alterations at the transcriptional and/or translational levels do not contribute significantly to this phenomenon. This conclusion is further supported by experiments carried out by Pitha et al. (20) who unequivocally showed that in "late passage" human fibroblasts in culture, no misincorporation could be found in the progeny of infecting virus particles. On the other hand, Linn et al. (21) reported decreased fidelity of DNA polymerase activity in human fibroblasts as a function of passage time in culture. It should be pointed out, though, that synthetic polynucleotides which differ substantially from the natural template were used as templates for the polymerase. For instance, initiation of transcription on these synthetic templates is not restricted to specific points but can occur at random. The degree of infidelity reported in this study would be so catastrophic to cells that it would not allow any appreciable viability. No such catastrophic conditions have been observed *in vivo.*

Posttranslational modifications of proteins that occur under physiological conditions have been characterized and found in a variety of protein and enzyme systems. Some of these may be relevant to the observed process of accumulation of inactive enzyme molecules in aging organisms. Among these postsynthetic modifications are phosphorylation, deamidation, glycosylation, acetylation, methylation, adenylylation, carbamylation, peptide cleavage on either the amino or carboxyl termini, and oxidation of SH groups. The lack of discernible charge differences among proteins from young and old animals (13,19) make most of these unlikely as universal modes of modification in aging, but do not rule out several of them as minor contributors to the phenomenon.

Oxidation of the SH group and of methionine does not involve changes in the charge of proteins. These may involve proteins some of whose SH groups and methionines are exposed and susceptible to oxidation. Although oxidation of methionine to methionine sulfoxide had not been considered possible under physiological conditions, it has recently been shown to occur in considerable amounts in human lens proteins during senile cataract formation (22). Methionine sulfoxide and cystine can be the result of oxidation with oxygen and peroxide in cells. Preliminary work in our laboratory (Reiss, *unpublished results*) demonstrates the existence in liver cells of an enzymatic system that reduces methionine sulfoxide to methionine. This activity decreases considerably with age and may thus provide less protection to proteins in aging cells. These should probably be proteins of relatively long half-lives. Further chemical analysis of the proteins and their hydrolysis products is necessary to unequivocally determine the significance of this posttranslational modification in the phenomenon of age-dependent alterations of proteins.

POSSIBLE PHYSIOLOGICAL CONSEQUENCES OF THE ACCUMULATION OF ALTERED PROTEIN FORMS IN CELLS OF AGING ORGANISMS

The decrease in specific activity per antigenic unit in a variety of enzymes probably leads to a reduction in efficiency of various biochemical functions in aging cells. It is well known that in structural genes, many mutations that cause the formation of altered enzyme molecules severely affect essential physiological functions of cells, thus limiting their viability. In these cases, a large proportion of the molecules of the specific enzyme are affected by the mutation and show lowered activity or none at all. In aging cells, the reported reduction in specific activities per mg purified enzyme ranges from 30 to 70%. In most cases studied, the number of enzyme molecules per cell (as judged by the amount of antigen per microgram DNA) remains approximately constant throughout life despite the loss in activity. However, it is conceivable that there are cases analogous to the one described by us for SOD in the brain and heart wherein total activity per cell was maintained unchanged throughout life by virtue of a compensatory increase in the number of molecules per cell (6). This undoubtedly requires either an increase in synthesis or a decrease in degradation of the enzyme molecules, or both, in cells of aging animals. Such a change probably entails additional expenditure of energy and use of cellular components that might be utilized for other functions.

Nevertheless, as stated above, most of the affected enzymes studied so far have shown declines of from 30 to 70% in specific activity per antigenic unit. Are such declines in a large number of enzymes in a cell detrimental? There is hardly any information regarding this question. In other words, it is not known whether and to what extent cells can tolerate such a degree of "noise." Initial information regarding this question has been obtained by Edelmann and Gallant (23) and by Reznick and Gershon (23a). Edelmann and Gallant (23) used the simple model system of *Escherichia coli* in which errors in translation were stimulated by streptomycin. By monitoring the amount of cysteine incorporated into flagellin, which is normally devoid of this amino acid, these authors found that the bacteria can tolerate surprising degrees of altered proteins. This work must be viewed cautiously, since other proteins were not tested in the streptomycin-treated cells. Flagellin is a secreted protein and it is conceivable that altered intracellular proteins are degraded rapidly (24) or, alternatively, exert more detrimental effect on the cell. Also, the prokaryotic system may be very different from the eukaryotic system in its tolerance to altered proteins and reduced fidelity.

Reznick and Gershon (23a) studied the same question in a nematode system, utilizing the amino acid analogs canavanine and 6-fluorotryptophan to induce the formation of altered proteins in young animals. Under controlled conditions, it was possible to obtain the same levels of altered proteins in young animals as were found naturally in aging animals. These levels were highly detrimental to the young animals, and resulted in early mortality. However, our conclusions that such is the situation in aging animals must be a qualified one since the altered proteins formed by analog incorporation into proteins differ in some properties from "native" altered proteins of aging animals.

The protein degradation system may play a very important role in the accumulation of altered protein molecules in cells of aging animals. This accumulation may be due to reduced efficiency of the degradation system. The following scheme of events in enzyme synthesis and degradation is proposed for animals of all ages (see also ref. 2):

protein synthesis —(a)→ active enzyme (unaltered proteins) —(b)→

modified enzyme form(s) (reduced activity or none at all) —(c)→

modified form(s) (no activity; better substrates for proteases) —(d)→ degradation (1. specific 2. nonspecific)

Our work and that of others has shown that with age, there is a considerable slowdown in protein synthesis [step (a) in the scheme]. Steps (b) and (c), which also take place in aging animals at an undetermined rate, produce enzyme molecules "earmarked" for degradation. Steps (b) and (c) are at least partially nonenzymatic. Step (d) apparently requires two classes of proteases: firstly, specific proteases, which are probably neutral and perform an initial specific cleavage (25) followed by rapid, nonspecific hydrolysis of the resulting peptides; and secondly, nonspecific proteases, which may be contained in the lysosomes. Like other enzymes, the proteases involved in both stages of hydrolysis are probably altered in old animals thus leading to impaired disposal of altered proteins. As in many other examples described in the literature, we have observed in old animals the accumulation of swollen, distended lysosomes that contained lipofuscin (26). This damage may well be the result of incomplete degradation of proteins. In addition, we have recently found that the disposal of analog-containing proteins is vary much impaired in older animals (23a).

Protein degradation has very important physiological functions whose impairment with age may have serious consequences on cellular performance. These functions are a) supply of amino acids during starvation; b) contribution to the control of levels of inducible and constitutive enzymes; and c) disposal of altered proteins by discriminately rapid hydrolysis.

CONCLUSION

All the studies described here have been carried out with large, heterogeneous populations of cells, and the results therefore constitute an overall average of each parameter studied. Even small average changes in the proportion of inactive enzymes may, for a small number of cells, constitute large changes that render them inviable or severely incapacitated. In some tissues, the elimination of a small percentage of cells may have a very severe effect on total tissue function.

Further studies on the chemical nature of protein modifications in aging animals are necessary to facilitate the development of methods to counter this phenomenon.

Also, a better link must be created between aging studies at the biochemical-subcellular level and at the physiological organ level to enable evaluation of the contribution of molecular alterations to the decline in viability of the whole animal.

ACKNOWLEDGMENTS

The work reported here was supported by NIH Grant no. RO1-AG-00459-04 and a grant from Deutches Forschungsgemeinschaft.

REFERENCES

1. Gershon, D. (1978): Current status of age-altered enzymes: alternative mechanisms. *Mechanisms of Aging and Development (in press).*
2. Rothstein, M. (1977): Recent developments in the age-related alteration of enzymes: a review. *Mechanisms of Aging and Development,* 6:241–257.
3. Gershon, H., and Gershon, D. (1973): Altered enzyme molecules in senescent organisms: mouse muscle aldolase. *Mechanisms of Aging and Development,* 2:33–41.
4. Gershon, H., and Gershon, D. (1973): Inactive enzyme molecules in aging mice: liver aldolase. *Proc. Natl. Acad. Sci. USA,* 70:909–913.
5. Reiss, U., and Gershon, D. (1976): Rat liver superoxide dismutase. Purification and age-related modifications. *Eur. J. Biochem.,* 63:617–623.

6. Reiss, U., and Gershon, D. (1976): Comparison of cytoplasmic superoxide dismutase in liver, heart and brain of aging rats and mice. *Biochem. Biophys. Res. Commun.,* 73:255–261.
7. Dresnick, J., and Gershon, D. (1978): *(in preparation).*
8. Jacobus, S., and Gershon, D. (1978): *(in preparation).*
9. Schapira, F., Weber, A., and Gregori, C. (1975): Vieillisement de la lacticodeshydrogenase hepatique du rat et renouvellement cellulaire. *Compt. Rend. Acad. Sci. (Paris),* 280:1161–1163.
10. Gershon, H., and Gershon, D. (1970): Detection of inactive enzyme molecules in aging organisms. *Nature,* 227:1214–1217.
11. Reiss, U., and Rothstein, M. (1975): Age-related changes in isocitrate lysate from the free-living nemate, *Turbatrix aceti. J. Biol. Chem.,* 250:826–830.
12. Zeelon, P., Gershon, H., and Gershon, D. (1973): Inactive enzyme molecules in aging organisms. Nematode fructose–1,6–diphosphate aldolase. *Biochem.,* 12:1743–1750.
13. Sharma, H. K., Gupta, S. K., and Rothstein, M. (1976): Age-related alteration of enolase in the free-living nematode *Turbatrix aceti. Arch. Biochem. Biophys.,* 174:324–332.
14. Gupta, S. K., and Rothstein, M. (1976): Phosphoglycerate kinase from young and old *Turbatrix aceti. Biochim. Biophys. Acta,* 445:632–644.
15. Bolla, R., and Brot, N. (1975): Age-dependent changes in enzymes involved in macromolecular synthesis in *Turbatrix aceti. Arch. Biochem. Biophys.,* 169:227–236.
16. Gupta, S. K., and Rothstein, M. (1976): Triosephosphate isomerase from young and old *Turbatrix aceti. Arch. Biochem. Biophys.,* 174:333–338.
17. Obenrader, M., and Prouty, W. F. (1977): Detection of multiple forms of rat liver ornithine decarboxylase. *J. Biol. Chem.,* 252:2860–2865.
18. Obenrader, M., and Prouty, W. F. (1977): Production of monospecific antibodies to rat liver ornithine decarboxylase and their use in turnover studies. *J. Biol. Chem.,* 252:2866–2872.
19. Gorin, P., Reznick, A. Z., Reiss, U., and Gershon, D. (1977): Isoelectric properties of nematode aldolase and rat liver superoxide dismutase from young and old animals. *FEBS Lett.,* 84:83–86.
20. Pitha, J., Stork, E., and Wimmer, E. (1975): Protein synthesis during aging of human cells in culture. *Exp. Cell Res.,* 94:310–314.
21. Linn, S., Kairis, M., and Holliday, R. (1976): Decreased fidelity of DNA polymerase activity isolated from aging human fibroblasts. *Proc. Natl. Acad. Sci. USA,* 73:2818–2822.
22. Truscott, R. J. W., and Augusteyn, R. C. (1977): Oxidative changes in human lens proteins during senile nuclear cataract formation. *Biochim. Biophys. Acta,* 492:43–52.
23. Edelmann, P., and Gallant, J. (1977): On the translational error theory of aging. *Proc. Natl. Acad. Sci. USA,* 74:3396–3398.
23a. Reznick, A. and Gershon, D. (1978): *(submitted for publication).*
24. Goldberg, A., and Dice, J. F. (1974): Intracellular protein degradation in mammalian and bacterial cells. *Ann. Rev. Biochem.,* 43:835–869.
25. Etlinger, J. D., and Goldberg, A. L. (1977): A soluble ATP-dependent proteolytic system responsible for the degradation of abnormal proteins in reticulocytes. *Proc. Natl. Acad. Sci. USA,* 74:54–58.
26. Epstein, J., and Gershon, D. (1972): Studies on aging in nematodes. IV. The effect of antioxidants on cellular damage and life span. *Mech. Aging and Devel.,* 1:257–267.

Physiology and Cell Biology of Aging
(Aging, Volume 8), edited by A. Cherkin, et al.
Raven Press, New York © 1979.

Cell Changes Associated with Declining Immune Function

Marguerite M. B. Kay and Lynn S. Baker

Geriatric Research, Education and Clinical Center, V.A. Wadsworth Hospital Center, Los Angeles, California 90073; and Department of Medicine, University of California at Los Angeles, Los Angeles, California 90024

Aging is characterized by the declining ability of an individual to adapt to environmental stress. Because more and more people are surviving to an older age, aging is potentially the most critical issue facing mankind—socially, economically, and biomedically (Fig. 1). Enhanced interest in the diseases and disabilities associated with aging has led to the careful investigation of many biologic systems with the hope that methods will be found for delaying the onset of aging, lessening its severity, or perhaps preventing some of its pathologic changes.

The immune system is perhaps the most productive field in which to conduct such studies for the following reasons:

1. Certain immune functions normally tend to decline with age in humans (81), pigs (9), hamsters (65), rats (13), mice (64), dogs (52), and rabbits (69,82,97) (Fig. 2), although the onset, magnitude, and rate of decline vary with the type of immune function and with the species.
2. The immune system is "organismal" in that it is in continuous contact with most, if not all, cell, tissue, and organ systems within the body. Thus, any alteration of the immune system would be expected to affect all other systems. If one views aging as a perturbation of homeostasis, then a mobile, dynamic system such as the immune system is perhaps the ideal one in which to study such perturbations and their consequences.
3. As immunologic vigor decreases, the incidence of infections, autoimmune and immune complex diseases, and cancer increases (31,35,59,60) as in the case of immunodeficient newborns and immunosuppressed adults (26,27,71). A strong piece of evidence linking decreased immunologic vigor to disease is that the occurrence of immunodeficiency, wasting disease, amyloidosis, and autoimmunity in neonatally thymectomized and genetically susceptible mice can be prevented, or at times even reversed, by reconstituting them with young, but not old, syngeneic thymus or spleen grafts (24).

Present evidence suggests that the decline in immune functions that accompanies aging is due primarily to changes in the T cell component of the immune

This is publication number 17 from V.A. Wadsworth GRECC.

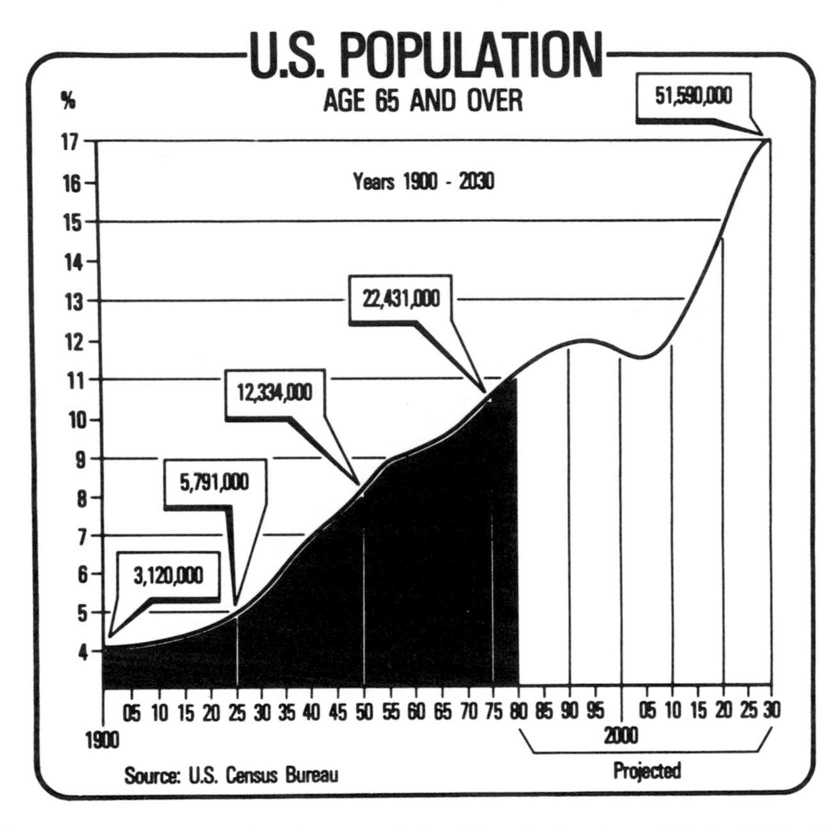

FIG. 1. Percentage of the American population 65 and older from 1900 to 1975 with the projected percentage for 1980 to 2030.

system, and that B cell changes are minimal or secondary to those of the T cells (55). In this presentation, therefore, an overview of the rapidly developing area of the aging of the immune system will be presented by first discussing the age-related changes in T cell activities, and then by focusing on the possible mechanisms that may be responsible for these changes. However, before doing so, we will briefly review the major components of the immune system.

THE IMMUNE SYSTEM

The immune system is an organismal one, as its cells are distributed throughout the body and circulate within the blood and lymph. The major lymphoid tissues are the thymus, lymph nodes, spleen, and bone marrow (Fig. 3). The stem cells, found primarily in the bone marrow, can differentiate into either T cells, B cells, or macrophages. B cells and macrophages are generated in the bone marrow, but T cells originate from pre-T cells under the influence of the thymus.

Obviously, anything which affects the stem cells will affect the other compo-

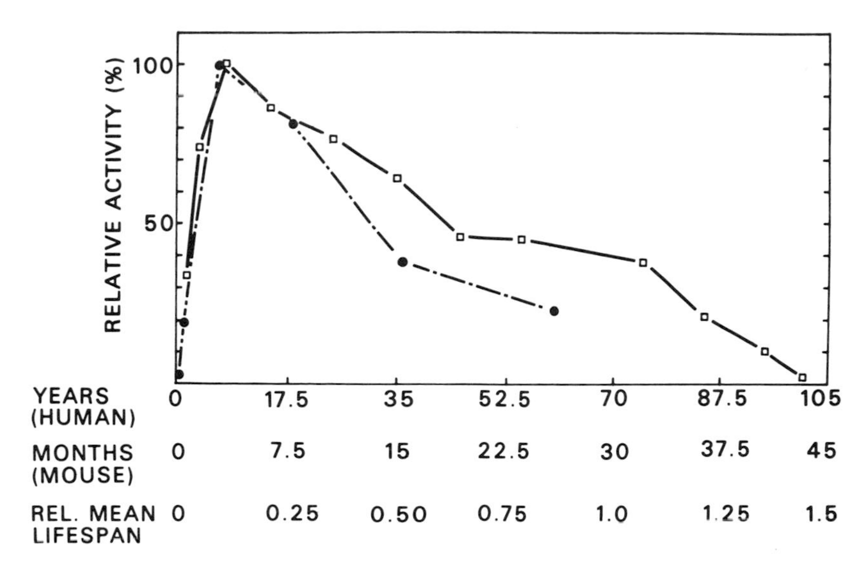

FIG. 2. Effects of age on serum agglutinin titers in humans and mice. Natural serum anti-A isoagglutinin titers in the human (97). Peak serum agglutinin response titer to RBC stimulation by intact long-lived mice (64). (Reprinted from Makinodan and Adler, ref. 61.)

nents of the immune system. Unlike stem cells passaged *in vivo* (41,42,58,87) whose self-replicating ability can be exhausted, stem cells can self-replicate *in situ* throughout the natural lifespan of an individual. However, the ability of stem cells to expand clonally and their rate of division decreases with age (63,3), as does their ability to repair X-ray–induced DNA damage (20), and their

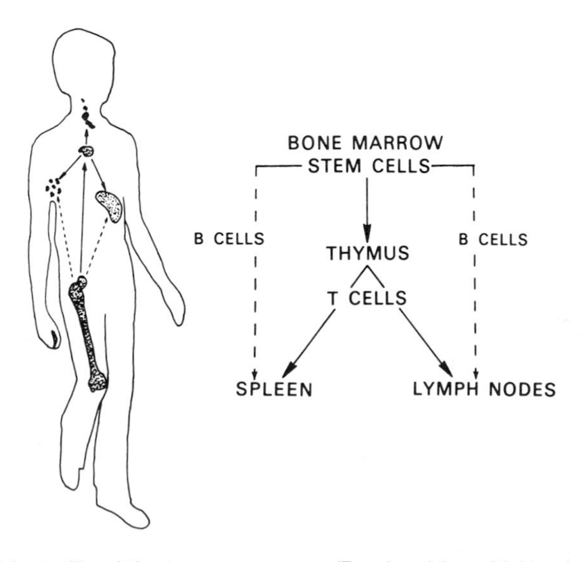

FIG. 3. Cellular traffic of the immune system. (Reprinted from Makinodan, ref. 60a.)

ability to home into the thymus (53,99). Interestingly, when attempts were made to reverse these age-related kinetic parameters by enabling "old" stem cells to self-replicate in syngeneic young recipients for an extended period, it was found that they still behaved as old stem cells kinetically. However, when "young" stem cells were allowed to self-replicate in syngeneic old recipients, they behaved like old stem cells kinetically. These results indicate that the milieu of stem cells induces subtle but stable irreversible changes that affect their responsiveness to differentiation-homeostatic factors.

T cells are responsible for protecting the body against viruses, fungi, and certain types of bacteria; for preventing the growth of certain neoplasms; and for regulating B cell antibody production to a large number of antigens. Human T cells are identified morphologically by their ability to form rosettes with sheep red blood cells (RBC) (Fig. 4), and mouse T cells by the presence of the θ-antigen on their surface. Assays for T cell activity include responsiveness to plant mitogenic lectins such as phytohemagglutinin (PHA), concanavalin A (Con A), participation in delayed-type hypersensitivity (i.e., skin test) reactions, ability to mount graft-versus-host (GVH) reactions, and ability to help or suppress the responses of other cells.

B cells can be identified by the presence of immunoglobulin (Ig) on their

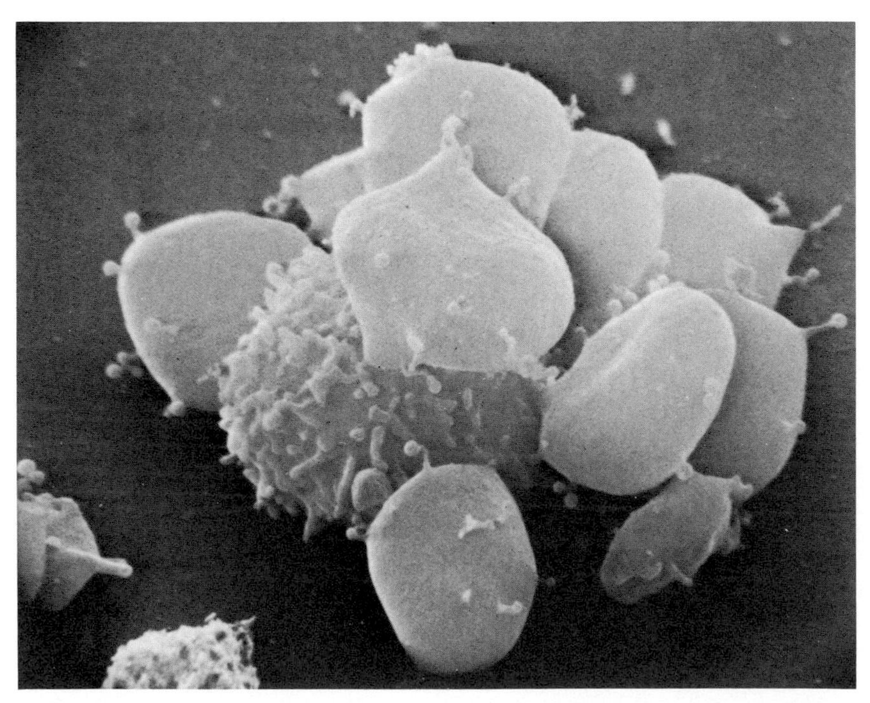

FIG. 4. Scanning electron micrograph of an unstimulated human T lymphocyte. The ability of human lymphocytes from nonimmunized individuals to bind sheep red blood cells without antibodies or complement being present has been shown to be a marker for T cells.

FIG. 5. Human B cells are usually identified by the binding of fluoroscene-labeled antihuman immunoglobulin to the immunoglobulins on their surface. When exposed to UV light, the fluorescene labeled molecules fluoresce bright green. This B cell has immunoglobulin represented by "Y" on its surface to which fluoroscene-labeled antiimmunoglobulin antibodies represented by "Y*" have bound.

surface. The Ig can be detected by electron microscopy, or more commonly by immunofluorescent staining (Fig. 5). Generally, B cells cannot produce antibody without the direction of T cells, unless the antigens consist of repetitive subunits as are found in most carbohydrates or polysaccharides. Assays for B cell activity include responsiveness to certain mitogens such as lipopolysaccharide (LPS), and antibody production in response to T cell-independent and -dependent antigens.

Another cell that participates in the immune response is the macrophage (Fig. 6). Macrophages are derived from peripheral blood monocytes, which are derived from a bone marrow precursor cell (100). Macrophages cooperate with T and B cells in many immune responses and they phagocytize bacteria, some viruses, and senescent and damaged cells. Since macrophages usually confront antigens before the T and B cells participate in most immunologic activities, many of the earlier studies on the mechanism of loss in immunologic vigor with age were focused on macrophages. These studies showed that aging does not adversely affect macrophages in their handling of antigens during both the induction of immune responses and phagocytosis (5,6,44,96).

AGE-RELATED CHANGES IN T CELL FUNCTION

Morphology

The first hint that normal T cell functions may be declining with age came from the findings of classical morphologists. They showed that the thymic lymphatic mass decreased with age, primarily as a result of atrophy of the cortex. The onset of this decrease coincided with the attainment of sexual maturity, and was found in both laboratory animals and humans (4,83). Subsequently, atrophy of the epithelial cells and decreased levels of thymus hormone(s) have been observed. Histologically, the cortex of an involuted thymus is sparsely populated with lymphocytes, which are replaced by numerous macrophages

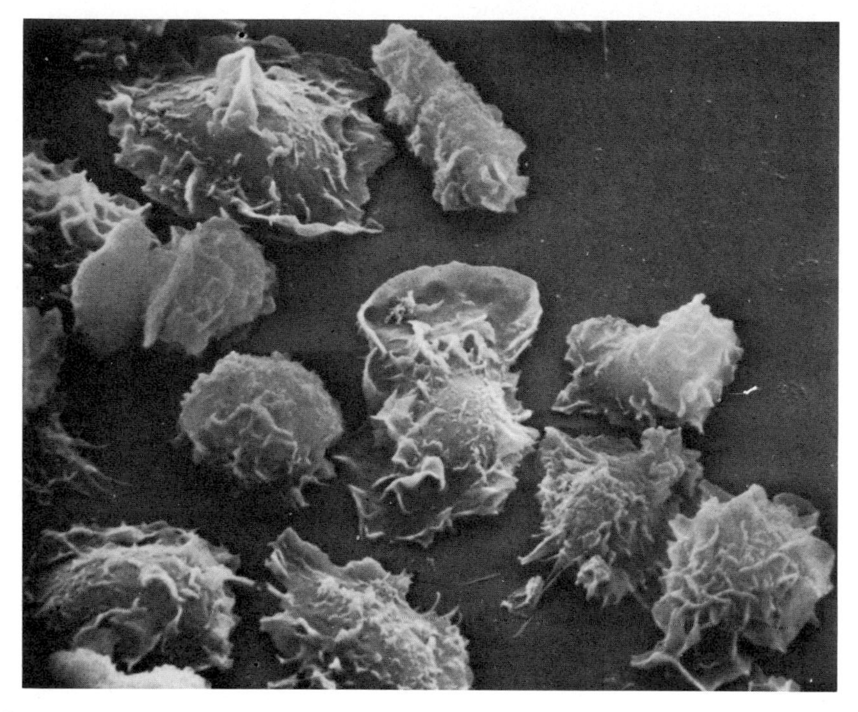

FIG. 6. Peritoneal macrophages crawling on a glass coverslip. The surfaces of these cells are covered with lamellae (ruffles).

filled with lipoid granules (48). In addition, infiltration of plasma and mast cells can be observed in the medulla as well as in the cortex. (48). Although the size of lymph nodes and of spleen remains about the same after adulthood in individuals without lymphatic neoplasia (4,83), the cellular composition of these tissues shifts so that there are diminished numbers of germinal centers and increased numbers of plasma cells and macrophages, as well as an increase in the amount of connective tissue (21,73). In humans, the number of circulating T cells has been reported either to remain the same or to decrease progressively after adulthood to a level at which, by the sixth decade of life, it is 70% that of a young adult (7,18,22).

Cellular Changes

In Vivo

In general, T cell-dependent cell-mediated functions decline with age. However, there are conflicting reports in the literature concerning certain functions in aging humans and animals (9,31,34,35,36,70). For example, some investigators report a decrease in delayed skin hypersensitivity to common test antigens to which individuals have been previously sensitized (such as purified protein derivative or tuberculosis, streptokinase-streptodornase, monilia, or trichophyton). Others report no decrease except in those elderly persons with acute illness. It could be argued that any decrease in delayed hypersensitivity seen in the elderly reflects aging of the skin rather than of the immune system. However, results of tumor cell rejection tests in aging mice were comparable, and since tumor cells were injected intraperitoneally in these tests, aging of the skin was not the limiting factor (34). It is more likely that the conflicting results could be attributed to (a) utilization of only one skin test antigen to assess T cell function; (b) selection of skin test antigens (e.g., since only a fraction of the U.S. population has been immunized to tuberculosis, a negative result cannot be interpreted as being reflective of defective immunological memory); and (c) selection of the population samples (i.e., some studies used hospitalized patients and medical clinic populations). Based on these considerations, it is recommended that a battery of common test antigens be utilized for the assessment of secondary delayed skin hypersensitivity reactions.

In contrast to its controversial role in secondary delayed skin reactions, T cell function declines with age in primary delayed skin reactions in response to antigens to which the individuals have *not* been sensitized previously, such as dinitrochlorobenzene (DNCB). Studies performed *in vivo* with mice indicate that various T cell dependent functions decline with age (34,72,93,95). Thus, cells from old mice have a decreased ability to mount a GVH reaction (72,103) even when enriched T cell populations are utilized to compensate for the possibility that old animals may have fewer T cells (103). Resistance to challenge with syngeneic and allogeneic tumor cells *in vivo* decreases dramatically with age (72,90). In these studies, it is significant that the decline in response to PHA measured *in vitro* approximated the decline in GVH and tumor cell challenge measured *in vivo* (72).

In Vitro

The findings from experiments performed *in vitro* show that the proliferative capacity of T cells of humans and rodents declines with age in response to PHA, Con A, and allogeneic target cells (1,25,37,45,49,54,56,57,65,72,76,81,51). This decline is most striking in mice, regardless of their life-span.

The decline in the cell-mediated lymphocytotoxicity (CML) index of T cells of long-lived mice has been reported to be moderate against allogeneic tumor cells and not readily apparent against certain syngeneic tumor cells (34). The discrepancy between results of tumor cytotoxicity tests performed *in vivo* and *in vitro* may be due in part to inadequate culture conditions in the latter, as indicated by the demonstration that a significant decline with age in mixed lymphocyte culture (MLC) reaction can be detected with improved culture methods (46).

There has been one report that the response to PHA in mice shows only a minimal decline with age (86). However, in this report, "middle-aged" rather than old C57Bl/6J mice were used [i.e., 15- to 20-month-old mice were used when the mean life-span (MLS) of the mice in that laboratory was 24 months, which is equivalent to assaying 44- to 58-year-old humans whose MLS is 70 years]. This directs attention to an important point: The age at which a mouse is "old" varies with the strain. For example, the MLS of DBA/2 males is 86 weeks in one laboratory, whereas the MLS of C67Bl/6J males is 130 weeks (57,88). Furthermore the MLS of a strain can vary between laboratories depending upon the animal housing conditions [e.g., C57B1/6J which have a MLS of 120 weeks in some laboratories may have a mean life-span of 97 weeks in others (91,101)]. Thus, a 90-week-old C57B1/6J could be considered "old" in one laboratory but "middle-aged" in another.

It should be emphasized, furthermore, that young *adult* mice (3 to 10 months old) rather than "young" mice (≤2 months old) should be used as a reference point for aging studies. The latter may be "immature" immunologically, depending on the immunologic index. For example, it was shown in one laboratory that the antisheep red blood cell response of $BC3F_1$ mice (MLS, 30 months) does not mature until 6 months of age (64), while their response to PHA peaked at 8 months of age (49). A further complication encountered when using immature young mice is that variability between individuals is greater than in young adults. Therefore, the description of mice as "old" or "young" should be based upon the survival and developmental pattern of each individual strain. This would reduce the number of conflicting reports regarding the effect of age on immune functions.

There are conflicting reports in the literature on the effect of age on the MLC reaction. Some investigators reported a marked decrease with age (1,49). Others reported that cells from old mice are equally, if not more, efficient than cells from young mice, as both responding and stimulating cells in the MLC reaction. However, the same cells showed a decreased GVH index (103). Some of these discrepancies could be avoided if investigators would not use mitomycin-treated cells as the stimulating cells in the MLC reaction. Mitomycin may leak from the stimulating cells, and responding cells from old animals may be more susceptible to this drug than are cells from young animals. In studying the effects of age on cell-mediated immunity, a sounder approach to the utilization of the MLC index would be to use either X-rayed or hybrid cells from donors of one age group as the stimulating cells.

MECHANISMS OF THE AGE-RELATED DECLINE IN NORMAL IMMUNE FUNCTIONS

One of the hallmarks of immunosenescence is the increase in variability of immune indices (56,62). If more than one factor is responsible for the increase in variability, it would not be surprising to find that the decline in immune capacity of aging mice results from changes in the immune cells or from changes in their milieu, or both. To differentiate between the influences of cells and their milieu, researchers employed the cell transfer method, which assesses immunocompetent cells from young and old mice in immunologically inert old and young syngeneic recipients, respectively (2,34,77,78). The results revealed that both types of changes affect the immune response and that about 10% of the normal age-related decline can be attributed to changes in the cellular milieu, whereas 90% of the decline can be attributed to changes intrinsic to the old cells (77,78).

Cellular Milieu

The responsible factor(s) in the cellular environment was shown to be systemic and noncellular (78). Spleen cells from young mice were cultured with the test antigen either in the young (or old) recipient's spleen by the cell transfer method, or in the recipient's peritoneal cavity by the diffusion chamber method (33). A twofold difference in response was observed between young and old recipients at both sites, indicating that the factor(s) is systemic. The fact that the effect was observed in cells grown in cell-impermeable diffusion chambers further indicates that a noncellular factor is involved. A comparable twofold difference was also observed when bone marrow stem cells were assessed in the spleens of young and old syngeneic recipients (19), indicating that the systemic, noncellular factor(s) influences both lympho- and hematopoietic processes.

The factor(s) could be a deleterious substance of molecular or viral nature, or it could be an essential substance that is deficient in old mice. Factors of both types probably change with age, and several factors of each type may exist.

Unfortunately, this area of research has not progressed as rapidly as was anticipated, because a simple, sensitive *in vitro* assay to analyze mouse sera has not yet been perfected. Thus, for example, it is unclear why normal adult mouse serum is toxic for mouse immunocompetent cells grown *in vitro.*

Cellular Changes

Three types of cellular changes could cause a decline in normal immune functions: (a) an absolute decrease in cell number through death, which may be caused by autoimmune cells, (b) a relative decrease in cell number as a result of an increase in the number of "suppressor" cells, and (c) a decrease in functional efficiency caused, perhaps, by somatic mutation. One approach in resolving this problem is first to estimate the frequency of old mice that exhibit these cellular changes. This can be done by assessing, in the presence of immune cells from old individuals, the activity of reference immune cells from adult individuals. A response of young-old cell mixtures that was less than the sum of the responses given by pure young and pure old cells would indicate that the decreased response of old individuals was due to an increase in suppressor cells. If the response of the mixture was comparable, it would indicate that the decreased response of old individuals was caused either by a decrease in their functional efficiency or by a general loss of immune cells. If the response of the mixture was higher, it would indicate that the decreased response of old individuals was caused by a selective loss of one type of immune cell that exists in excess in young individuals. The results showed that all three types of interactions can occur. This supports the contention that although there may be only one underlying mechanism responsible for the loss of immunologic vigor with age, it is expressed differently by aging individuals, and this contributes to the increased variability in immunologic performance with age.

Since involution of the thymus precedes the age-related decline in T cell function, a "cause and effect" relationship was suspected (i.e., thymic involution results in a decreased capacity of the system to generate functional T cells). Despite the obvious importance of firmly establishing such a relationship, this relationship was subjected to rigorous experimental proof only recently. Previous experiments had shown that adult thymectomy accelerates the decline of immune responsiveness by demonstrating: (a) a decreased hemagglutinin response to sheep RBC, particularly in the early response (19S) phase (68), (b) a marked decrease in the GVH reaction, (c) the poor general condition of the mice, and (d) a reduced antibody response to bovine serum albumin (BSA) (131). The effect on the latter response was less pronounced than the effect on the GVH reaction. Recently, Hirokawa and Makinodan (47) transplanted thymus lobes from mice 1 day to 33 months old into young adult, T cell-deprived recipients, and assessed kinetically the emergence of T cells (Fig. 7). They found that

→

FIG. 7. Hirokawa and Makinodan performed an experiment to establish the extent to which age-related involution of the thymus affects its capacity to transform pre-T cells into T cells. The experimental design was as follows: 1. T cell deficient mice were prepared first by taking out the thymus so that the mice could not make more T cells, and then by X-irradiating the mice to destroy the T cells that they already had. Mice were then given stem cells from the bone marrow so that they could produce all blood cells. (Remember, they didn't have a thymus so they could not produce T cells from the pre-T cells produced by the bone marrow.) Mice prepared in this manner are called TXB mice ("T" for thymectomized, "X" for X-irradiated, and "B" for bone marrow reconstituted). 2. TXB (recipient) mice were given a thymus transplant (graft) from donor mice ranging in age from 1 day to 33 months. The thymus graft was placed in the capsule of one of the kidneys to provide it with a good blood supply. In this situation, pre-T cells from the bone marrow of recipient mice migrate into the thymus graft and are transformed into T cells if the thymus has the ability to induce the transformation. 3. The recipient TXB mice were then examined for (a) repopulation of the paracortical (T cell dependent) areas of the lymph nodes; (b) number of cells with the theta (T cell) antigen on their surface; (c) assay response to the T cell mitogens PHA and succinyl (s)-Con A; (d) T cell-dependent antibody responses to sheep red blood cells using a plaque assay; and (e) mitogenic reactivity of splenic cells to allogeneic lymphocytes (MLC reaction). The results showed that thymic tissues lose the following influences: first, the influence on lymphocyte repopulation of T cell-dependent areas of lymph nodes (compare "A" in Fig. 7**a** and **b**); second, the influence on mitogen responsiveness of splenic T cells to PHA and s-Con A (see "B" and "C" of Fig. 7**a** and **b**); third, the influence on the number of spleen cells with the theta (T cell) antigen (refer to "B" in Fig. 7**a** and **b**) and on splenic T cell-dependent antibody response (see "D" of Fig. 7**a** and **b**); and fourth, the influence on the mitogenic reactivity of spleen cells to allogeneic lymphocytes.

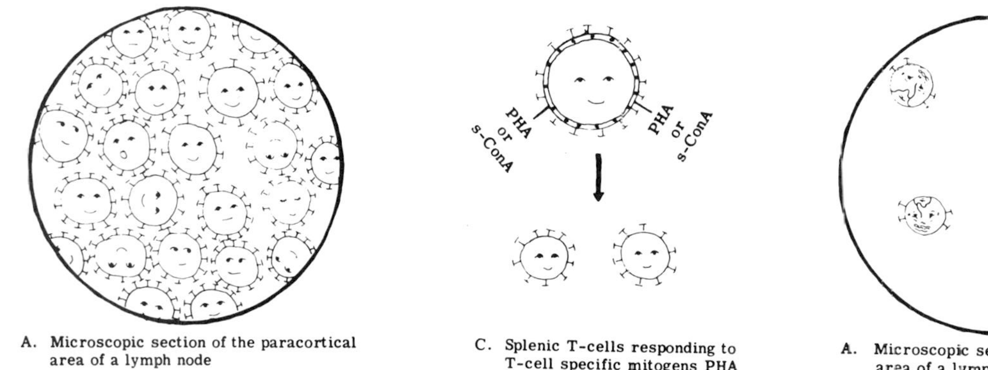

A. Microscopic section of the paracortical area of a lymph node

or

B. Microscopic section of the spleen

C. Splenic T-cells responding to T-cell specific mitogens PHA or s-ConA

TXB mouse with thymic graft from a 1-day old donor

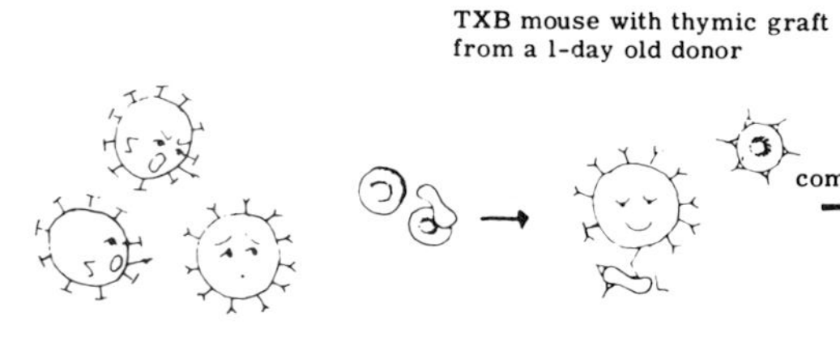

+ complement → Plaque Formation

D. Splenic T-cells helping an uncertain B-cell mount an antibody response to sheep red blood cells

(a)

l. s. baker, md

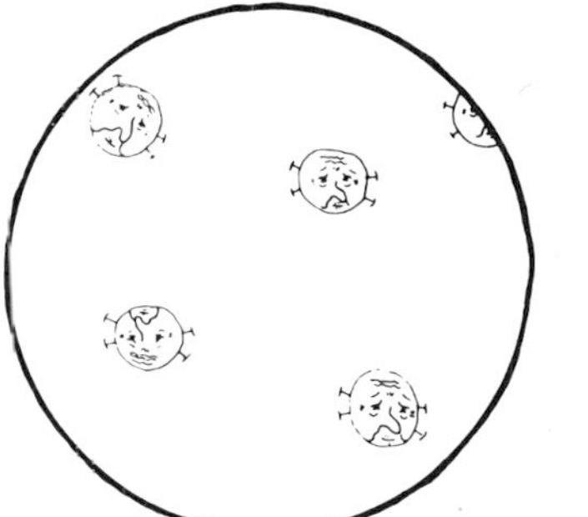

PHA or s-ConA

PHA or s-ConA

No Action

A. Microscopic section of the paracortical area of a lymph node

or

B. Microscopic section of the spleen

C. Splenic T-cell failing to respond to T-cell specific mitogens PHA or s-ConA

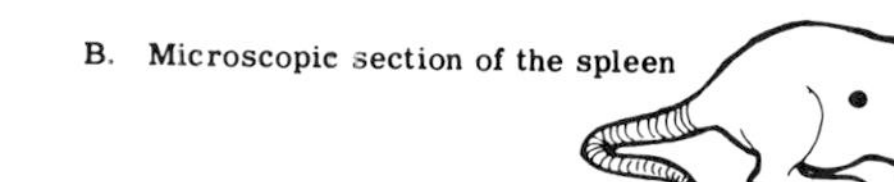

TXB mouse with thymic graft from a 33-month old donor

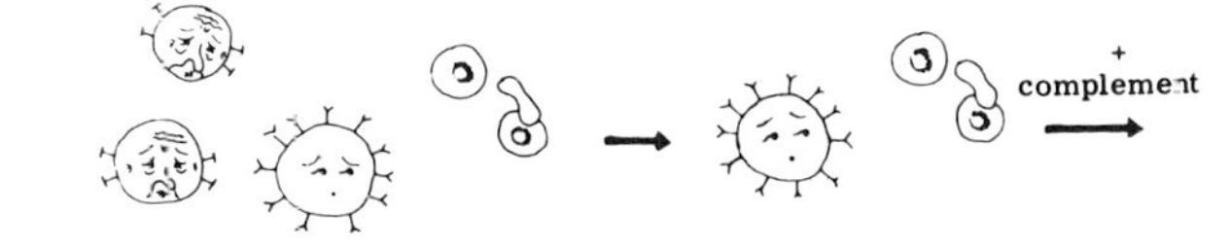

+ complement → No Action

D. Splenic T-cells not being very helpful to an uncertain B-cell, who then leaves sheep red blood cells alone

(b)

l. s. baker, md

the ability of the thymus tissue to influence the differentiation or maturation of precursor cells into functional T cells decreases with age, and that the various T cell activities exhibit differential susceptibility to thymus involution (see ref. 47).

Recent experiments performed in my laboratory utilizing chromosome markers and parabiotic mice indicate that after 6 to 10 weeks of age, the movement of stem cells into the thymus seems to cease (56). Thus, at least during this growth phase, it appears that the cells within the thymus are self-perpetuating. This may also be true of peripheral T cells. The cessation of stem cell traffic into the thymus prior to adolescence could be due to (a) a defect in the stem cells, (b) a defect in the thymus, or (c) a defective regulatory mechanism that is in neither the bone marrow nor the thymus. It is significant that the cessation

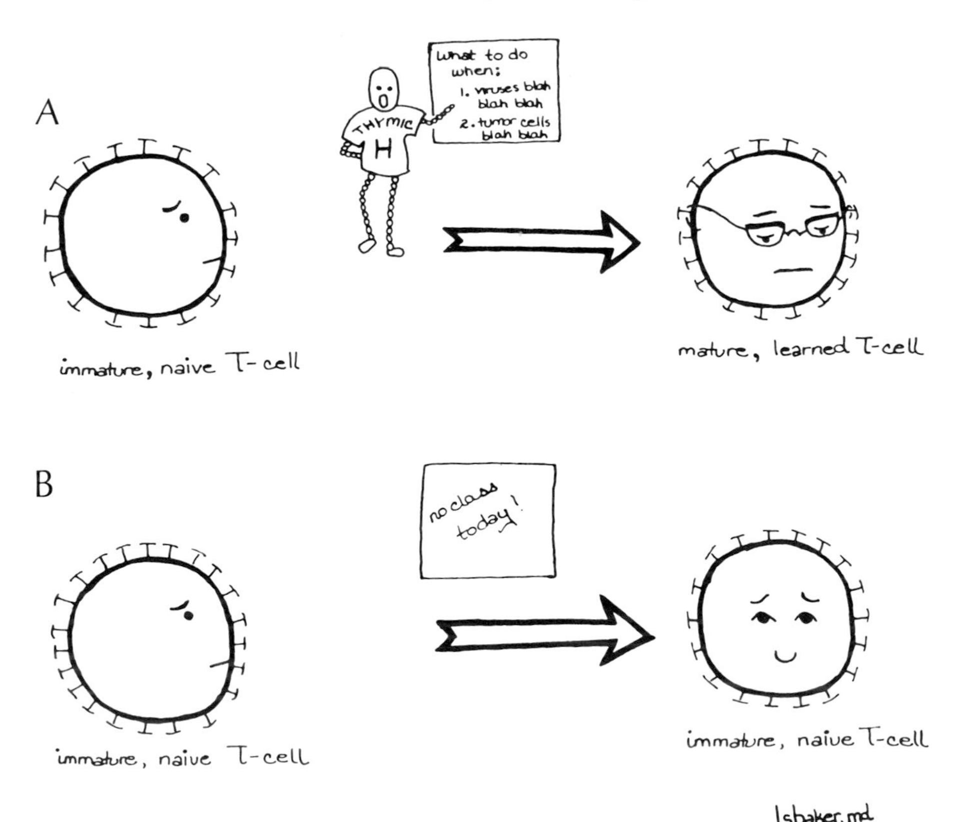

FIG. 8. Present evidence suggests that thymic involution leads to a decrease in normal peripheral effector functions by altering the proportion of T cell subsets. It seems that some T cell subsets, which are necessary for initiating certain immunologic activities, decrease with age. It has been hypothesized that certain T cells, under the influence of thymic hormone(s) (shown here as the teacher, thymic H), differentiate into mature T cells. (**A**). With age, the level of the thymic hormone(s) decreases. Thus, some T cell subsets do not become "educated" and are not effective in the process of life maintenance (**B**).

of stem-cell-to-T-cell traffic occurs during a growth phase of the individual, and before detectable thymic involution or a decrease in the number of T cells in the thymus, lymph nodes, or spleen is observed (54).

These results indicate that thymic involution precedes, and is responsible for, the age-dependent decline in the ability of the immune system to generate functional T cells. What is the mechanism by which thymic involution leads to a decrease in normal peripheral effector cell functions? It appears that involution of the thymus alters the proportion of T cells, possibly through decreased synthesis of differentiation hormones or through synthesis of factors that can alter cells, or both (Fig. 8). The following observations support this view:

1. The proportion of mouse lymphocytes bearing the θ-antigen decreases with age, as does the amount of θ-antigen on cell surfaces. Yet there is no compensatory increase in B lymphocytes, nor does the lymphocyte number change significantly (Fig. 9). This suggests an increase in T cells that do not carry detectable θ-receptors on their surfaces (15).

2. Although PHA-induced blastogenesis of cells from old mice is significantly reduced (see Fig. 4), these cells bind ^{125}I-labeled PHA just as well as cells from young mice (50). Because there seems to be no significant decrease in binding affinities or receptor sites for PHA on cells of old mice, the defect cannot be in their membrane receptors.

3. The cyclic nucleotide 3′,5′-guanosine monophosphate, which has been shown to increase when T cells are stimulated by mitogens, is found in relatively

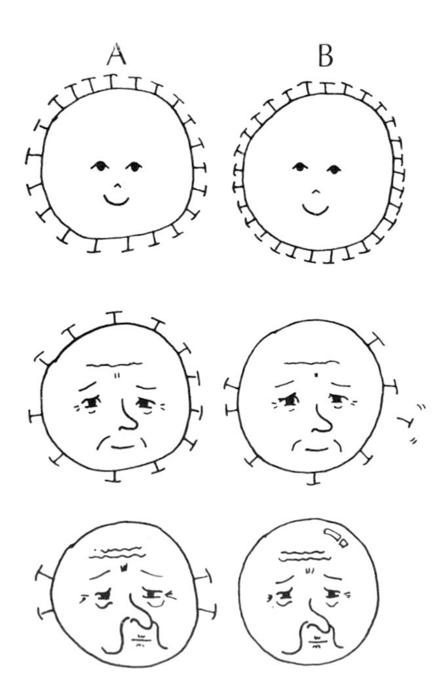

FIG. 9. This figure provides the rare opportunity of seeing a T cell age before your eyes. **A**: depicts T cells which have decreased amounts of theta or T cell antigen on their surfaces. **B**: depicts T cells which lose theta as they age.

low concentrations in mitogen-stimulated T cells of old mice (Fig. 10). It is known that the levels of cyclic nucleotides are hormone-dependent (43). This suggests that the defect in T cells is intracellular and may be hormone-dependent.

4. The life-span of hypopituitary dwarf mice, which are T cell-deficient, can be extended from 4 months to 12 months by a single intraperitoneal injection of 150 × 10^6 lymph node cells rich in mature T cells; but one injection of an equal number of cells from the thymus deficient in mature T cells, or 50 ×

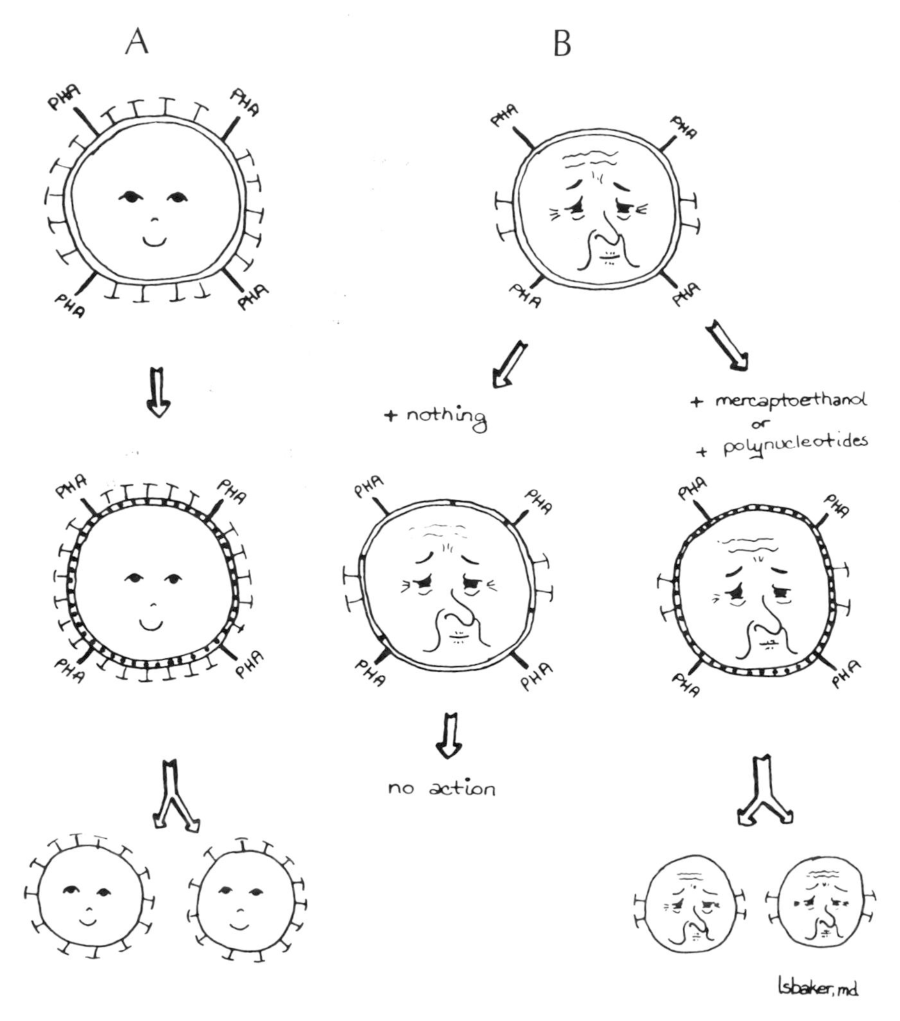

FIG. 10. A: When T cells from young individuals are stimulated with PHA, the level of the cyclic nucleotide 3′,5′-guanosine monophosphate (shown here as black dots within the membrane) increases. **B:** When T cells from old individuals are stimulated with PHA, the cyclic nucleotide level and the amount of cell proliferation remain relatively low. However, addition of 2-mercaptoethanol or polynucleotides to the culture significantly increases the blastogenic response of old cells to PHA.

10^6 bone marrow cells, is ineffective (Fig. 11) (24,74). Comparable life prolongation was demonstrated by injection of growth factors and thyroxine in untreated, but not in thymectomized, dwarf mice (24,23). This suggests that mature T cells are required to approximate homeostasis, but that their activities cannot be sustained throughout the normal life-span of the species without replenishment from a precursor pool and, perhaps, favorable hormonal environment.

5. Transfer of 5×10^7 to 5×10^8 thymus cells from immunologically mature congeneic donors into athymic mice reconstitutes the recipients' capacity to reject allogeneic skin grafts, whereas transfer of an equal number of thymus cells from newborn mice will not (75). Presumably, the ability of thymus cell suspensions from adult mice to reconstitute graft rejections depends on the

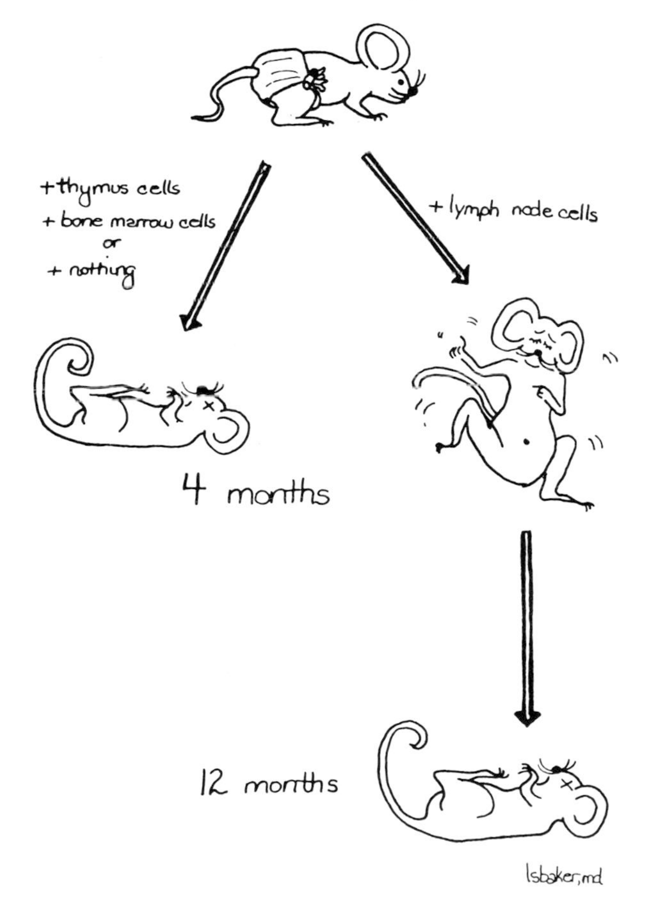

FIG. 11. Hypopituitary dwarf mice in the laboratory of Fabris, Pierpaoli, and Sorkin. Besides being dwarves, these mice were also T cell deficient, as a result of which they died at 4 months of age. The doctors tried to find a way to prolong the lives of their mice. They injected thymus or bone marrow cell suspensions, but these treatments were futile because the cell suspensions lacked mature T cells. Finally, they injected lymph node cells which contain mature T cells. The dwarf mice lived to be 12 months old (i.e., three times their normal life-span).

number of "competent" lymphocytes resident in the thymic medulla or those that come from the recirculating pool and are trapped in the thymus cell suspensions.

6. Studies of "synergy" between subpopulations of lymphocytes in the MLC reaction (28,66) suggest that lymph node "T_2" (amplifier) cells display a greater functional decline with age than do spleen and thymus "T_1" lymphocytes (the precursors of T_2 cells). This suggests that the mature T cells, which are required for the recruitment of precursor T cells and for the MLC reaction, decline preferentially with age.

7. Treatment of thymus cell suspensions with anti-TL serum and complement (which spares the more mature TL^- thymocytes, which have undergone differentiation within the thymus and are about to migrate to the periphery, and kills the less mature TL^+ thymocytes) does not reduce the GVH reactivity of thymocyte suspensions (98). However, the capacity of thymocytes, when combined with peripheral blood leukocytes, to give superadditive GVH reactions was decreased by treatment with anti-TL. These results indicate that synergy among subpopulations of T cells in GVH reactions involves the participation of "immature" T cells under the influence of mature T cells.

8. Density distribution analysis, using BSA and Ficoll discontinuous gradients, shows that the frequency of less dense cells (ρ 1.06–1.08) increases at the expense of the more dense cells (ρ 1.10–1.12) (61). This density shift within the lymphocyte population can also be seen in young mice shortly after they are immunized with foreign red cells or allogeneic lymphocytes, and in tumor-bearing mice. In these latter cases, however, the spleen cell number increases, whereas in unimmunized old mice it does not. This suggests that there is a relative increase with age in immature T_1 cells at the expense of mature T_2 cells.

9. Shortly after the thymus begins to involute and atrophy, the level of serum thymic hormones(s) decreases with age (8). It would seem reasonable to assume that this hormone(s) is necessary for terminal differentiation of T cells. This could lead to a decrease in certain normal effector cell functions and an increase in suppressor function, and thus to a deficit in overall T cell responsiveness (see Fig. 6).

10. The PHA response of T cells from old mice can be significantly increased *in vitro* by addition of certain chemicals (e.g., mercaptoethanol and polynucleotides) to the cultures, as can the antibody response to sheep RBC (see Fig. 8A and B; refs. 14,38,63). This supports the view that old mice are deficient in a hormone(s) that is essential for the differentiation of T cells to become responsive to a mitogen, and suggests that simple chemicals can substitute for this hormone.

How these regulatory T cells function as suppressors and helpers is not known. Therefore, there have not been extensive studies on age-related changes in T cells involved in the regulation of B cell immune responses. This is due in part to the complexities of the system. The recent development of antisera directed against allelic T cell determinants (Ly differentiation antigens), which allows identification of three T cell subsets, should greatly facilitate research in the area of age effects on T cell suppressor function (16,17). It seems that T cells expressing the Ly 1 antigen are responsible for helper and delayed-type hypersensitivity effects, whereas cells expressing Ly 2,3 are responsible for suppressor effects and for cell-mediated cytolysis (16,17). No specific function has yet been assigned to Ly 1,2,3 cells (for a complete review of suppressor cells in aging, see ref. 29).

The evidence that suppressor T cell activity declines with age was derived from studies of short-lived NZB and related mice (39). C-type virus infection, however, plays a major role in the pathogenesis of the disease of NZB. The causes and mechanisms of the decline in normal immune functions, and of immunodeficient diseases in these animal models, may be different from those occurring in long-lived mice and elderly humans. We may be observing phenotypic caricatures of old-age immune deficiency that are analogous to the phenotypic features of accelerated aging seen in progeric humans. Thus, although a decrease of suppressor T cell activity with age could account for the emergence of autoantibodies in the older short-lived mice, it cannot account for the emergence of autoantibodies in the older long-lived mice. In the later case, the relative number of suppressor T cells *increases* slightly with age (36,85,77,78).

By reducing either the number of antigen responsive IU or the magnitude of IBS, increases in numbers of regulator T cells with suppressor activity can interfere with the B cell response to antigenic stimulation. The notion that the number of regulator cells with inhibitory activities increases with age was initially tested by assessing the antisheep RBC response of spleen cells from young mice in the presence or absence of spleen cells from old mice (63). The ratio of the observed response to that which would be expected if the response of young and old cells were additive was <1. These results indicate that spleens of old mice contain regulator cells that can interfere with the immunologic activities of spleen cells from young mice. It should be noted that the proliferative response of T cells from young mice to PHA and allogeneic target cells was also assessed in a similar manner. These initial studies have since been confirmed and extended with cells from both aging mice and humans (84,85).

The helper function of T cells declines with age. This has been demonstrated in intact animals as well as in assays performed both *in vivo* and *in vitro* (39,44, 77,78). The assays employed are based on the ability of T cells to promote antibody response to T cell-dependent antigens. Many of the assays have utilized foreign red blood cells as a T-dependent antigen. Many of the descriptive studies have been completed, and since it is now necessary to elucidate cellular and molecular mechanisms of age-related alteration, use of antigenetically undefined red blood cells as a test antigen should be discouraged. They contain many antigens, some of which are probably T-independent, whereas others are T-dependent. Instead, defined hapten antigens, such as dinitrophenol-bovine serum albumin (DNP-BSA), should be used for T-dependent antigen studies, and antigens such as DNP-Ficoll or pneumococcus SIII should be used for T-independent antigen studies.

Finally, it should be noted that T cells may also regulate hematopoiesis (32), for it was shown that hematopoiesis of parental stem cells in heavily irradiated F_1 recipients can be augmented in the presence of parental T cells.

Thymus

All the relevant studies to date indicate that the process of involution and atrophy of the thymus is the key to the aging of the immune system. It follows, then, that the search for the cause(s) and mechanism(s) of aging of the immune system should be centered on the thymus. The causes can be either extrinsic or intrinsic to the thymus. The most likely extrinsic cause is a possible regulatory breakdown in the hypothalamus-pituitary neuroendocrine axis in relation to thymus (10,23,24,74). Intrinsic causes might be found at either the DNA level or the non-DNA level (for molecular theories of aging, see, for example, reference 92).

With regard to a neuroendocrine axis in relation to the thymus, it has been found that stress and other psychological factors can affect the immune response, as measured by skin graft rejection and by primary and secondary antibody responses. Lesions in the anterior basal hypothalamus, but not in the median or posterior hypothalamus, can depress delayed hypersensitivity and reduce the severity of anaphylactic reactions (for review, see ref. 89). Growth hormone and insulin have been shown to act preferentially on T-dependent immune functions. Thyroxine and sex hormones influence both T and B cell responses (for review, see refs. 23,74). On the other hand, there is evidence that the thymus can modulate hormone levels (12,23). The area of thymic–neuroendocrine interactions should be an exciting and fruitful one for future research on aging.

Considering intrinsic causes, three possible mechanisms of involution and atrophy can be proposed. One is clonal exhaustion (42) (i.e., thymus cells might have a genetically programmed clock mechanism to self-destruct and die after undergoing a fixed number of divisions) (55). The mechanism would be similar to that of the Hayflick phenomenon, which is seen *in vitro* in human fibroblasts. It would require either the thymus to count the cells leaving it or the cells to count the divisions they have undergone, or both. Another possible mechanism is an alteration of thymus cell DNA, either randomly or through viral infection (80). Various stable alterations of DNA can occur, including cross-linking and strand breaks (for review, see, for example, ref. 73). The third possible mechanism is a stable molecular alteration at the non-DNA level through subtle error-accumulating mechanisms.

At the level of the genes, there is mounting evidence indicating that genes of the H-2 system in mice and the HLA region in humans influence immune responsiveness and disease susceptibility (for details, see refs. 11,67,102).

CONCLUSIONS

An attempt has been made both to summarize present knowledge and to show the potential for future progress in the areas of age-related changes in T cell immune function and mechanisms of the age-related decline.

Normal immune functions can begin to decline shortly after an individual reaches sexual maturity. Although changes in the environment of the cells are partially responsible, the decline is due primarily to changes in the cells. Foremost among the cellular changes are those seen in the stem cells, as reflected in their growth properties, and in T cells, where a shift in subsets may be occurring with age. It appears that the process(es) regulating involution and atrophy of the thymus could be the key to immunosenescence. Future studies on the mechanism(s) responsible for decreased function are expected, therefore, to focus as much on the area of the neuroendocrine-thymic axis as they do on possible intrinsic mechanisms. The increasing size of the aged population is rapidly becoming one of the most critical socioeconomic issues on this planet. The diseases associated with age, and with the loss of immunologic vigor, are numerous. As understanding of the aging immune system increases, it is anticipated that methods will be developed to predict, minimize, delay, and prevent the debilitative processes associated with aging.

REFERENCES

1. Adler, W., Takiguchi, T., and Smith, R. T. (1971): Effect of age upon primary alloantigen recognition by mouse spleen cells. *J. Immunol.*, 107:1357–1362.
2. Albright, J. F., and Makinodan, T. (1966): Growth and senescence of antibody-forming cells. *J. Cell Physiol.*, 67:(Suppl. 1)185.
3. Albright, J., and Makinodan, T. (1976): Decline in the growth potential of spleen-colonizing bone marrow stem cells of long-lived aging mice. *J. Exp. Med.*, 144:1204–1213.
4. Andrew, W., editor (1952): *Cellular Changes With Age.* Charles C Thomas, Springfield, Illinois.
5. Aoki, T., and Teller, M. N. (1966): Aging and cancerigenesis. III. Effect of age on isoantibody formation. *Cancer Res.*, 26:1648–1652.
6. Aoki, T., Teller, M. N., and Robitaille, M. L. (1965): Aging and cancerigenesis. II. Effect of age on phagocytic activity of the reticuloendothelial system and on tumor growth. *J. Natl. Cancer Inst.*, 34:255–264.
7. Augener, W., Cohnen, G., Reuter, A., and Brittinger, G. (1974): Decrease of T lymphocytes during aging. *Lancet*, 1:1164.
8. Bach, J. F., Dardenne, M., and Salomon, J. C. (1973): Studies on thymus products. IV. Absence of serum "thymic activity" in adult NZB and (NZB × NZW)F_1 mice. *Clin. Exp. Immunol.*, 14:247–256.
9. Baer, H., and Bowser, R. T. (1963): Antibody production and development of contact skin sensitivity in guinea pigs of various ages. *Science*, 140:1211–1212.
10. Bearn, J. G. (1968): The thymus and the pituitary adrenal axis in anencephaly. *Br. J. Exp. Pathol.*, 49:136–144.
11. Benacerraf, B., and McDevitt, O. H. (1972): Histocompatibility-linked immune response genes. *Science*, 175:273–279.
12. Besedovsky, H., and Sorkin, E. (1975): Changes in blood hormone levels during the immune response. *Proc. Soc. Exp. Biol. Med.*, 150:466–470.
13. Bilder, G. E., and Denckla, W. D. (1977): Restoration of ability to reject xenografts and clear carbon after hypophysectomy of adult rats. *Mech. Ageing Dev.*, 6:153–163.
14. Braun, W., Yajima, Y., and Ishizuka, M. (1970): Synthetic polynucleotides as restorers of normal antibody forming capacities in aged mice. *J. Reticuloendothel. Soc.*, 7:418–424.

15. Brennan, P. C., and Jaroslow, B. N. (1975): Age-associated decline in theta antigen on spleen thymus-derived lymphocytes of B6CF1 mice. *Cell Immunol.,* 15:51–56.
16. Cantor, H., and Boyse, E. A. (1975): Functional subclasses of T lymphocytes bearing different Ly antigens. I. The generation of functionally distinct T-cell subclasses is a differentiative process independent of antigen. *J. Exp. Med.,* 141:1376–1389.
17. Cantor, H., and Boyse, E. A. (1975): Functional subclasses of T lymphocytes bearing different Ly antigens. II. Cooperation between subclasses of Ly^+ cells in the generation of killer activity. *J. Exp. Med.,* 141:1390–1399.
18. Carosella, E. D., Monchanko, K., and Braun, M. (1974): Rosette-forming T cells in human peripheral blood at different ages. *Cell. Immunol.,* 12:323–325.
19. Chen, M. G. (1971): Age-related changes in hematopoietic stem cell populations of a long-lived hybrid mouse. *J. Cell Physiol.,* 78:225–232.
20. Chen, M. G. (1974): Impaired Elkind recovery in hematopoietic colony-forming cells of aged mice. *Proc. Soc. Exp. Biol. Med.,* 145:1181–1186.
21. Chino, F., Makinodan, T., Lever, W. H., and Peterson, W. J. (1971): The immune system of mice reared in clean and dirty conventional laboratory farms. I. Life expectancy and pathology of mice with long life spans. *J. Gerontol.,* 26:497–507.
22. Diaz-Jouanen, E., Williams, R. C., Jr., and Strickland, R. G. (1974): Age-related changes in T and B cells. *Lancet,* 1:688–689.
23. Fabris, N. (1977): Hormones and aging. In: *Immunity and Aging,* edited by T. Makinodan and E. Yunis, pp. 73–89. Plenum Press, New York.
24. Fabris, N., Pierpaoli, W., and Sorkin, E. (1972): Lymphocytes, hormones and aging. *Nature,* 240:557–559.
25. Fernandez, L. A., MacSween, J. M., and Langley, G. R. (1976): Lymphocyte responses to phytohaemagglutinin: Age-related effects. *Immunology,* 31:583–587.
26. Fudenberg, H. H. (1971): Genetically determined immune deficiency as the predisposing cause of "autoimmunity" and lymphoid neoplasia. *Amer. J. Med.,* 51:295–298.
27. Fudenberg, H. H., Good, R. A., Goodman, H. C., Hitzig, W., Kundel, H. G., Roitt, I. M., Rosen, F. S., Rowe, D. S., Selligmann, M., and Soothill, J. R. (1971): Primary immunodeficiencies. *Bull. WHO,* 45:125–142.
28. Gerbase-DeLima, M., Meredith, P., and Walford, R. (1975): Age-related changes, including synergy and suppression, in the mixed lymphocyte reaction in long-lived mice. *Fed. Proc.,* 34:159–161.
29. Gershon, R. K., and Metzler, C. M. (1977): Suppressor cells in aging. In: *Immunity and Aging,* edited by T. Makinodan and E. Yunis, pp. 103–110. Plenum Press, New York.
30. Goidl, E. A., Innes, J. B., and Weksler, M. E. (1976): Immunological studies of aging. II. Loss of IgG and high avidity plaque-forming cells and increased suppressor cell activity in aging mice. *J. Exp. Med.,* 144:1037–1048.
31. Good, R. A., and Yunis, E. J. (1974): Association of autoimmunity, immunodeficiency and aging in man, rabbits and mice. *Fed. Proc.,* 33:2040–2050.
32. Goodman, J. W., and Shinpock, S. G. (1968): Influence of thymus cells on erythropoiesis of parental marrow in irradiated hybrid mice. *Proc. Soc. Exp. Bio. Med.,* 129:417.
33. Goodman, S. A., Chen, M. G., and Makinodan, T. (1972): An improved primary response from mouse spleen cells cultured *in vivo* in diffusion chambers. *J. Immunol.,* 108:1387–1399.
34. Goodman, S. A., and Makinodan, T. (1975): Effect of age on cell-mediated immunity in long-lived mice. *Clin. Exp. Immunol.,* 19:533–542.
35. Gross, L. (1965): Immunologic defect in aged population and its relation to cancer. Cancer, 18:201–204.
36. Grossman, J., Baum, J., Fusner, J., and Condemi, J. (1975): The effect of aging and acute illness on delayed hypersensitivity. *J. Allergy Clin. Immunol.,* 55:268–275.
37. Hallgren, H. M., Buckley, E. C., III, Gilbertsen, V. A., and Yunis, E. J. (1973): Lymphocyte phytohemagglutinin responsiveness, immunoglobulins and autoantibodies in aging humans. *J. Immunol.,* 111:1101–1107.
38. Han, I. H., and Johnson, A. G. (1976): Regulation of the immune system by synthetic polynucleotides. VI. Amplification of the immune response in young and aging mice. *J. Immunol.,* 117:423–427.
39. Hardin, J. A., Chuseo, T. M., and Steinberg, A. D. (1973): Suppressor cells in the graft vs. host reaction. *J. Immunol.,* 111:650–651.

40. Harrison, D. E. and Doubleday, J. W. (1975): Normal function of immunologic stem cells from aged mice. *J. Immunol.,* 114:1314–1317.
41. Harrison, D. E., Astle, C. M., and Doubleday, J. W. (1977): Stem cell lines from old immunodeficient donors give normal response in young recipients. *J. Immunol.,* 118:1223–1227.
42. Hayflick, L. (1965): The limited *in vitro* lifetime of human diploid cell strains. *Exp. Cell Res.,* 37:614–636.
43. Heidrick, M. L. (1973): Imbalanced cyclic-AMP and cyclic-GMP levels in concanavalin-A stimulated spleen cells from aged mice. *J. Cell Biol.,* 57:139a.
44. Heidrick, M. L., and Makinodan, T. (1973): Presence of impairment of humoral immunity in nonadherent spleen cells of old mice. *J. Immunol.,* 111:1502–1506.
45. Heine, K. M. (1971): Die reaktionsfahigkeit der lymphozyten im alter. *Folia Haematol.,* 96:29–33.
46. Hirano, T., and Nordin, A. (1976): Age-associated decline in the *in vitro* development of cytotoxic lymphocytes in NZB mice. *J. Immunol.,* 117:1093–1098.
47. Hirokawa, K., and Makinodan, T. (1975): Thymic involution: effect on T cell differentiation. *J. Immunol.,* 114:1659–1664.
48. Hirokawa, K. (1977): The thymus and aging. In: *Immunity and Aging,* edited by T. Makinodan and E. Yunis, pp. 51–72. Plenum Press, New York.
49. Hori, Y. Perkins, E. H., and Halsall, M. K. (1973): Decline in phytohemagglutinin responsiveness of spleen cells from aging mice. *Proc. Soc. Exp. Biol. Med.,* 144:48–53.
50. Hung, C.-Y. Perkins, E. H., and Yang, W.-K. (1975b): Age-related refractoriness of PHA-induced lymphocyte transformation. II. ^{125}I-PHA binding to spleen cells from young and old mice. *Mech. Ageing Dev.,* 4:103–112.
51. Inkeles, B., Innes, J. B., Kuntz, M. M., Kadish, A. S., and Weksler, M. E. (1977): Immunological studies of aging. III. Cytokinetic basis for the impaired response of lymphocytes from aged humans to plant lectins. *J. Exp. Med.,* 145:1158–1168.
52. Jaroslow, B. N., Suhrbier, K. M., and Fritz, T. E. (1974): Decline and restoration of antibody forming capacity in aging beagle dogs. *J. Immunol.,* 112:1467–1476.
53. Kay, M. M. B. (1978): Effect of age on T cell differentiation. *Fed. Proc.,* 37:1241–1244.
54. Kay, M. M. B. (1978): Immunological aging patterns: Effect of parainfluenza type I virus infection on aging mice of 8 strains and hybrids. In: *Genetic Effects on Aging,* edited by D. Bergsma and D. Harrison, pp. 213–240. Alan R. Liss, Inc., New York.
55. Kay, M. M. B. and Makinodan, T. (1976): Immunobiology of aging: evaluation of current status. *Clin. Immunol. Immunopathol.,* 6:394–413.
56. Kay, M. M. B., Mendoza, J., Denton, T., Union, N., and Lajiness, M. (1978): Age-related changes in the immune system of 8 medium and long-lived strain and hybrids I: Weight, cellular and activity changes. *Mech. Ageing Dev. (in press).*
57. Konen, T. G., Smith, G. S., and Walford, R. L. (1973): Decline in mixed lymphocyte reactivity of spleen cells from aged mice of a long-lived strain. *J. Immunol.,* 110:1216–1221.
58. Lajtha, L. J., and Schofield, R. (1971): Regulation of stem cell renewal and differentiation: possible significance in aging. *Adv. Gerontol. Res.,* 3:131–146.
59. Mackay, I. R. (1972): Ageing and immunological function in man. *Gerontologia,* 18:285–304.
60. Mackay, I. R., Whittingham, S. F., and Mathews, J. D. (1977): The immunoepidemiology of aging, In: *Immunity and Aging,* edited by T. Makinodan and E. Yunis, pp. 35–49. Plenum Press, New York.
60a. Makinodan, T. (1976): Immunity and aging. In: *Handbook of the Biology of Aging,* edited by C. E. Finch and L. Hayflick, pp. 379–408. Van Nostrand-Reinhold Co., New York.
61. Makinodan, T., and Adler, W. (1975): The effects of aging on the differentiation and proliferation potentials of cells of the immune system. *Fed. Proc.,* 34:153–158.
62. Makinodan, T., Chino, F., Lever, W. E., and Brewen, B. S. (1971): The immune systems of mice reared in clean and dirty conventional laboratory farms. II. Primary antibody-forming activity of young and old mice with long life-spans. *J. Gerontol.,* 26:508–514.
63. Makinodan, T., Deitchman, J. W., Stoltzner, G. H., Kay, M. M. B., and Hirokawa, K. (1975): Restoration of the declining normal immune functions of aging mice. *Proc. 10th Internatl. Cong. Gerontol.,* 2:23.
64. Makinodan, T., and Peterson, W. J. (1962): Relative antibody-forming capacity of spleen cells as a function of age. *Proc. Natl. Acad. Sci. USA,* 48:234–238.
65. Mathies, M., Lipps, L., Smith, G. S., and Walford, R. L. (1973): Age-related decline in response

to phytohemagglutinin and pokeweed mitogen by spleen cells from hamsters and a long-lived mouse strain. *J. Gerontol.,* 28:425–430.
66. Meredith, P., Tittor, W., Gerbase-DeLima, M., and Walford, R. (1975): Age-related changes in the cellular immune response of lymph node and thymus cells in long-lived mice. *Cell Immunol.,* 18:324–330.
67. Meredith, P., and Walford, R. L. (1977): Effect of age on the response to T and B cell mitogens in mice congenic at the H-2 locus. *Immunogenetics (in press).*
68. Metcalf, D. (1965): Delayed effect of thymectomy in adult life on immunological competence. *Nature,* 208:1336.
69. Nomaguchi, T. A., Okuma-Sakurai, Y., and Kimura, I. (1976): Changes in immunological potential between juvenile and presenile rabbits. *Mech. Ageing Dev.,* 5:409–417.
70. Novick, A., Novick, I., and Potoker, S. (1972): Tuberculin skin testing in a chronically sick aged population. *J. Am. Geriatr. Soc.,* 20:455–458.
71. Penn, I., and Starzl, T. E. (1972): Malignant tumors arising *de novo* in immunosuppressed organ transplant recipients. *Transplantation,* 14:407–417.
72. Perkins, E. H., and Cacheiro, L. H. (1977): A multiple-parameter comparison of immunocompetence and tumor resistance in aged Balb/c mice. *Mech. Ageing Dev. (in press).*
73. Peter, C. P. (1973): Possible immune origin of age-related pathological changes in long-lived mice. *J. Gerontol.,* 28:265–275.
74. Piantanelli, L., and Fabris, N. (1977): Hypopituitary dwarf and athymic nude mice to the study of the relationships among thymus, hormones, and aging. In: *Genetic Effects on Aging,* edited by D. Harrison, pp. 315–333. Alan R. Liss, Inc., New York.
75. Pierpaoli, W. (1975): Inability of thymus cells from newborn donors to restore transplantation immunity in athymic mice. *Immunology* 29:465–468.
76. Pisciotta, A. V., Westring, D. W., Deprey, C., and Walsh, B. (1967): Mitogenic effect of phytohaemagglutinin at different ages. *Nature,* 215:193–194.
77. Price, G. B., and Makinodan, T. (1972): Immunologic deficiencies in senescence. I. Characterization of intrinsic deficiencies. *J. Immunol.,* 108:403–412.
78. Price, G. B., and Makinodan, T. (1972): Immunologic deficiencies in senescence. II. Characterization of extrinsic deficiencies. *J. Immunol.,* 108:413–417.
79. Price, G. B., and Makinodan, T. (1973): Aging: alterations of DNA-protein information. *Gerontologia,* 19:58–70.
80. Proffitt, M. R., Hirsch, M. S., and Black, P. H. (1973): Murine leukemia: a virus-induced autoimmune disease? *Science,* 182:821–823.
81. Roberts-Thomson, I., Whittingham, S., Youngchaiyud, U., and Mackay, I. R. (1974): Ageing, immune response, and mortality. *Lancet,* 2:368–370.
82. Rowley, D. A., Fitch, F. W., Stuart, F. P., Köhler, H., and Cosenza, H. (1973): Specific suppression of immune responses. *Science,* 181:1133–1141.
83. Santisteban, G. A. (1960): The growth and involution of lymphatic tissue and its interrelationships to aging and to the growth of the adrenal glands and sex organs in CBA mice. *Anat. Rec.,* 136:117–126.
84. Segre, D., and Segre, M. (1976): Humoral immunity in aged mice. I. Age-related decline in the secondary response to DNP of spleen cells propagated in diffusion chambers. *J. Immunol.,* 116:731–734.
85. Segre, D., and Segre, M. (1976): Humoral immunity in aged mice. II. Increased suppressor T cell activity in immunologically deficient old mice. *J. Immunol.,* 116:735–738.
86. Shigemoto, S., Kishimoto, S., and Yamamura, Y. (1975): Change of cell-mediated cytotoxicity with aging. *J. Immunol.,* 115:307–309.
87. Siminovitch, L., Till, J. E., and McCulloch, E. A. (1964): Decline in colony-forming ability of marrow cells subjected to serial transplantation into irradiated mice. *J. Cell. Physiol.,* 64:23–31.
88. Smith, G. S., and Walford, R. L. (1977): Influence of the H-2 and H-1 histocompatibility systems upon lifespan and spontaneous cancer incidence in congenic mice. In: *Genetic Effects On Aging,* edited by D. Bergsma and D. Harrison, pp. 281–312. Alan R. Liss, Inc., New York.
89. Stein, M., Schiavi, R. C., Camerino, M. (1976): Influence of brain and behavior on the immune system. *Science,* 191:435–440.
90. Stjernsward, J. (1966): Age dependent tumor-host barrier and effect of carcinogen initiated immune depression of rejection of siografted methylcholanthrene induced sarcoma cells. *J. Nat. Cancer Inst.,* 37:505–512.

91. Storer, J. B. (1966): Longevity and gross pathology at death in 22 inbred mouse strains. *J. Gerontol.*, 21:404–409.
92. Strehler, B., Hirsch, G., Gusseck, D., Johnson, R., and Bick, M. (1971): Codon-restriction theory of aging and development. *J. Theoret. Biol.*, 33:429–474.
93. Stutman, O., Yunis, E. J., and Good, R. A. (1968): Deficient immunologic functions of NZB mice. *Proc. Soc. Exp. Biol. Med.*, 127:1204–1207.
94. Taylor, R. B. (1965): Decay of immunological responsiveness after thymectomy in adult life. *Nature*, 208:1334–1335.
95. Teague, P. O., Yunis, E. J., Rodey, G., Fish, A. J., Stutman, O., and Good, R. A. (1970): Autoimmune phenomena and renal disease of mice. Role of thymectomy, aging, and involution of immunologic capacity. *Lab. Invest.*, 22:121–130.
96. Teller, M. N. (1972): Age changes and immune resistance to cancer. *Adv. Gerontol. Res.*, 4:25–43.
97. Thomsen, O., and Kettel, K. (1929): Die stärke der menschlichen isoagglutinine und entsprechenden blutkörperchenrezeptoren in verschiedenen lebensaltern. *Z. Immunitätsforsch.*, 63:67–93.
98. Tigelaar, R. E., Gershon, R. K., and Asofsky, R. (1975): Graft-versus-host reactivity of mouse thymocytes: Effect in *in vitro* treatment with anti-TL serum. *Cell. Immunol.*, 19:58–64.
99. Tyan, M. L. (1977): Age-related decrease in mouse T-cell progenitors. *J. Immunol.*, 118:846–851.
100. Volkman, A., and Gowans, J. L. (1965): The origin of macrophages from bone marrow in the rat. *Br. J. Exp. Pathol.*, 46:62–70.
101. Walford, R. L. (1976): When is a mouse "old"? *J. Immunol.*, 117:352–353.
102. Walford, R. L. (1976): Human B cell alloantigenic systems: Their medical and biological significance. In: *Proceeding of the International Congress of HLA System.* Elsevier, North Holland *(in press).*
103. Walters, C. S., and Claman, H. N. (1975): Age related changes in cell mediated immunity of Balb/c mouse. *J. Immunol.*, 115:1438–1443.

Physiology and Cell Biology of Aging
(Aging, Volume 8), edited by A. Cherkin, et al.
Raven Press, New York © 1979.

Thymosin: The Endocrine Thymus and Its Role in the Aging Process

Allan L. Goldstein,* Gary B. Thurman,* Teresa L. K. Low,* Glenn E. Trivers,* and Jeffrey L. Rossio**

Department of Biochemistry, George Washington University Medical Center, Washington, D.C. 20037

The thymus gland controls, in part, the development and function of the immune system by secreting a family of polypeptide hormone factors which we have named thymosin (8,10,13). Thymosin, and perhaps other factors such as thymic humoral factor (THF) (30), serum thymic factors like factor of thymic serum (FTS) (3) and serum factor (SF) (2), and thymopoietin (27), regulate the formation of subpopulations of T cells and may play a major role in maintenance of immune balance. Recently, thymosin has proved to have significant clinical value in treatment of children with a variety of primary immunodeficiency diseases (11,31–33). Furthermore, in phase I (6,22–25) and phase II (4,5) trials, thymosin has appeared promising in the treatment of cancer when given in combination with intensive chemotherapy or radiotherapy.

In this paper, we will review the progress we are making in the chemical characterization of the polypeptides we have isolated from thymosin fraction 5, and discuss the potential application of thymosin in dealing with the diseases of aging and the senescence of the immune response.

SENESCENCE OF THYMIC-DEPENDENT IMMUNITY WITH AGE

The thymus gland is vital to the normal development and function of the immune system and is involved directly in the aging process. It has been known since the time of Galen (7) that the thymus is the first gland in the body to begin to atrophy (Fig. 1). The two most critical periods of life with regard to susceptibility to disease occur shortly after birth and after the sixth decade of life (Fig. 1). After puberty, both cellular (1) and humoral (18,19) immunity gradually deteriorate with age. There is increasing support for the concept that the deterioration of thymic-dependent, cell-mediated immunity is causally related

* Visiting scientist from the National Cancer Institute, University of Texas Medical Branch, Galveston, Texas.
** Current address: Department of Microbiology & Immunology, Wright State University College of Medicine, Dayton, Ohio

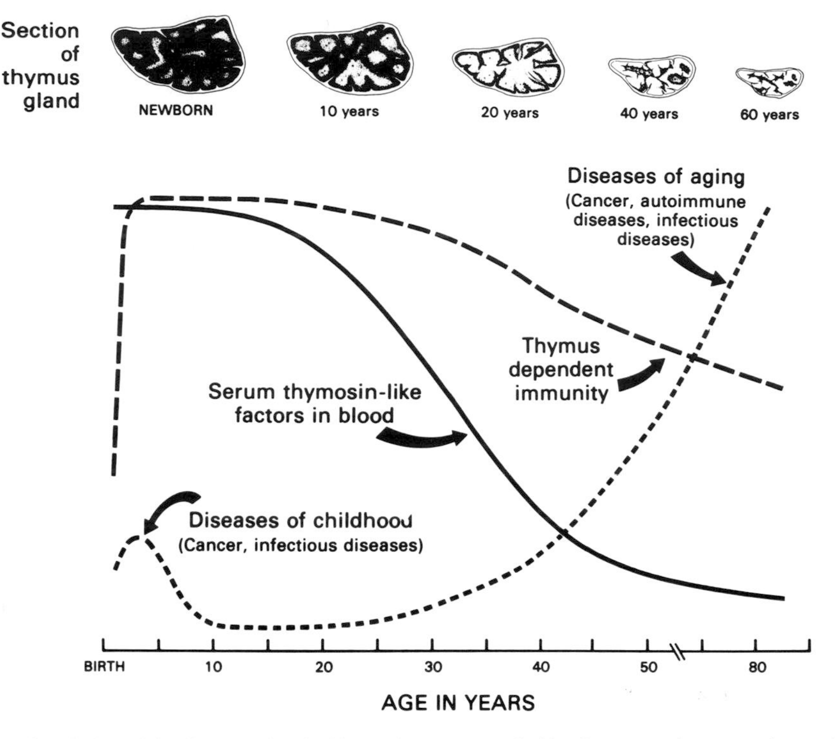

FIG. 1. Involution of the thymus gland with age is accompanied by decreases in serum thymosin-like activity and thymus-dependent immunity. As these decline, diseases of aging increase.

to the observations that the incidence of malignancies (29) and autoimmune disorders (17) increases with age.

The involution of the thymus gland begins in man and most other mammals just prior to puberty and primarily involves a depletion of the lymphoid elements of the cortex and medulla (12). As involution progresses, thymic lymphoid cells are gradually replaced by adipose tissue; however, the medullary epithelial cells (the hormone-secreting cells) remain (20). This explains why serum thymosin and thymosin-like activity remain demonstrable by both bioassay (2,16) and radioimmunoassay (28) with advancing age, albeit at markedly lower levels. It had not been established whether thymic involution is itself a consequence of aging or is an etiologic factor in the process. Studies are in progress to determine if the decrease in the endocrine role of the thymus with age results in an immune imbalance. Our ongoing clinical studies have demonstrated that the failure of the thymus gland to function properly is a major factor in a number of debilitating and often fatal diseases in children and adults. This immune imbalance, as illustrated in Fig. 2, is associated with a significant increase in the incidence of autoimmune diseases, cancer, and a number of infectious diseases.

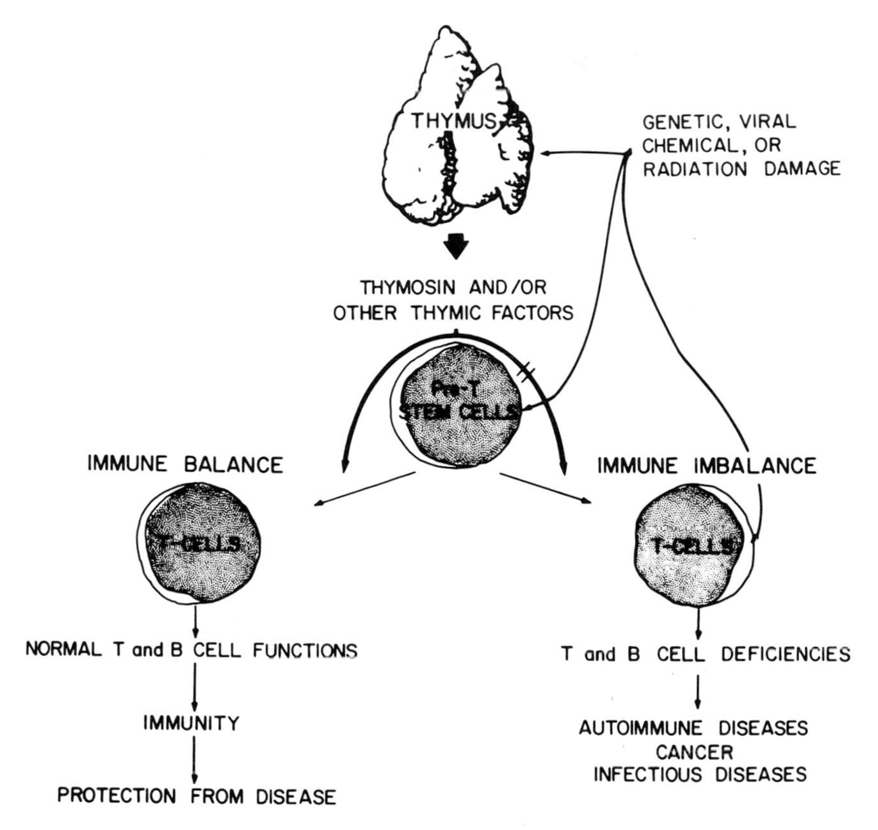

FIG. 2. Failure of the thymus gland to produce thymosin or other thymic factors can result in defects in immune balance, which may lead to the appearance of a number of disease states.

We know specifically that thymosin or thymosin-like activities in the blood are very high in young people, and that between the ages of 25 and 45 years, the levels decrease very significantly in the blood and continue to fall with age (9,28). In light of these observations, it would now be of interest to try to determine whether the very early fall in thymosin-like factors in the blood is related to the decline of immunity and susceptibility to disease that occur later in life. We think there is a relationship, but how strong this association might be has yet to be established experimentally.

CHEMISTRY OF THYMOSIN POLYPEPTIDES

Figure 3 shows an isoelectric focusing gel of the partially purified thymosin fraction 5 that has been used in the clinical trials. Thymosin fraction 5 consists of a family of small heat-stable acidic polypeptides with molecular weights ranging from 1,000 to 15,000 (10). Ongoing studies suggest that the biological potency

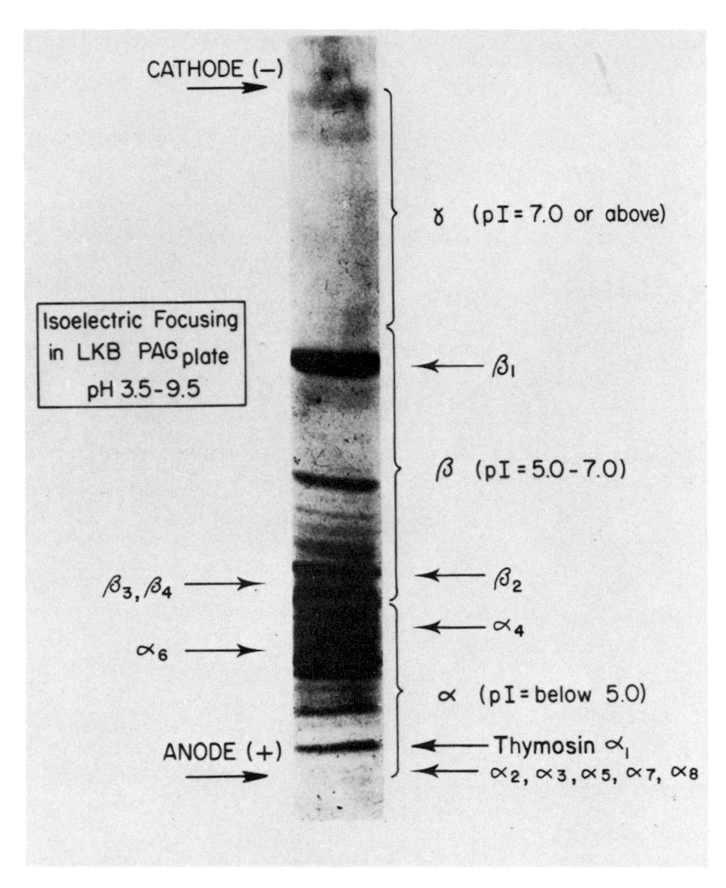

FIG. 3. Suggested nomenclature for thymosin polypeptides. Isoelectric focusing gel of thymosin fraction 5. The α (most acidic), β (acidic to neutral) and γ (basic) regions are indicated. α_1 to α_8 and β_1 to β_4 are individual polypeptides which have been isolated from thymosin fraction 5 (see text).

of fraction 5 may be due to one or more of these polypeptides acting either in concert or individually on various subpopulations of T cells. We have suggested a nomenclature for each of these polypeptides based upon their isoelectric points, as indicated in Fig. 3 (10). To date we have isolated eight polypeptides from the α region and four from the β region. One of the biologically active polypeptides in fraction 5 termed thymosin α_1 has now been sequenced (10). The sequence of thymosin α_1 is shown in Fig. 4. It consists of 28 amino acids and has a molecular weight of 3,108. Thymosin α_1 is 10 to 1,000 times as active as thymosin fraction 5 in inducing the activation and function of selected subpopulations of T cells. The observation that it does not work in all our bioassay systems gave us the first indication that there might be more than one biologically active molecule present in thymosin fraction 5. Of particular interest in addition

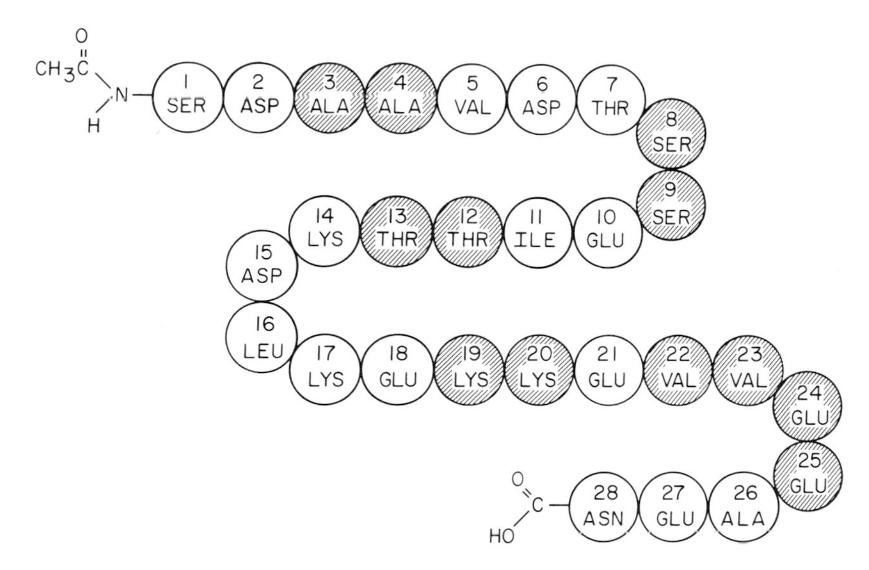

FIG. 4. The complete amino acid sequence of thymosin α_1.

to thymosin α_1 is thymosin α_7, which induces suppressor function (S. Horowitz and A. Ahmed, *personal communication*) and thymosin fraction 5 and thymosin β_3 and β_4, which specifically induce the expression of terminal deoxynucleotidyl transferase (TdT) (21; N. Pazmino and J. Ihle, *personal communication*) in immature lymphoid cell populations.

CLINICAL TRIALS WITH THYMOSIN

Primary Immunodeficiency Diseases

Over 50 children have received thymosin for a variety of primary immunodeficiency diseases (11,31–33). These patients have been treated with injections of thymosin up to 400 mg/M^2 for periods of over 48 months (usually daily for 2 to 4 weeks, then once per week). Most of the patients have received 50 mg/M^2 thymosin by subcutaneous injection. To date this group has shown no evidence of central nervous system (CNS), liver, kidney, or bone marrow toxicity due to thymosin administration. More than 80% of the pediatric patients who have responded *in vitro* in the E-rosette assay have also responded *in vivo.* We have seen a significant increase in T-lymphocyte number and function in over 50% of the patients studied, as well as significant clinical improvement in some cases. Six of 7 children with severe combined immunodeficiency disease who did not respond *in vitro* to thymosin have similarly not responded *in vivo.* Results in responding patients include such objective clinical improvements as decreased incidence of infection and weight gain.

Phase I and Phase II Cancer Trials

More than 150 cancer patients have been treated according to phase I or phase II protocols. The route of administration has been either intramuscular or by subcutaneous injection. Following an initial course of thymosin, some of the patients in the phase I study have been placed on maintenance therapy which, depending upon the protocol used, generally consists of one injection per week at approximately 50 mg/M^2. Cancer patients have been treated for periods of up to 40 months. As with pediatric patients, no major side effects have been seen in the majority of patients. About 20% of the cancer patients have experienced mild local skin reactions, and about 10% have experienced moderate skin reactions or systemic reactions, including urticarial rashes and low-grade fevers. None of the patients has shown any evidence of liver, CNS, kidney, or bone marrow toxicity due to the administration of thymosin. More than 50% of the cancer patients in the phase I trial have shown a significant increase in percentage and absolute number of E-rosettes.

The first efficacy trial of thymosin has now been completed in nonresectable small cell carcinoma of the lungs by Chretien and his associates at the National Cancer Institute (4,5). In this trial, thymosin fraction 5, when given in conjunction with intensive chemotherapy, was found to significantly prolong the survival of cancer patients. Mean survival time was increased from 225 days with chemotherapy alone to 450 days with chemotherapy plus 60 mg/M^2 thymosin twice per week for the first six weeks of the chemotherapy induction period.

Thymosin and Autoimmune Disease

To date, 5 patients with autoimmune diseases have been treated with thymosin fraction 5 for periods ranging from 4 to 16 months (15). Four of the patients had systemic lupus erythematosus (SLE), and the fifth had rheumatoid arthritis and Sjogren's syndrome. In this preliminary open study, there have been no ill effects related to the administration of thymosin, and in 3 patients there has been objective immunological and clinical improvement. Based upon these encouraging findings, a phase II randomized trial is planned to determine the efficacy of thymosin in SLE.

Although the mechanism of thymosin's reconstitutive action is not as yet defined, it may be related to the restoration of a subpopulation of thymosin-activated T cells termed suppressor or regulator cells (14,26). Results in NZB mice suggest that an endocrine disturbance may underlie many immunologic abnormalities associated with autoimmune disease. *In vitro* human studies by Scheinberg et al. (26) and Horowitz et al. (14) have confirmed that in the blood of many patients with SLE, there is a decrease in the absolute number of T cells and suppressor cells, a situation that is correctable by the addition of thymosin.

CONCLUSIONS

Phase I trials and the first phase II trial with thymosin have been successfully completed. No evidence of toxicity has appeared, and we have observed encouraging immunologic changes in cancer patients and in children with primary immunodeficiency diseases. Based upon these studies, additional Phase II trials are under way to determine the efficacy of thymosin in patients with primary immunodeficiency diseases, cancer, and autoimmune diseases. The range of thymosin activity in enhancing immune reconstitution, as illustrated in Figure 5, appears to be broader than immunoenhancing agents such as bacillus Calmette-Guerin (BCG). Most immunotherapeutic agents are effective only if the patient has

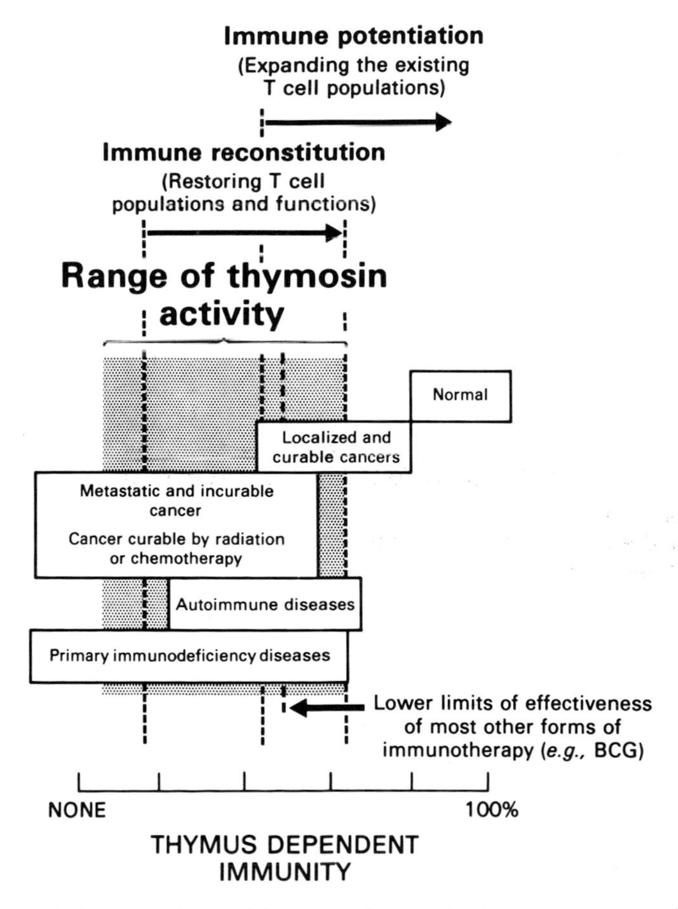

FIG. 5. A conceptual representation of the use of thymosin in the reconstitution and potentiation of immune responses. Thymosin may be useful in reconstituting disease states associated with severely depressed immune responses.

functional T cells. Thymosin appears to require only the presence of stem cells, and thus has a broader range of activity.

To date, the studies suggest that thymosin may be helpful in restoring immune balance by augmenting specific lymphocyte functions in children with hypothymic function. In immunosuppressed cancer patients, thymosin may increase T cell function, induce endogenous antitumor activity, and increase host resistance to pathogens, thus reducing the high incidence of infection that often accompanies treatment of cancer with conventional modalities. Of particular interest is the observation of Chretien and his colleagues that in small cell carcinoma of the lung, thymosin is very effective in increasing survival in severely immunosuppressed patients (4,5). The capacity of thymosin to increase *in vitro* the number of T cells (26) and suppressor cells (14) in patients with active SLE points to a potential role for thymosin in these disorders as well. The marked clinical improvement observed following thymosin treatment of primary immunodeficiency diseases in children, including resolution of chronic fungal or viral diseases (11,31–33), points to a potential role for thymosin in the treatment of other infectious diseases not associated with primary thymic abnormalities, and in particular, treatment of many of the diseases of aging which are associated with decreased thymic function.

In summary, our studies support the hypothesis that many diseases of the aged may result from an inability of the thymus gland to produce normal amounts of thymosin, thus resulting in decreased resistance to infection.

The results of our studies raise the following questions, which may now be approached and answered because of the availability of several of the purified thymosin polypeptides.

1. Does the maintenance of "normal" thymosin levels in the blood retard the senescence of the T cell system and thus prevent, or delay, the onset of specific age-related diseases?
2. How does thymosin control the ontogenesis of the T cell and the maintenance of immune function at the cellular and molecular levels?
3. What is the relationship between hyper- and hypothymosin secretion and immunodeficiency diseases?
4. What stimulates thymosin release into the circulation, and is this process subject to feedback controls?
5. Can determination of thymosin levels by radioimmunoassay be used as an early diagnostic test to predict loss of normal thymic function?

Our studies indicate that the thymus affects the maturation of T cells by an endocrine mechanism. Therefore, the contention that the sole mechanism of T cell maturation requires a sojourn through the thymus gland may no longer be tenable. We propose that the thymus gland secretes several thymic hormones that can act at sites distant from the thymus to influence the ontogenesis, function, and perhaps the senescence of lymphoid cells involved in cell-mediated immunity.

ACKNOWLEDGMENTS

This study was supported in part by grants from the National Cancer Institute (CA14108 and CA16964), the John Hartford Foundation, Inc., and Hoffmann-LaRoche, Inc., Nutley, New Jersey.

REFERENCES

1. Albright, J. F., and Makinodan, T. (1966): Growth and senescence of antibody-forming cells. *J. Cell. Comp. Physiol.* (Suppl.), 67:185.
2. Astaldi, A., Astaldi, G. C. B., Wijermans, P., Groenewoud, M., Schellekens, P. Th. A., and Eijsvoogel, V. P. (1977): Thymosin-induced human serum factor increasing cyclic AMP. *J. Immunol.,* 119:1106.
3. Bach, J. F., and Dardenne, M. (1963): Studies of thymus products. II. Demonstration and characterization of a circulating thymic hormone. *Immunology,* 25:353.
4. Chretien, P. (1978): Thymosin as adjuvant in treatment of human carcinomas. In: *Proceedings of 4th Conference on Immune Modulation and Control of Neoplasia by Adjuvant Therapy,* ed. M. A. Chirigos, *(in press).*
5. Cohen, M. H., Chretien, P. B., Ihle, D. C., Fassieck, B. E., Jr., Bunn, P. A., Kennedy, D. E., Lipson, S. D., and Minna, J. D. (1978): *Cancer Res.,* 19:117.
6. Costanzi, J. J., Gagliano, R. G., Loukas, D., Delaney, F., Sakai, H., Harris, N. S., Thurman, G. B., and Goldstein, A. L. (1977): The effect of thymosin on patients with disseminated malignancies: A Phase I study. *Cancer,* 40:14–19.
7. Duckworth, W. L. (1962): In: *Galen on Anatomical Procedures,* edited by M. C. Lyans and B. Towers, p. 160. Cambridge University Press, London and New York.
8. Goldstein, A. L., Slater, F. D., and White, A. (1966): Effects of the thymus lymphocytopoietic factor. *Ann. N. Y. Acad. Sci.,* 135:485–495.
9. Goldstein, A. L., Hooper, J. A., Schulof, R. S., Cohen, G. H., Thurman, G. B., McDaniel, M. C., White, A., and Dardenne, M. (1974): Thymosin and the immunopathology of aging. *Fed. Proc.,* 33(9):2053–2056.
10. Goldstein, A. L., Low, T. L. K., McAdoo, M., McClure, J., Thurman, G. B., Rossio, J. L., Lai, C-Y., Chang, D., Wang, S-S., Harvey, C., Ramel, A. H., and Meienhofer, J. (1977): Thymosin α_1: Isolation and sequence analysis of an immunologically active thymic polypeptide. *Proc. Natl. Acad. Sci.,* 24:725–729.
11. Goldstein, A. L., Wara, D. W., Ammann, A. J., Sakai, H., Harris, N. S., Thurman, G. B., Hooper, J. A., Cohen, G. H., Goldman, A. S., Constanzi, J. J., and McDaniel, M. C. (1975): First clinical trial with thymosin: Reconstitution of T cells in patients with cellular immunodeficiency diseases. *Transplant. Proc.,* 7(1):681–686.
12. Hammer, J. A. (1966): In: *Die Normal—Morphologische Thymus Forschung in Letzten Vierteijahrlundert.* Barton, Leipzig.
13. Hooper, J. A., McDaniel, M. C., Thurman, G. B., Cohen, G. H., Schulof, R. S., and Goldstein, A. L. (1975): The purification and properties of bovine thymosin. *Ann. N. Y. Acad. Sci.,* 249:125–144.
14. Horowitz, S., Borcherding, W., Moorthy, A. V., Chesney, R., Schulte-Wisserman, H., Hong, R., and Goldstein, A. L. (1977): Induction of suppressor T cells in systemic lupus erythematosus by thymosin and cultured thymic epithelium. *Science,* 197(4307):999–1001.
15. Lavastida, M. T., Goldstein, A. L., Rossio, J. L., and Daniels, J. C. (1978): Thymosin treatment in autoimmune disorders. *(Submitted for publication.)*
16. LeBrand, H. (1972): Evidence for a serum-factor secreted by the human thymus. *Lancet,* 2:1056.
17. MacKay, I. R., Masel, M., and Burnett, F. M. (1964): Thymic abnormality in systemic lupus erythematosus. *Anst. Ann. Med.,* 13:5.
18. Makinodan, T. (1966): Secondary antibody-forming potential of mice in relation to age—its significance in senescence. *Dev. Biol.,* 14:96.
19. Makinodan, T., and Peterson, W. J., (1964): Growth and senescence of the primary antibody-forming potential of the spleen. *J. Immunol.,* 93:886.
20. Metcalf, D. (1966): The thymus, its role in immune responses, leukemia, development and

carcinogenesis. In: *Recent Results in Cancer Research,* edited by P. Rentchnick. Springer-Verlag, Berlin.
21. Pazmino, N. H., Ihle, J. N., and Goldstein, A. L. (1978): Induction *in vivo* and *in vitro* of terminal deoxynucleotidyl transferase by thymosin in bone marrow cells from athymic mice. *J. Exp. Med.,* 147:708–718.
22. Rossio, J. L., and Goldstein, A. L. (1977): Immunotherapy of cancer with thymosin. *World J. Surg.,* 1:605–616.
23. Schafer, L. A., Goldstein, A. L., Gutterman, J. U., and Hersh, E. M. (1976): *In vitro* and *in vivo* studies with thymosin in cancer patients. *Ann. N.Y. Acad. Sci.,* 227:609–620.
24. Schafer, L. A., Gutterman, J. U., Hersh, E. M., Mavligit, G. M., Dandridge, K., Cohen, G., and Goldstein, A. L. (1976): Partial restoration by *in vivo* thymosin of E-rosettes and delayed-type-hypersensitivity reactions in immunodeficient cancer patients. *Cancer Immunol. Immunother.,* 1:259–264.
25. Schafer, L. A., Gutterman, J. U., Hersh, E. M., Mavligit, G. M., and Goldstein, A. L. (1977): *In vitro* and *in vivo* studies of thymosin activity in cancer patients. In: *Progress in Cancer Research and Therapy,* Vol. 2, *Control of Neoplasia by Modulation of the Immune System,* edited by M. A. Chirigos, pp. 329–346. Raven Press, New York.
26. Scheinberg, M. A., Cathcart, E. S., and Goldstein, A. L. (1975): Thymosin-induced reduction of "null cells" in peripheral-blood lymphocytes of patients with systemic lupus erythematosus. *Lancet,* i:424–429.
27. Schlesinger, D. H., and Goldstein, G. (1975): The amino acid sequence of thymopoietin II. *Cell,* 5:361–365.
28. Schulof, R. S. Ph.D. Thesis: Radioimmunological Studies of Thymosin.
29. Teller, M. N., Stohr, G., Cartlett, W., Kubisek, M. L., and Curtis, D. (1969): Aging and cancerigenesis. I. Immunity to tumor and skin grafts. *J. Natl. Cancer Inst.,* 33:649.
30. Trainin, N. (1974): Thymic hormones and immune response. *Physiol. Rev.,* 54:272.
31. Wara, D. W., and Ammann, A. J. (1978): Thymosin treatment of children with primary immunodeficiency disease. *Transplant. Proc. (in press).*
32. Wara, D. W., Goldstein, A. L., Doyle, N., and Ammann, A. J. (1975): Thymosin activity in patients with cellular immunodeficiency. *N. Engl. J. Med.,* 292:70–74.
33. Wara, D. W., Johnson, A. C., and Ammann, A. J. (1978): In: *Proceedings 3rd Conference on Modulation of Heart Immune Resistance in the Presentation and Treatment of Neoplasia,* ed. M. A. Chirigos *(in press).*

Physiology and Cell Biology of Aging
(Aging, Volume 8), edited by A. Cherkin, et al.
Raven Press, New York © 1979.

Prevention and Restoration of Age-Associated Impaired Normal Immune Functions

Takashi Makinodan

Geriatric Research, Education, and Clinical Center, V.A. Wadsworth Hospital Center, Los Angeles, California 90073; and Department of Medicine, University of California at Los Angeles, Los Angeles, California, 90024

As emphasized throughout this volume, certain normal immune functions decline with advancing age, and associated with the decline is an increase in susceptibility to certain types of infection, autoimmune and immune complex diseases, and cancer (27,29). Because these two inversely related, age-associated events appear to be causally related (27) and because such diseases and infections contribute to morbidity and mortality of the elderly, there has been a growing interest in methods of intervention.

I will attempt to review methods that appear most promising, either preventively or restoratively, in controlling immunologic abnormalities associated with aging. Emphasis will be given to methods that appear likely to succeed in humans and to those that could serve as probes to understand the biochemical mechanism(s) responsible for the decline in normal immune functions with age. To date, there have been four preventive and three restorative types of methods attempted with some success, and one of these methods has been used for both purposes. The preventive methods are (a) internal body temperature control, (b) tissue ablation, (c) genetic manipulation, and (d) dietary manipulation. The restorative methods are (a) tissue ablation, (b) cell grafting, and (c) chemical therapy.

PREVENTIVE METHODS

Internal Body Temperature Control

Walford and Liu demonstrated that the life-span of an annual fresh-water fish can be extended significantly by subjecting it to mild hypothermia during the last half of its life (23,41). The significance of the effect of hypothermia on the relationship between immunity and disease-pattern with respect to life-span must await further studies. Regardless, this method is not practical in humans at this time, for only a few individuals are proficient in lowering their internal body temperature through yoga and related practices.

Tissue Ablation

One approach in controlling a disease that can disrupt normal immune functions and shorten the life-span is to remove either the tissue from which a disease originates or the one that promotes the disease, provided that the tissue is not essential. This approach was first employed most successfully by Furth over 30 years ago (16). Using a short-lived (mean life-span, 38 weeks), thymoma-susceptible strain of mice called AKR, he thymectomized them at young adulthood. As a consequence, the incidence of leukemia decreased drastically, and the mice lived significantly longer. This approach was also used successfully with long-lived hybrid mice (mean life-span, 130 weeks) (3). Preliminary studies indicated that at old age many of these mice were dying with reticulum cell sarcoma originating in the spleen at about 100 weeks of age, an observation which has since been confirmed and extended (8). These mice, therefore, were either splenectomized or sham-splenectomized at about 100 weeks of age. The results revealed that the remaining survival time of splenectomized mice can be nearly doubled over that of sham-splenectomized mice.

Although the results of these experiments are impressive, tissue ablation is not a practical method in humans. The method, however, could be used in experimental animals in resolving the type of diseases that emerge late in life and that can compromise general and specific immune functions.

Genetic Manipulation

Genetic manipulation has been performed in long-lived and short-lived strains of mice. In studies with long-lived mice, it was reasoned that aging is influenced genetically by only a limited number of regulatory genes (9), that the immune system plays a major role in aging (40), and that the major histocompatibility complex (MHC) system represents such a "super-regulatory gene complex system" of the immune system (4), i.e., the H-2 region of chromosome 17 in the mouse and the HLA region of chromosome 6 in the human. Mice differing only at the H-2 region in several inbred strains were therefore assessed for their age-related immune functions, age-specific diseases, and life-spans (29,36). These mice, which are genetically identical, except for those in the MHC, are commonly referred to as congenic mice within a genetic strain. The results revealed that variation in these parameters between these congenic mice within a given inbred strain was as great as that observed between different inbred strains of mice. If the MHC system in the mouse did not exert a significant influence upon aging, age-related immunologic abnormalities, and life-span, one would have expected a greater uniformity in life-span, immunologically associated disease pattern, and immunologic activities between congenic sets of mice within an inbred strain than between inbred strains. It can be argued, therefore, that the MHC system plays a major role in age-associated immunologic abnormalities and life-span in mice.

With short-lived mice, the focus has been on susceptibility to autoimmune and immune complex manifestations (12,37). The studies revealed that susceptibility to autoimmune and immune complex diseases involves more than one gene. An example is reflected in the life span of different inbred strains of mice and their hybrids. The mean life span of autoimmune-susceptible NZB, NZW, (NZB × NZW)F_1, (NZB × CBA)F_1, and nonautoimmune-susceptible CBA mice were 12, 21, 12, 29, and 28 months.

Obviously genetic manipulation is not a practical method in humans. However, it can be used to dissociate the relative role genetic and environmental factors play in aging, from which practical methods could arise.

Dietary Manipulation

In the 1930s, McCay, et al. (28) first discovered that the life-span of rats can be extended significantly by restricting their caloric intake during growth. Recently a more exhaustive study was carried out, also in rats, by Ross and Bras (32,33), which confirmed and extended the classical work. Thus, they found that early caloric restriction decelerates the aging rate, as judged by age-related biochemical and pathological changes. Walford, et al. (42) then showed that the immune system of long-lived mice, subjected to the life-extending, calorically restricted but nutritionally supplemented diet, matures more slowly and begins to age later in life. Fernandes, et al. (14,15) showed that a diet high in fat and relatively low in protein, which favors reproduction in experimental rodents, significantly decreased cell-mediated autoimmune manifestations and shortened the life expectancy of short-lived, autoimmune-susceptible mice. In contrast, a diet low in fat and relatively high in protein, which is less favorable for reproduction, decreased autoimmune manifestations and prolonged the life expectancy of these mice. They further showed that the life-span of these short-lived, autoimmune-susceptible mice can be dramatically extended by restricting their caloric intake (13).

These results are encouraging. They suggest that through appropriate dietary manipulation (a) the aging rate of mice with long life-spans can be retarded by preserving their immunologic vigor, (b) the life-span of short-lived mice can be extended significantly by delaying the expression of life-shortening autoimmune and immune complex diseases, and (c) the developing T cell arm of the immune system and the neuroendocrine system may be the targets of the life-prolonging dietary regimen.

RESTORATIVE METHODS

Tissue Ablation

The tissue ablation approach, as mentioned earlier, has been used as a preventive method. Recently, however, it has been used effectively by Bilder and Denckla in an attempt to restore normal immune functions (5). In their study, rats were used to test the hypothesis by Denckla that hypothyroidism in old individuals could be caused by a substance(s), secreted by the pituitary, that competes with thyroid hormones for the same receptors of target cells, including those of the immune system, and that secretion of the substance begins shortly after sexual maturity (10). Bilder and Denckla reasoned that if the hypothesis were correct, hypophysectomy of aging individuals supplemented with the standard hormones should have a beneficial effect. Accordingly, they hypophysectomized old rats and, for control, young rats. The results revealed that hypophysectomy had a pronounced immunorestorative effect on old, but not young rats.

This method is not practical in humans. However, hypophysectomized old rats could be used to resolve the problem of the relationship between the immune system's "aging clock" and its extrinsic "pacemaker." Moreover, if the pituitary substance can be purified, a specific antiserum reagent could be prepared and used clinically. Thus, it would appear that this type of tissue ablation study has both basic and clinical importance.

Cell Grafting

Grafting of thymus, spleen, lymph nodes, and bone marrow has been attempted individually or in combinations in genetically compatible old recipients with varying success in terms of immunologic restoration and extension of life span.

Fabris et al. demonstrated most impressively that the life-spans of short-lived, growth-hormone–deficient, hypopituitary dwarf mice can be extended as much as 3- to 4-fold by injecting large doses of lymph node cells (11). A comparable life-prolonging effect was obtained by injecting growth hormone and thyroxin into these dwarf mice with intact thymus, but not into dwarf mice whose thymuses had been removed beforehand. These results indicate that the pituitary "turns on" the immune system through the thymus, and the T cells turn on the endocrine system through the pituitary. Injection of young spleen or thymus cells into old, autoimmune-susceptible, short-lived mice has had less spectacular results. Thus, it did delay but did not permanently prevent the appearance of certain types of autoantibodies (38), and it had a minimal life-prolonging effect (43). Similarly, multiple thymus grafts did extend their life expectancy, but only by one month (21). Furthermore, the age-associated pathological changes were unaltered by these treatments (44).

The first attempts to graft young thymus or bone marrow cells into old nonautoimmune-susceptible, long-lived mice were also not too encouraging, as the life-span was not extended appreciably (1,30). Subsequent studies on the mechanism of decline with age in normal immune functions indicated why these earlier cell grafting attempts were not successful. These studies revealed that the loss of immunologic vigor is due in part to changes in the T cell population (25), in part to the reduced rate at which stem cells can self-generate and generate progeny cells (2), and in part to the inability of involuted thymus to transform precursor cells to T cells efficiently (19). Therefore, we grafted both newborn thymus and young adult bone marrow stem cells into long-lived old mice, and succeeded in restoring their immune functions to levels approaching those of

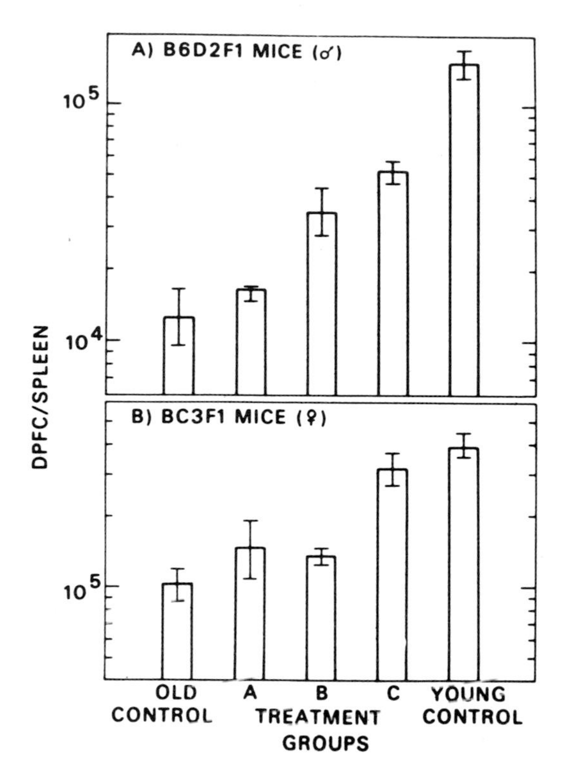

FIG. 1. Effect of bone marrow and thymus grafts on the antisheep RBC response of aging hybrid mice. B6D2F_1 mice were assessed 4.5 months after grafting when they were 19.5 months old, and BC3F_1 mice were assessed 3 months after grafting when they were 29 months old. Treatment groups: **A,** 850 R + 5 × 10^6 bone marrow cells from 6–8 week old syngeneic donors; **B,** newborn thymus graft; **C,** 850 R + 5 × 10^6 bone marrow cells + newborn thymus graft; young control mice 3–4 months old; vertical bars, 1 standard error of the mean. From Hirokawa, et al. (18).

adult mice (Fig. 1) (18). The restorative effect was observed for at least six months after graft treatment in mice with a normal mean life-span of 27 months (an equivalent to 0.22 of a mean mouse life-span, or about 15 human years). Current immunorestorative studies by cell grafting should resolve what effect, if any, grafting will have on the frequency and severity of diseases of the aged, and if cell grafting can alter the life expectancy of short-lived and long-lived mice.

Studies by Perkins, et al. (31) on susceptibility to infection have also generated encouraging data. These showed that old mice can be made to resist lethal doses of virulent *Salmonella typhimurium* by prior injection with spleen cells from young genetically compatible mice which had been immunized with the vaccine. The findings also indicated that sensitized spleen cells can persist in the recipients for a long time after injection and that spleen cells can be stored cryogenically for an extended period of time without loss of immunologic activity. This means that the method of Perkins, et al. (31) has a practical application in humans and in other species with noninbred individuals. The reason is that graft rejection due to incompatibility between donor and recipient, the major problem in transplantation involving noninbred individuals, can be bypassed. Thus, during adulthood, when individuals are healthy, they can be systematically immunized against an array of microbial agents to which elderly individuals are highly susceptible. Blood can then be drawn at appropriate intervals, and their immune cells (stem cells, T cells, B cells, and monocytes) can be fractionated and stored in ampules at liquid nitrogen temperature (−190°). Later in life, as individuals approach their seventies, eighties, and nineties, they can reclaim their own immune cells which had been primed to combat various infective agents.

Chemical Therapy

Only a few chemical agents have been shown to possess immunorestorative activity. These include thymosin (a thymic immunorestorative hormone), double stranded polynucleotide, and mercaptoethanol.

As reported by Goldstein *(this volume),* repeated injection of thymosin can alleviate many of the symptoms in mice and humans manifesting immunodeficiency diseases. The exciting possibility exists that repeated injection of thymosin may also have immunorestorative effect on elderly individuals manifesting reduced T cell-dependent immune functions. The basis for this optimism comes from a recent preliminary report showing that the number of T cells of old individuals can be increased by exposing their white blood cells to thymosin *in vitro* (34), a predictive index with a high probability of clinical success (35). However, before assessing the effect of thymosin on elderly individuals, it would seem prudent to carry out animal studies, i.e., to assess the effects of repeated injection into aging short-lived and long-lived mice on their normal immune functions, disease patterns, and life-spans.

Braun et al. (7) were the first to demonstrate that double-stranded polynucleotides (e.g., polyadenlic-polyuridylic acid complexes) can restore the T cell-dependent antibody response of middle-aged long-lived mice to that of young adult mice. Han and Johnson (17) not only confirmed this observation but proceeded to demonstrate that the supernatant of cultures of thymocytes treated with the polynucleotides is equally effective as an immunorestorative agent. This would suggest that by acting on the T cells, the polynucleotide restores the immunologic vigor of aging mice. Further studies are required to determine its mechanism of action on T cells and on the disease pattern and life-span.

By exposing the cells to mercaptoethanol, we have found that the reduced *in vitro* T cell proliferation and T cell-dependent antibody forming activities of old long-lived mice can be restored to levels approaching those of young adult mice (Table 1) (24,26). Because mercaptoethanol can influence mitotic and other cellular functions by affecting various activities including cyclic nucleotide and prostaglandin metabolisms (6,20,22), it would appear to be a very

TABLE 1. *Effect of 2-mercaptoethanol* (5×10^{-5}M) *on* in vitro *anti-sheep RBC by* 10^7 *spleen cells of young (3–4 months) and old (24–26 months) BC3F1 mice*[a]

	Sample no.	(A)/(B)
I. Young BC3F1	1	1.0
	2	1.3
	3	1.1
	4	1.3
	5	1.2
	6	1.5
	X ± SEM	1.2 ± 0.07
II. Old BC3F1	1	2.1
	2	1.6
	3	1.5
	4	2.8
	5	6.3
	6	16.1
	7	17.9
	X ± SEM	6.9 ± 2.69

[a]Antibody response measured in terms of number of direct plaque-forming cells (DPFC) per culture; Ag, antigen; X ± SEM, mean ± standard error of the mean; A, number of DPFC in Ag + 2-ME culture; B, number of DPFC in Ag culture. From Makinodan, et al. (26).

promising chemical probe in understanding the nature of the cellular defect at the molecular level. Appropriate studies are therefore in progress in our laboratory to determine its mechanism of action(s).

The possibility exists that mercaptoethanol is an effective immunorestorative agent *in vitro* only, as are many chemical ingredients in tissue culture media which promote cells to grow better in artificial media. If so, it may not have a practical application. We have therefore been assessing the effectiveness in intact old mice of various genetic strains and hybrids of strains. Our preliminary results are encouraging, because with appropriate doses of mercaptoethanol, we have succeeded in restoring the T cell-dependent antibody responding capacity of long-lived old mice to that of young mice (24,26). These results would suggest that mercaptoethanol and related chemicals may have practical applications. Current studies are therefore focused on its long-term effect on immune functions, disease patterns, and life-span of aging short-lived and long-lived mice.

CONCLUSION

Internal body temperature, tissue ablation, dietary manipulation, genetic manipulation, cell grafting, and chemical therapy have been employed to prevent and restore the decline in normal immune functions with age and to extend the life expectancy. Many of these studies are very preliminary, but overall, the findings are most encouraging.

Of the six model approaches, genetic manipulations and chemical therapy appear to be most promising in serving as probes to understand the biochemical nature and mechanism(s) of the decline. Genetic manipulations should enable one to determine which gene(s) are primarily responsible for the decline in normal immune functions with age and the diseases associated with it. Chemical therapy should enable one to determine the cell types(s) (stem cells, T cells, B cells, macrophages) most severely affected functionally and the nature of changes associated with it at the subcellular level.

In terms of practical application, dietary manipulation appears to be the most promising in preventing the decline. Chemical therapy could also serve as an effective preventive method. Cell grafting and chemical therapy appear to be most promising in restoring immunologic vigor. I believe that the most effective approach in controlling loss of immunologic vigor with age would be a combination of these three methods (dietary manipulation, chemical therapy, and cell grafting).

In view of these considerations, current studies are focused not only on the effectiveness of these immunomanipulative approaches on immune functions of aging individuals, but also on the diseases associated with death and lifespans.

ACKNOWLEDGMENTS

This is publication number 016 from V. A. Wadsworth GRECC, supported in part by VA Merit Review #5444 and by The Intra-Science Research Foundation.

REFERENCES

1. Albright, J. F., and Makinodan, T. (1966): Growth and senescence of antibody-forming cells. *J. Cell Comp. Physiol.,* 67(Suppl. 1):185–206.
2. Albright, J. W., and Makinodan, T. (1976): Decline in the growth potential of spleen-colonizing bone marrow stem cells of long-lived aging mice. *J. Exp. Med.,* 144:1204–1213.
3. Albright, J. W., Makinodan, T., and Deitchman, J. W. (1969): Presence of life-shortening factors in spleens of aged mice of long lifespan and extension of life expectancy by splenectomy. *Exp. Gerontol.,* 4:267–276.
4. Benacerraf, B., ed. (1975): *Immunogenetics and Immunodeficiency.* University Park Press, Baltimore.
5. Bilder, G. E., and Denckla, W. D. (1977): Restoration of ability to reject xenografts and clear carbon after hypophysectomy of adult rats. *Mech. Ageing Develop.,* 6:153–163.
6. Braun, W., Lichtenstein, W. M., and Parker, C., eds. (1974): *Cyclic AMP, Cell Growth and the Immune Response,* Springer-Verlag, New York.
7. Braun, W., Yajima, Y., and Ishizuka, M. (1970): Synthetic polynucleotides as restorers of normal antibody formation capacities in aged mice. *J. Reticuloendothel. Soc.,* 7:418–424.
8. Chino, F., Makinodan, T., Lever, W. E., and Peterson, W. J. (1971): The immune systems of mice reared in clean and in dirty conventional laboratory farms. I. Life expectancy and pathology of mice with long life spans. *J. Gerontol.,* 26:497–507.
9. Cutler, R. G. (1975): Evolution of human longevity and the genetic complexity governing aging rats. *Proc. Natl. Acad. Sci. USA,* 72:4664–4668.

10. Denckla, W. D. (1974): Role of pituitary and thyroid glands in the decline of minimal O_2 consumption with age. *J. Clin. Invest.,* 53:572–581.
11. Fabris, N., Pierpaoli, W., and Sorkin, E. (1972): Lymphocytes, hormones and aging. *Nature,* 240:557–559.
12. Fernandes, G., Good, R. A., and Yunis, E. J. (1977): Attempts to correct age-related immunodeficiency and autoimmunity by cellular and dietary manipulation in inbred mice. In: *Immunology and Aging,* edited by T. Makinodan and E. Yunis, pp. 111–133. Plenum Press, New York.
13. Fernandes, G., Yunis, E. J., and Good, R. A. (1976): Influence of diet on survival of mice. *Proc. Natl. Acad. Sci. USA,* 73:1279–1283.
14. Fernandes, G., Yunis, E. J., Jose, D. G., and Good, R. A. (1973): Dietary influence on antinuclear antibodies and cell-mediated immunity in NZB mice. *Int. Arch. Allergy Appl. Immunol.,* 44:770–782.
15. Fernandes, B., Yumis, E. J., Smith, J., and Good, R. A. (1972): Dietary influence on breeding behavior, hemolytic anemia, and longevity in NZB mice. *Proc. Soc. Exp. Biol. Med.,* 139:1189–1196.
16. Furth, J. (1946): Prolongation of life with prevention of leukemia by thymectomy in mice. *J. Gerontol.,* 1:46–52.
17. Han, I. H., and Johnson, A. G. (1976): Regulation of the immune system by synthetic polynucleotides. VII. Amplification of the immune response in young and aged mice. *J. Immunol.,* 117:423–427.
18. Hirokawa, K., Albright, J. W., and Makinodan, T. (1976): Restoration of impaired immune functions in aging animals. I. Effect of syngeneic thymus and bone marrow cells. *Clin. Immunol. Immunopathol.,* 5:371–376.
19. Hirokawa, K., and Makinodan, T. (1975): Thymic involution: Effect on T cell differentiation. *J. Immunol.,* 114:1659–1664.
20. Johnson, N., Jessup, R., and Ramwell, P. W. (1974): The significance of protein disulfide and sulfphenyl groups in prostaglandin action. *Prostaglandins,* 5:125–136.
21. Kysela, S., and Steinberg, A. D. (1973): Increased survival of NZB/W mice given multiple syngeneic young thymus grafts. *Clin. Immunol. Immunopathol.,* 2:133–136.
22. Lands, W., Lee, R., and Smith, W. (1971): Factors regulating the biosynthesis of various prostaglandins. *Ann. N.Y. Acad. Sci.,* 180:107–122.
23. Liu, R. K., and Walford, R. L. (1975): Mid-life temperature-transfer effects on lifespan of annual fish, *J. Gerontol.,* 30:129–131.
24. Makinodan, T. (1978): Control of immunologic abnormalities associated with aging. *Mech. Ageing Dev. (in press.)*
25. Makinodan, T., and Adler, W. H. (1975): The effects of aging on the differentiation and proliferation potentials of cells of the immune system. *Fed. Proc.,* 34:153–158.
26. Makinodan, T., Deitchman, J. W., Stoltzner, G. H., Kay, M. M., and Hirokawa, K. (1975): Restoration of the declining normal immune functions of aging mice. *Proc. 10th Intl. Cong. Gerontol.,* 2:23.
27. Makinodan, T., and Yunis, E., eds. (1977): *Immunology and Aging.* Plenum Medical Book Company, New York.
28. McCay, C. M., Crowell, M. F., and Maynard, L. A. (1935): The effect of retarded growth upon the length of life span and upon the ultimate body size. *J. Nutr.,* 10:63–79.
29. Meredith, P. J., and Walford, R. L. (1978): Effect of age on response to T and B cell mitogens in mice congenic at the H-2 locus. *Immunogenetics (in press).*
30. Metcalf, D., Moulds, R., and Pike, B. (1966): Influence of the spleen and thymus on immune responses in aging mice. *Clin. Exp. Immunol.,* 2:109–120.
31. Perkins, E. H., Makinodan, T., and Seibert, C. (1972): Model approach to immunological rejuvenation of the aged. *Infect. Immunity,* 6:518–524.
32. Ross, M. H. (1969): Aging, nutrition, and hepatic enzyme activity patterns in the rat. *J. Nutr.,* Suppl. 1, part 2, 97:565–601.
33. Ross, M. H., and Bras, G. (1971): Lasting influence of early caloric restriction on prevalence of neoplasia in the rat. *J. Nat. Cancer Inst.,* 47:1095–1113.
34. Rovensky, J., Goldstein, A. L., Holt, P. J. L., Pwkarek, J., and Mistina, T. (1977): Obnova funkcie T lymfocytor tymosinom u klinicky zdravych asob vyssieho veku. *Cas. Lek. Ces.,* 116:1063–1065.
35. Schafer, L. A., Gutterman, J., Hersh, E. M., Mavligit, G. M., Dandridge, K., Cohen, G., and Goldstein, A. L. (1975): Partial restoration by thymosin of E-rosettes and delayed-type-

hypersensitivity reactions in immunodeficient patients. *Cancer Immunol. and Immunother. (In press.)*
36. Smith, G. S., and Walford, R. L. (1977): Influence of the main histocompatibility complex on aging in mice. *Nature,* 270:727–729.
37. Talal, N., and Steinberg, A. D. (1974): The pathogenesis of autoimmunity in New Zealand black mice. *Curr. Top. Microbiol. Immunol.,* 64:79–103.
38. Teague, P. O., and Friou, G. J. (1969): Antinuclear antibodies in mice. II. Transformation with spleen cells, inhibition or prevention with thymus or spleen cells. *Immunology,* 17:665–675.
39. Walford, R. L. (1969): *The Immunologic Theory of Aging.* Munksgaard, Copenhagen.
40. Walford, R. L. (1974): The immunologic theory of aging, current status. *Fed. Proc.,* 33:2020–2027.
41. Walford, R. L., and Liu, R. K. (1965): Husbandry, life span, and growth rate of the annual fish Cynolebias Adloffi. *Exp. Gerontol.,* 1:161–171.
42. Walford, R. L., Liu, R. K., Mathies, M., Gerbase-DeLima, M., and Smith, G. S. (1974): Response to sheep red blood cells and to mitogenic agents. *Mech. Ageing and Develop.,* 2:447–454.
43. Yunis, E. J., and Greenberg, L. J. (1974): Immunopathology of aging. *Fed. Proc.,* 33:2017–2019.
44. Yunis, E. J., Fernandes, G., and Stutman, O. (1971): Susceptibility to involution of the thymus-dependent lymphoid system and autoimmunity. *Am. J. Clin. Pathol.,* 56:280–292.

Physiology and Cell Biology of Aging
(Aging, Volume 8), edited by A. Cherkin, et al.
Raven Press, New York © 1979.

Studies on Hormonal Regulation and Target Cell Response in the Aging C57BL/6J Mouse

Caleb E. Finch

Andrus Gerontology Center, University of Southern California, University Park, Los Angeles, California 90007

The life-span is a species characteristic in mammals, and varies over a 30-fold range (58). Yet the succession of events during postnatal development and aging follows a remarkably similar course in most mammals when the timing of the changes is expressed as a fraction of the maximum life-span (18). The highly predictable age-related changes of each species are doubtless under genetic control.

A major question, which at present can only be crudely considered, is, which level of the mammalian organism contains the pacemakers regulating the expression of the genetic program for aging? Two extreme hypothetical cases may be considered: (a) the pacemakers of aging are intrinsic or localized in each cell—therefore, each cell ages independently of others; or (b) there is a limited population of cells that serve as pacemakers of aging and regulate aging elsewhere in the body by means of physiological controls, such as extracellular factors, hormones and other circulating molecules, and neural factors (18).

An approach to evaluating the extent of intrinsic cellular aging is to examine the response of various target cells to directly acting hormonal stimuli. This approach is illustrated by studies of hepatic enzyme regulation.

HEPATIC ENZYMES, HORMONES, AND HORMONE RECEPTORS

The liver enzyme tyrosine aminotransferase (TAT) is rapidly induced in rodents by perfused (or injected) glucocorticoids (31) or insulin (35). The induction is blocked by inhibitors of RNA synthesis (32,35), and involves a selective increase in the rate of synthesis of TAT polypeptide chains (35,44). It has been shown repeatedly that the induction of TAT by directly acting hormones is not impaired during the average life-span (24–30 months) of laboratory rats (3,6,7,33) and mice (16). These results are consistent with the maintenance of hepatic cytosol glucocorticoid receptors in aging mice (46) (Figure 1, Table 1) and rats (59), and hepatic plasma membrane insulin receptors in aging rats

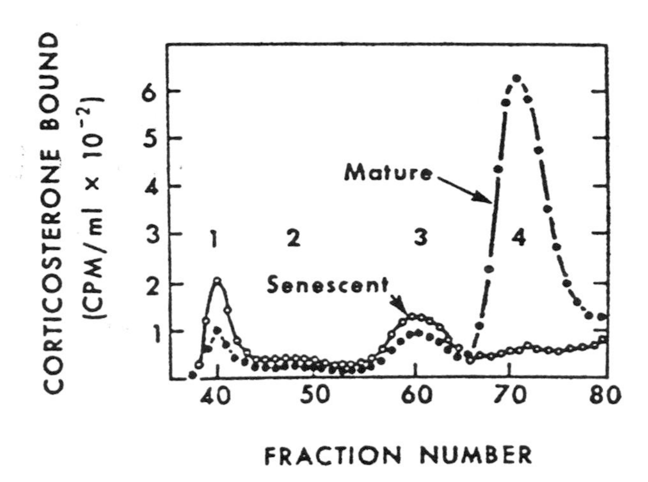

FIG. 1. Hepatic cytosol corticosterone binders from C57BL/6J male mice aged 8–12 months (mature) and 28–32 months (senescent). Cytosols from adrenalectomized mice were incubated for 2 hr with ^{3}H-corticosterone and fractionated on Sephadex G-100 in 0.1 M sodium phosphate, pH 6.9. Free ^{3}H-corticosterone elutes after fraction 85. The peak at fraction 60 ("3") represents the glucocorticoid receptor which can translocate to the nucleus. The cytosols from senescent mice contained slightly more ^{3}H-corticosterone in peak 3. The peak at fraction 70 ("4") shows major decreases with age. It represents an unknown steroid binder, which is distinct from the nuclear receptor by its small size (29,000 daltons) and by the ability of progesterone to compete with ^{3}H-corticosterone binding. For details, see ref. 46.

(27)[1]. As shown in Figure 1 and Table 1, there are at least several binders of glucocorticoids that can be distinguished by molecular weight and steroid affinities. The analysis of steroid binding by unfractionated cytosols thus may obscure significant changes. [By these procedures, no age-related changes in glucocorticoid binding were found in three brain regions of aging C57BL/6J mice—hippocampus, hypothalamus, or cerebral cortex (in this region a slight but nonsignificant decrease was observed) (52).] Although these reports may be interpreted as evidence that the key control mechanisms in genomic regulation of TAT are unimpaired during aging, a more complete understanding will require examination of the mRNA for TAT and for the types of TAT polypeptides synthesized. The possibility of heterogenous polypeptides, some of which may be biologically inactive, is shown by the unusual properties of liver aldolase from aging mice, which includes immunologically active but *enzymatically inactive* forms (29). Such forms appear to include enzymes with abnormal conformations (30). Because not all enzymes in aging rodents show altered forms (54) and some cases may represent changes in isozymes (70), each case must be characterized in detail. Such information has not yet been published for TAT.

[1] In these and subsequently cited papers, the comparisons refer to 8- to 14-month-old adult rodents and 24- to 30-month-old senescent rodents; age-related changes are often reported during the period of growth, which generally approaches completion by 1 year. However, it seems reasonable to consider these periods of life as being physiologically distinct, although a number of functions may change continuously throughout.

TABLE 1. *Hepatic cytosol glucocorticoid binders*

Component[a]	Mol. wt. (daltons)	Total binding (%)	Age change[b] (%)	Characteristics
1	130,000	12	+26	?
2	95,000	6	+ 8	?
3	45,000	14	+12	Glucocorticoid receptor which can translocate to nucleus
4	29,000	68	−83	Distinct from component 3 by the ability of progesterone to compete with ^{3}H-corticosterone

[a] See Figure 1 for Sephadex chromatographic profile and details (46).
[b] Average of three studies.

In contrast to the absence of age-related impairment to TAT induction by directly acting hormones, major delays of TAT induction were observed in aging mice during cold stress (16). Delayed TAT induction was also observed in aging rats if ACTH (rather than glucocorticoids) was injected (2). Thus, age-related impairments of enzyme induction can be revealed by circumstances involving indirectly acting stimuli: cold stress, which alters a variety of hormones; and ACTH, which stimulates adrenal corticosteroid secretions directly and pancreatic and other hormones indirectly. Although the cold-stress activated induction of TAT is dependent on intact adrenals (17), the particular hormones acting on the liver are not yet known. As with TAT, the induction of hepatic glucokinase in aging rats shows no age-related impairment when insulin is injected (a directly acting hormone), but shows extensive delays when fasting is used as the inducer (1). In this case, the delay may be attributed to delayed secretion of insulin in the fasted aging rat (Adelman, *personal communication).* Additionally, fasting-induced corticosterone secretion is delayed in the aging mouse (6).

AGING IN THE FEMALE REPRODUCTIVE SYSTEM

Recently we have begun an extensive analysis of female reproductive aging, since major irreversible aging changes are known to occur in the ovary. The contributions of the aging ovary (with its dwindling output of hormones) to aging in sex steroid target cells thus provide a useful model to study aging in a physiologically interacting system in which at least one of the components undergoes definitive aging changes.

The Ovary

In all mammals, the stock of primordial follicles and oocytes is finite and declines irreversibly after birth (43,68). In turn, the number of primary follicles in the ovary which differentiate into steroid-producing follicles and corpora lutea during the ovulatory cycle is also finite. In humans, the postmenopausal state is associated with a major decrease in plasma estradiol and progesterone, closely approximating plasma levels found in ovariectomized women (reviewed in ref. 19). Impaired secretion of estradiol and progesterone is also observed in premenopausal women who still retain menstrual cycles (64,65). Corresponding endocrine data are not yet available for the aging rats or mice at ages when regular estrous cycles occur less frequently, i.e., 8–14 months. However, since there is an age-related reduction of plasma progesterone during early pregnancy in C57BL/6J mice (41), it is likely that deficiencies of sex steroids also occur in aging rodents during the onset of reproductive senescence.

A major question concerns the relationship of altered ovarian function (sex steroid and ova production) to the major losses of fertility observed in aging rodents. Although the stock of oocytes is finite, there is no significant reduction in the number of ova shed at ovulation during the time when fertility sharply declines, e.g., in the 11- to 12-month-old C57BL/6J mouse (36). The decline of fertility is not uniform in 11- to 12-month-old mice, and several subgroups are suggested in preliminary analysis of recent data (Table 2). Because there is only a slight decrease of implantation sites during early gestation of 11- to 12-month-old mice, the cause of small litters thus derives from fetal death or from increased failure of implantation, rather than from a decreased number of ova at ovulation in this age group. Although if cycles continue the ovary must ultimately exhaust its finite store of oocytes, substantial numbers of undifferentiated oocytes remain in the ovary at the time fertility is lost in most strains of mice (43) and humans (10). Because age changes in uterine function are potentially important to the increased resorption frequency and fetal mortality during aging, we have undertaken studies of uterine functions, particularly regarding response to estradiol and progesterone.

TABLE 2. *Aging and fertility in C57BL/6J mice*

	Mean ± SEM (No. Mice)		
	3–7 months[a]	11–12 months[b]	Signif.
A. Implantation sites:[c]			
no. sites/mouse, d7-10 gestation	8.50 ± .07 (55)	7.41 ± .36 (79)	$p<.01$
B. Resorption frequency, d13-17			
0–15% resorptions/litter	71 ± 2.7% (50)	33 ± 3.4% (21)	
16–66%	36 ± 6.8% (24)	54 ± 5.3% (48)	$p<.01$
67–100%	0	13 ± 2.3% (12)	
C. Litter size at birth	7.79 ± .40	4.62 ± .32	$p<.001$

Data from Holinka et al. (41).
[a]All mice had a previous litter.
[b]Obtained as retired breeders.
[c]After breeding with C3Heb/FeJ male mice of established fertility.

The Uterine Response to Deciduogenic Stimulae

In laboratory mice, the embryo remains free in the uterine cavity during the third to the fourth day after mating. Then, during the fifth day, the zona pellucida of the blastocyst is shed, and the embryo begins to penetrate or implant into the uterine wall. During the next four days there is a characteristic proliferation of the endometrial stromal cells, which is accompanied by major increases in alkaline phosphatase and glycogen (23,25,63). These parameters are very useful indices of decidualization, as shown below.

The induction of the decidual response in the nonpregnant rabbit, discovered by Loeb (47), provides a model for studying uterine cell functions during the decidual response in the absence of the embryo in an ovariectomized rodent with steroid replacements. The hormonal requirements for response to deciduogenic stimuli in mice and rats are reasonably well defined (21,63), and the young ovariectomized, hormonally primed mouse's response to intrauterine oil shares many characteristics of the decidua resulting from normal implantation (25).

Previously many reports showed major impairments of the decidual response [e.g., in transplanted fertilized ova from young donors to aging host mice (67) and hamsters (5); or as induced experimentally by deciduogenic stimuli such as oil droplets or thread in the uterine lumen of the hormonally primed, ovariectomized mouse (22,62)]. However, in these studies, the reproductive history (degree of parity or time elapsed since the last litter) could have influenced the decidual response. Therefore, we studied virgin C57BL/6J mice (38).

The study compared 4- and 10-month-old virgin mice, which were in three experimental groups: (I) ovariectomized (2 wk); (II) ovariectomized and hormonally primed, with a sequence of estradiol and progesterone injections to stimulate early pregnancy; and (III) ovariectomized and primed as in group II, but given in addition a supramaximal deciduogenic stimulus of Planter's Peanut Oil (arachis oil). In 10-month-old mice, there was a major impairment in decidual response (Figure 2), which can now be characterized as an age-related change, rather than as a variable related to reproductive history. Impairments occurred in the incidence of decidual swellings (74% of 4-month-old mice showed visible decidua, whereas only 8% of 10-month-old mice did so). Net increases of uterine weight, glycogen, and alkaline phosphatase were also much smaller in the older mice (Figure 2). The increase of DNA was equivalent to the formation of 38 $\times$ 10^6 additional uterine cells in the 4-month-old mice, but only 10 $\times$ 10^6 additional uterine cells in the 10-month-old group. This calculation assumes that the average uterine cell has a DNA content of 6 pg (diploid value) (15).

The impaired induced decidual response may represent an exhaustion of proliferative capacity of uterine stromal cells, which are characterized by some as fibroblasts (55) and from which the decidual tissue is derived (28). Limited proliferation of diploid fibroblasts *in vitro* has been shown for explants from

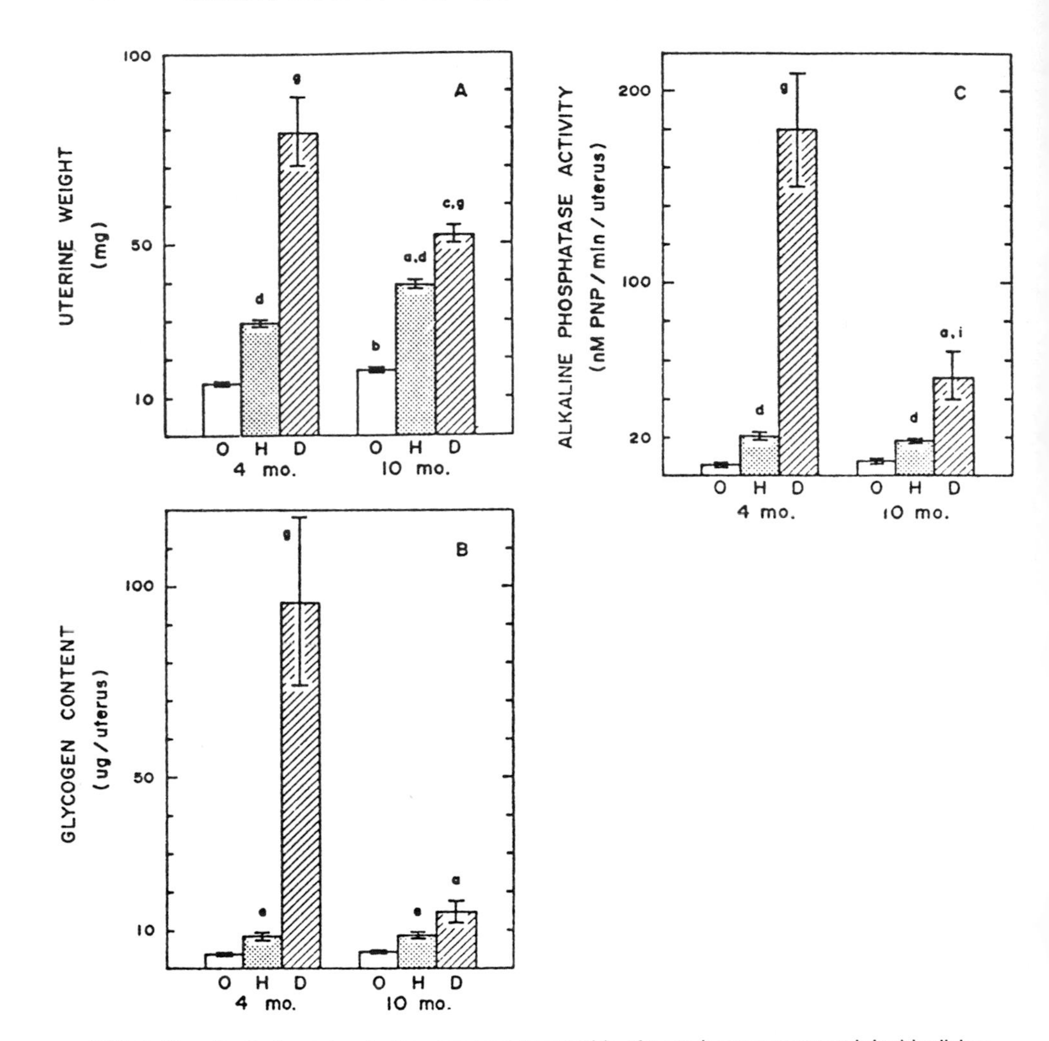

FIG. 2. The effect of age on uterine response to steroids of pseudopregnancy and decidualizing stimulae in virgin C57BL/6J mice. Data from ref. 38. Mice were ovariectomized and given a sequence of estradiol and progesterone injections to simulate pseudopregnancy. One group was further given intrauterine oil and killed at a time equivalent to day 8–9 of implantation. See also Table 3. In the above figure, "O" signifies ovariectomized controls; "H" signifies ovariectomized and steroid injected controls; and "D" signifies mice given steroids and decidualized.

human lung (37) and skin (49,60), and for mammary epithelial cells *in vivo* (11). The repeated waves of mitosis in the uterus during the estrous cycle (48) may exhaust a limited proliferative capacity of some uterine cell types. However, very recent data (described below) suggest that the decidual response may not be impaired during action implantation in aging mice.

Additionally, the young decidualized mice showed a markedly asymmetric distribution of decidua in favor of the right uterine horn—1.25:1 ($p < .025$).

Similarly, the distribution of implantation sites on day 8–9 of pregnancy favored the right uterine horn by 1.5:1 (40). Whereas the previously observed asymmetry in distribution of implantation sites in mice and other species was attributed to an ovarian asymmetry in ovulation, these results show that the uterus itself may have an ovarian-independent asymmetry. These studies do not rule out asymmetry in ovarian influence (e.g., via blood flow) prior to ovariectomy. The existence of ovarian-independent uterine asymmetry in the decidual response may require reinterpretation of a number of earlier studies on decidualization in which one uterine horn was a control and the other was injected with oil. No such problems arise from the present experimental design, which included separate control groups for each treatment.

Although the 10-month-old mice responded comparatively poorly to the decidualizing stimulus, there was no corresponding general age-related impairment in the hormone-primed, but not decidualized controls (group II; Table 3). Biochemical parameters are expressed per μg DNA to represent content per cell. The only significant age difference in average cell composition in the hormone-primed group (II) was a 30% lower activity of alkaline phosphatase/mg DNA. In all other parameters expressed per cell, the 4- and 11-month-old hormone-primed mice were indistinguishable. This result indicates that uteri of both age groups received comparable stimulation from the injected steroids. Thus, the impaired decidual response in aging mice appears to result from an incomplete or insufficient response to decidualization by intrauterine oil.

Uterine Short-Term Response to Estradiol

Because the hormonal priming in the decidualization study has a complex impact on the uterus [e.g., involving stimulation of cell proliferation in both epithelial (mucosal) and endometrial (stromal) cell populations (24)], it was of interest to examine uterine response to a single injection of estradiol. Mice were ovariectomized, and after two days were injected with 0.2 μg estradiol or with the vehicle only, and were killed one day after the hormone injections. In agreement with the results of the previous study (38), significant increases of protein, alkaline phosphatase, and glycogen resulted from injections of estradiol in 5- and 11-month-old mice (Table 4) (39). There was clearly no age-related deficit in the average increase of protein, glycogen, or alkaline phosphatase per cell. In this study, cell proliferation (measured as uterine DNA content) did not occur because of the short period of stimulation, ca. 24 hr.

Measurements of estradiol binding by uterine cytosols when expressed/mg DNA also show no age-related trends (Table 5) (53). These data are consistent with the maintained biochemical responses to estradiol and to the progesterone and estradiol responses described above. A major ambiguity in the interpretation of these data of whole uteri is that there is no indication of which cell populations or cell types are involved. Deeper understanding of changes in uterine function with age will require detailed histological characterization of each cell type

TABLE 3. *The effect of age on the induced decidual response in virgin C57BL/6J mice*

Age group	4 months			10 months		
	(I) OVX	(II) OVX + HOR	(III) OVX + HOR + DECID	(I) OVX	(II) OVX + HOR	(III) OVX + HOR + DECID
Uterine weight, mg wet tissue	13.7 ± 0.3	29.3 ± 0.9 [b]	78.9 ± 9.9 [a]	17.1 ± 0.8 [c]	39.3 ± 1.2 [b,c]	52.6 ± 2.4 [a,c]
Uterine DNA mg/uterus	210 ± 26	277 ± 22	505 ± 41 [a]	234 ± 27	333 ± 11 [b,c]	396 ± 21 [a,c]
Protein/DNA mg/mg	4.7 ± 0.5	7.6 ± 0.5 [b]	10.9 ± 0.8 [a]	4.8 ± 0.3	8.0 ± 0.3 [b]	9.4 ± 0.4 [a]
Glycogen/DNA mg/mg	20.8 ± 3.9	28.9 ± 3.5	161.8 ± 26.9 [a]	18.0 ± 4.0	27.6 ± 4.0	37.1 ± 7.8 [a,c]
Alkaline phosphatase/DNA nM PNP/min/mg DNA	28.5 ± 3.2	77.6 ± 7.3 [b]	326.7 ± 36.1 [a]	32.8 ± 2.2	54.0 ± 3.6 [b,c]	129.9 ± 28.6 [a,c]

Data from Holinka (38).
[a] Differs significantly from hormone-treated control within age group.
[b] Differs significantly from OVX control.
[c] Differs significantly from 4-month group of equivalent treatment.

TABLE 4. *The effect of age on the uterine response to estradiol.*

		Mean ± SEM (no. mice)		
		6 months	11 months (I)	11 months (II)
Uterine weight,	Control	32.0 ± 1.1 (12)	28.2 ± 1.90 (9)	26.2 ± 0.6 (5)
mg wet tissue	E_2	57.3 ± 3.0 (9)	60.7 ± 3.60 (9)	60.9 ± 4.8 (5)
Uterine DNA	Control	0.331 ± .019	0.243 ± .025	0.164 ± .028
mg/uterus	E_2	0.320 ± .023	0.245 ± .023	0.135 ± .015
Glycogen/DNA	Control	90.77 ± 10.57	263.86 ± 36.07	289.76 ± 93.30
μg/mg	E_2	146 ± 25.70	355.80 ± 74.80	574.70 ± 150.50
Alkaline phosphatase/DNA	Control	137.7 ± 23.27	113.97 ± 12.40	194.78 ± 41.71
nM PNP/min/mg DNA	E_2	630.9 ± 130	565.59 ± 74.80	1076.70 ± 125.40

Data from Holinka, et al. (39).
Mice were previously parous (6 months) or retired breeders (11 months). After bilateral ovariectomy (46–52 hr), mice were injected with oil (controls) or 0.2 pg estradiol in oil (E_2).

and its biochemical response to hormones. Analysis combining quantitative biochemical and histological parameters have not yet been made for any sex steroid target tissue of aging animals.

Decidualization in the Aging Intact Mouse

Recent data during early pregnancy of 11- to 12-month-old mice suggest that the early phases of the decidual response induced by an implanting blastocyst may not be impaired. The net increase of uterine weight, i.e., the decidual tissue per implantation site in 3- to 7-month-old and 11- to 12-month-old mice, was similar during implantation (calculated for each mouse after subtraction of the average values of uterine wet weight day 3, the day just before implantation, when uterine weight approximates a nadir in both age groups) (41). These very surprising data (not shown here) suggest that if there are age-related changes in the response to artificial decidualizing stimuli (intraluminal oil or thread), then the change is in the threshold for decidualization rather than in the potential. Biochemical and histological characterization of the decidual response during

TABLE 5. *Estradiol binding by uterine cytosols of C57BL/6J mice during aging*

Age	pMol E_2/uterus	mgDNA/uterus	pMol E_2/mg DNA
3 months (uniparous)	1.02 ± .09	.308 ± .017	3.36 ± .32
8 months (retired breeders)	0.90 ± .10	.288 ± .011	3.14 ± .35
18 months	0.54 ± .08 [a]	.183 ± .010 [a]	3.06 ± .33

Mice were ovariectomized 48 hr; cytosol binding of estradiol was evaluated at 3×10^{-8}M ^{3}H-estradiol and competed with 3×10^{-6}M cold estradiol (53).
[a] Significantly different from 3-month values, $p < .01$.

implantation in both age groups is required before further conclusions can be drawn.

ENDOCRINE REGULATION IN THE AGING MALE MOUSE

Testicular Function

The general pattern of aging in male reproductive functions differs from the female in some mammals in that some individuals retain fertility after the age of mean life-span, e.g., C57BL/6J mice (26), CBF_1 mice (8), and even humans (61). In contrast, female fertility is almost universally lost by midlife in short- and long-lived mammals (68).

In studies on testicular function during aging, we found that health status is a major factor. Adult (8- to 12-month-old) and senescent (28- to 30-month-old) mice had indistinguishable plasma testosterone values and testicular weights if they were free of gross pathological lesions. However, tumors, lung disease, or other major health problems depressed both plasma testosterone and testicular weight (51). Unaltered plasma testosterone during the average life-span was also found in a sampling of DBA/2J and C57BL/6J male mice (14). It seems likely that the great heterogeneity observed in the human male plasma testosterone values (56,69) reflects the influence of age-related diseases.

Recently a novel, longitudinal study of CBF_1 mice revealed that dramatic declines of reproductive performance by 24 months were preceded by declining plasma testosterone and plasma LH levels (8). Various subpopulations of aging mice were distinguished. Sexually active, "robust-appearing" mice had normal distributions of plasma testosterone and LH, and a strong elevation of plasma LH after castration. Another group of aging mice retained a "robust appearance," but were sexually inactive. Although this group also had reduced testicular weights, plasma testosterone, and plasma LH, and only a slight increase of LH after castration, testicular sperm counts were comparable to sexually active mice. Other aging mice showed debilitation or weight loss and were not assayed. A difficulty in interpreting these data is that surveys for specific pathological conditions were not done. Because in our experience gross internal lesions are often not revealed by the external appearance, it is possible that the sexually inactive mice had impaired endocrine function as a correlate of disease, as observed by us and discussed above (51). The distinction between healthy and diseased mice during aging is more than an issue of philosophical debate. Not a single mammalian population with age-related increase of mortality is absolutely free of all deleterious or abnormal changes. The incidence of pathological lesions often increases exponentially and precedes the exponential increase of mortality by approximately 10% of the life-span (66). Clearly the burden is on the researcher to identify if particular disease-related changes account for the age-related phenomenon under study.

Pituitary Function in Aging C57BL/6J Mice

We have recently examined the pituitary response of an apparently healthy aging subpopulation (20). About 30% of 28-month-old mice are rejected by routine internal examination for gross lesions. As shown in Table 6, there was no age-related difference in basal plasma LH or FSH, or in LH after LRH. After castration, there tended to be a slightly smaller increase of LH in senescent mice in one of two studies. However, postcastration increases of plasma FSH were not age-related (Table 6). Basal levels of FSH, prolactin, TSH, and growth hormone also showed no age differences. We conclude that healthy subpopulations of C57BL/6J mice at the average life-span do not show impaired pituitary responses, within the limits imposed by the single time point plasma sampling possible in mice. These results are similar to those obtained in the CBF_1 subpopulation that was sexually active (8).

As a further test of testicular function, testes from "apparently healthy" 12- and 28-month-old mice were decapsulated and incubated with rat LH *in vitro* for 4 hr. No age-related impairment in testosterone release into the medium was detected over a wide range of LH concentrations (Figure 3). This result showing unimpaired response of an endocrine target cell to a directly acting hormone *in vitro* parallels the absence of an age effect on the response of adrenocortical cells to ACTH in rats (Adelman, *personal communication)* and the absence of age-related impairments in the induction of some rodent liver enzymes (see above).

Aging in male rats is consistently associated with substantial impairments in plasma pituitary and gonadal hormones and in the pituitary response to releasing hormones or castration (reviewed in refs. 50 and 57). It is possible that there are major species differences in the pattern of endocrine aging *or* in age-related lesions impinging on pituitary and testicular functions. The high incidence of pituitary (12,34,42) and testicular (9) tumors in some common rat strains is noteworthy and needs detailed consideration.

TABLE 6. *The effect of aging on plasma gonadotropins in C57BL/6J male mice*

	ng/ml plasma, Mean ± SEM			
	LH		FSH	
	12 months	28 months	12 months	28 months
A. Basal levels	19 ± 3	14 ± 10	1540 ± 125	1520 ± 215
B. After LRH				
300 ng/100 g body wt, I	292 ± 28	347 ± 41		
II	290 ± 37	303 ± 20		
C. After castration				
25 days	271 ± 45	172 ± 35[a]		
41 days	513 ± 40	382 ± 46	3570 ± 217	3240 ± 433

Data from Finch, et al. (20).
[a]Significantly lower than 12 month-old values, $p < .05$.

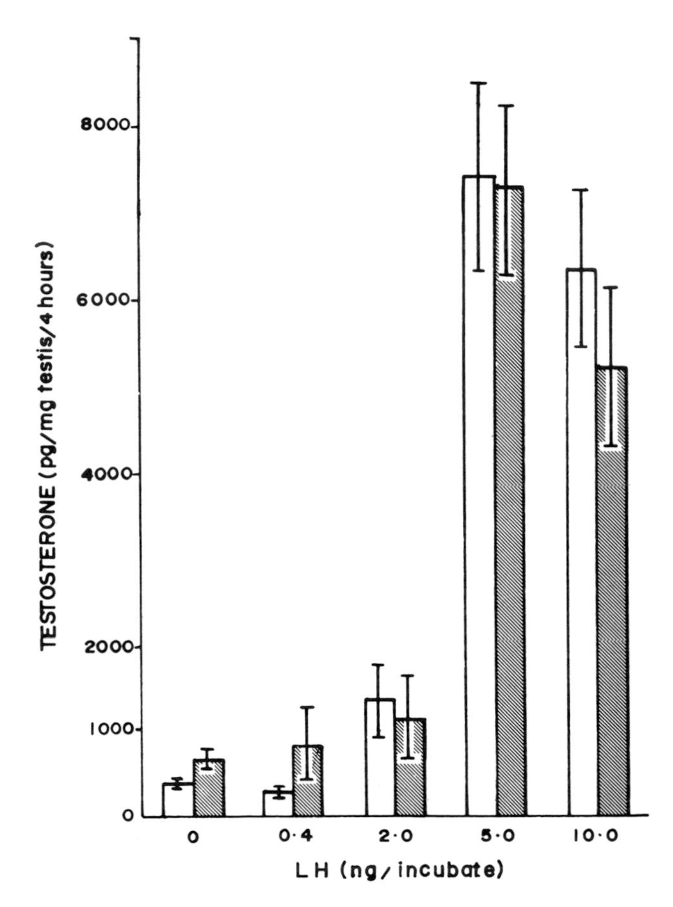

FIG. 3. The secretion of testosterone by decapsulated testes (C57BL/6J mice) during 4 hr incubation with varying concentrations of bovine LH. Data from ref. 20.

Finally, another possible confounding variable needs consideration. A number of labs have reported discrepancies between immunological and bioassays for peptide hormones—insulin (4), TSH (45), and LH (13). The discrepancies could represent "pre"- or "pro"-hormones, degradation products, or modified proteins similar to those observed by Gershon (30). Detailed biochemical characterization of the immunological and bioassayable pituitary peptides produced in the aging rodent could lead to a very different interpretation of cell function during aging.

SUMMARY

Detailed information of molecular events of aging at the level of amino acid and nucleotide sequence is not yet available. However, in at least some cases (e.g., induction of hepatic tyrosine transaminase by glucocorticoids, secretion of corticosterone by rat adrenal cells *in vitro* in response to ACTH, secretion of testosterone by mouse testes *in vitro* in response to LH), there is no indication of intrinsic aging in target cells. In other cases (uterine decidual response, secretion of rat pituitary hormones), the evidence is complex. In both types of examples, it will be valuable to characterize the key macromolecules involved to establish the contribution that molecular damage could make.

ACKNOWLEDGMENTS

This research was supported by grants from the NIH (AG 00446, AG 00560, and AG 00117) and the NSF (PCM 76–02168). This is contribution no. 34 from the Neurobiology Laboratory.

REFERENCES

1. Adelman, R. C. (1970): An age-dependent modification of enzyme regulation. *J. Biol. Chem.*, 245:1032–1036.
2. Adelman, R. C. (1971): Age-dependent effects in enzyme induction; a biochemical expression of aging. *Exp. Gerontol.*, 6:75–87.
3. Adelman, R. C., Freeman, C., and Cohen, B. S. (1972): Enzyme adaptation as a biochemical probe of development and aging. In: *Advances in Enzymology*, ed. G. Weber, pp. 365–382. Pergamon Press, New York.
4. Adelman, R. C. (1977): *This volume.*
5. Blaha, G. C. (1964): Effect of age of the donor and recipient on the development of the transferred golden hamster ova. *Anat. Rec.*, 150:413–416.
6. Britton, G. W., Rothenberg, S., and Adelman, R. C. (1976): Impaired regulation of corticosterone levels during fasting of aging rats. *Biochem. Biophys. Res. Commun.*, 64:184–188.
7. Britton, G. W., Britton, V. J., Gold, G., and Adelman, R. C. (1975): The capability for hormone-stimulated enzyme adaptation in liver cells isolated from aging rats. *Exp. Gerontol.*, 11:1–4.
8. Bronson, F. H., and Desjardins, C. (1977): Reproductive failure in aged CBF_1 male mice: Interrelationships between pituitary gonadotropic hormones, testicular function, and mating success. *Endocrinology*, 101:939–945.
9. Coleman, G. L., Barthold, S. W., Osbaldiston, G. W., Foster, S. J., and Jonas, A. M. (1977): Pathological changes during aging in barrier-reared Fischer 344 male rats. *J. Gerontol.*, 32:258–278.
10. Costoff, A., and Mahesh, V. B. (1975): Primordial follicles with normal oocytes in the ovaries of postmenopausal women. *J. Am. Gerontol. Soc.*, 23:193–196.
11. Daniel, C. W. (1977): Cell longevity in vivo. In: *Handbook of the Biology of Aging*, ed. C. E. Finch and L. Hayflick, pp. 122–158. Van Nostrand, New York.
12. Duchen, L. W., and Schurr, P. H. (1976): In: *Hypothalamus, Pituitary, and Aging*, ed. A. V. Everitt and J. A. Burgess, p. 137. Charles C Thomas, Springfield, Ill.
13. Dufau, M., Beitins, I., McArthur, J., and Catt, K. (1977): Bioassay of serum LH concentrations in normal and LHRH-stimulated human subjects. In: *The Testis in Normal and Infertile Men*, ed. P. Troen and H. R. Nankin, pp. 309–329. Raven Press, New York.
14. Eleftheriou, B. E. (1974): Changes with age in pituitary-adrenal responsiveness and reactivity to mild stress in mice. *Gerontologia*, 20:224–230.
15. Enesco, M., and Leblond, C. P. (1962): Increase in cell number as a factor in the growth of organs and tissues of the young male rat. *J. Embryol. Exp. Morphol.*, 10:530–562.
16. Finch, C. E., Foster, J. R., and Mirsky, A. E. (1969): Ageing and the regulation of cell activities during exposure to cold. *J. Gen. Physiol.*, 54:690–712.
17. Finch, C. E., Huberman, H. S., Mirsky, A. E. (1969): Regulation of liver tyrosine aminotransferase by endogenous factors in the mouse. *J. Gen. Physiol.*, 54:675–689.
18. Finch, C. E. (1976): The regulation of physiological changes during mammalian aging. *Q. Rev. Biol.*, 51:49–83.
19. Finch, C. E., and Flurkey, K. (1977): The molecular biology of estrogen replacement therapy. *Contemp. Ob. Gyn.*, 9:97–107.

20. Finch, C. E., Jonec, V., Wisner, J. R., Jr., Sinha, Y. N., de Vellis, J. S., and Swerdloff, R. S. (1977): Hormone production by the pituitary and testes of male C57BL/6J mice during aging. *Endocrinology,* 101:1310–1317.
21. Finn, C. A., and Hinchliffe, J. R. (1965): The reaction of the mouse uterus during implantation and deciduoma formation as demonstrated by changes in the distribution of alkaline phosphatase. *J. Reprod. Fertil.,* 8:331–338.
22. Finn, C. A. (1966): The initiation of the decidual cell reaction in the uterus of the aged mouse. *J. Reprod. Fertil.,* 11:423–428.
23. Finn, C. A. (1971): The biology of decidual cells. *Adv. Reprod. Physiol.,* 5:1–26.
24. Finn, C. A., and Martin, L. (1973): Endocrine control of gland proliferation in the mouse uterus. *Biol. Reprod.,* 8:585–588.
25. Finn, C. A., and Porter, D. G. (1975): In: *The Uterus.* Publishing Science Group, Inc., Acton, Mass.
26. Franks, L. M., and Payne, J. (1970): The influence of age on reproductive capacity in C57BL mice. *J. Reprod. Fertil.,* 21:563–565.
27. Freeman, C., Karoly, K., and Adelman, R. C. (1973): Impairments in availability of insulin to liver in vivo and in binding of insulin to purified hepatic plasma membrane during aging. *Biochem. Biophys. Res. Commun.,* 54:1573–1580.
28. Gelassi, L. (1968): Autoradiographic study of the decidual cell reaction in the rat. *Dev. Biol.,* 17:75–84.
29. Gershon, H., and Gershon, D. (1973): Inactive enzyme molecules in aging mice: liver aldolase. *Proc. Natl. Acad. Sci.,* 70:909–913.
30. Gershon, D. (1979): *(this volume).*
31. Goldstein, L., Stelle, E. J., and Knox, W. E. (1962): The effect of hydrocortisone on tyrosine α-ketoglutarate transaminase and tryptophan pyrrolase activities in the isolated, perfused rat liver. *J. Biol. Chem.,* 237:1723.
32. Greengard, O., and Acs, G. (1962): The effect of actinomycin on the substrate and hormonal induction of liver enzymes. *Biochim. Biophys. Acta,* 61:652.
33. Gregerman, R. I. (1959): Adaptive enzyme responses in the senescent rat: tryptophan peroxidase and tyrosine transaminase. *Am. J. Physiol.,* 197:63.
34. Greisbach, W. E. (1967): Basophil adenomata in the pituitary glands of 2-year-old male Long-Evans rats. *Canc. Res.,* 27:1813–1818.
35. Hager, C. B., and Kenney, F. T. (1968): Regulation of tyrosine-α-ketoglutarate transaminase in rat liver. VII. Hormonal effects on synthesis in the isolated, perfused liver. *J. Biol. Chem.,* 243:3296.
36. Harman, M. S., and Talbert, G. B. (1970): The effect of maternal age on ovulation, corpora lutea of pregnancy, and implantation failure in mice. *J. Reprod. Fertil.,* 23:33–39.
37. Hayflick, L. (1965): The limited in vitro lifetime of human diploid cell strains. *Exp. Cell Res.,* 37:614–636.
38. Holinka, C. F., and Finch, C. E. (1977): Age-related changes in the decidual response of the C57BL/6J mouse uterus. *Biol. Reprod.,* 16:385–393.
39. Holinka, C. F., Hetland, M. D., and Finch, C. E. (1977): The response to a single dose of estradiol in the uterus of ovariectomized C57BL/6J mice during aging. *Biol. Reprod.,* 17:262–264.
40. Holinka, C. F., and Finch, C. E. (1978): Dextral bias in the induced decidual response after ovariectomy and in implantation sites in the C57BL/6J mouse uterus. *Biol. Reprod.,* 18:418–420.
41. Holinka, C. F., Tseng, Y-C., and Finch, C. E. Reproductive aging in C57BL mice: plasma progesterone, viable embryos, and resorption frequency throughout pregnancy *(in preparation).*
42. Huang, H. H., Marshall, S., and Meites, J. (1976): Capacity of old versus young female rats to secrete LH, FSH, and prolactin. *Biol. Reprod.,* 14:538–543.
43. Jones, E. C., and Krohn, P. L. (1961): The relationships between age, numbers of oocytes and fertility in virgin and multiparous mice. *J. Endocrinol.,* 21:469–495.
44. Kenney, F. T. (1962): Induction of tyrosine-α-ketoglutarate transaminase in rat liver. IV. Evidence for an increase in rate of enzyme synthesis. *J. Biol. Chem.,* 237:3495–3498.
45. Klug, T. L., and Adelman, R. C. (1977): Evidence for a large thyrotropin and its accumulation during aging in rats. *Biochem. Biophys. Res. Commun.,* 77:1431–1437.
46. Latham, K., and Finch, C. E. (1976): Hepatic glucocorticoid binders in mature and senescent C57BL/6J male mice. *Endocrinology,* 98:1434–1443.

47. Loeb, L. (1907): Über die experimentelle Erzeugung von Knoten von Deciduagewebe in dem Uterus des Meerschweinchens nach stattgefundener Copulation. *Zentbl. allg. Path. path. Anat.*, 18:563.
48. Marcus, G. J. (1974): Mitosis in the rat uterus during the estrous cycle, early pregnancy, and early pseudopregnancy. *Biol. Reprod.*, 10:447–452.
49. Martin, G. M., Sprague, C. A., and Epstein, C. J. (1970): Replicative lifespan of cultivate human cells; effects of donor's age, tissue and genotype. *Lab. Invest.*, 23:86–92.
50. Meites, J. (1977): *(this volume).*
51. Nelson, J. F., Latham, K., and Finch, C. E. (1975): Plasma testosterone levels in C57BL/6J male mice: effects of age and disease. *Acta Endocrinol.(Kbh.)*, 80:744–750.
52. Nelson, J. F., Holinka, C. F., Latham, K. R., Allen, J. K., and Finch, C. E. (1976): Corticosterone binding in cytosols from brain regions of mature and senescent male C57BL/6J mice. *Brain Res.*, 115:345–351.
53. Nelson, J. F., Felicio, L. S., Holinka, C. F., and Finch, C. E.: *(in preparation).*
54. Oliviera, R. J., and Pfuderer, P. (1973): Test for missynthesis of lactic dehydrogenase in aging mice by use of a monospecific antibody. *Exp. Gerontol.*, 8:193–198.
55. Parkening, T. A. (1976): An ultrastructural study of implantation in the golden hamster. III. Initial formation and differentiation of decidual cells. *J. Anat.*, 3:485–498.
56. Pirke, K. M., and Doerr, P. (1973): Age-related changes and interrelations between plasma testosterone, oestradiol, and testosterone binding globulin in normal adult males. *Acta Endocrinol.(Kbh.)*, 74:792–800.
57. Riegle, G. D., and Miller, A. E. (1977): Aging effects on the hypothalamic hypophyseal-gonadal control system in the rat. In: *Aging and Reproduction*, ed. E. L. Schneider *(in press).*
58. Rockstein, M., Chesky, J. A., and Sussman, M. L. (1977): Comparative biology and evolution of aging. In: *Handbook of the Biology of Aging*, ed. C. E. Finch and L. Hayflick, pp. 3–34. Van Nostrand, New York.
59. Roth, G. S. (1974): Age-related changes in specific glucocorticoid binding by steroid responsive tissues of rats. *Endocrinology*, 94:82–90.
60. Schneider, E. L., and Mitsui, Y. (1976): The relationship between in vitro cellular aging and in vivo human age. *Proc. Natl. Acad. Sci.*, 73:3584–3588.
61. Seymour, F. I., Duffy, C., and Koerner, A. (1935): A case of authenticated fertility in a man of 94. *J. Am. Med. Assoc.*, 105:1423–1424.
62. Shapiro, M., and Talbert, G. B. (1974): The effect of maternal age on decidualization in the mouse. *J. Gerontol.*, 29:145–148.
63. Shelesnyak, M. C., Marcus, G. J., and Linder, H. R. (1970): Determinants of the decidual reaction. In: *Ovo-Implantation, Human Gonadotropins and Prolactin*, Karger, Basel.
64. Sherman, B. M., and Korenman, S. G. (1975): Hormonal characteristics of the human menstrual cycle throughout reproductive life. *J. Clin. Invest.*, 55:699–706.
65. Sherman, B. M., West, J. H., and Korenman, S. G. (1976): The menopausal transition: Analysis of LH, FSH, estradiol and progesterone concentrations during menstrual cycles of older women. *J. Clin. Endocrinol. Metab.*, 42:629–636.
66. Simms, H. S., and Berg, B. N. (1957): Longevity and the onset of lesions in male rats. *J. Gerontol.*, 12:244–252.
67. Talbert, G. B., and Krohn, P. L. (1966): Effect of maternal age on viability of ova and uterine support of pregnancy in mice. *J. Reprod. Fertil.*, 11:399–406.
68. Talbert, G. B. (1977): Aging of the reproductive system. In: *Handbook of the Biology of Aging*, ed. C. E. Finch and L. Hayflick, pp. 318–356. D. Van Nostrand Company, New York.
69. Vermeulen, A., Reubens, R., and Verdonck, L. (1972): Testosterone secretion and metabolism in male senescence. *J. Clin. Endocrinol. Metab.*, 34:730–735.
70. Wang, R. K. J., and Mays, L. L. (1978): Isozymes of glucose-6-phosphate dehydrogenase in livers of aging rats. *Age*, 1:2–7.

Physiology and Cell Biology of Aging (Aging, Volume 8), edited by A. Cherkin, et al.
Raven Press, New York © 1979.

The Relation of Hypothalamic Biogenic Amines to Secretion of Gonadotropins and Prolactin in the Aging Rat

Joseph Meites, J. W. Simpkins,* and H. H. Huang

Department of Physiology, Neuroendocrine Research Laboratory, Michigan State University, East Lansing, Michigan 48824

Hypothalamic biogenic amines have been shown to have an important role in regulating secretion of anterior pituitary hormones (14). In general, norepinephrine has been shown to stimulate release of gonadotropins, whereas dopamine inhibits release of prolactin. Many reports have indicated that serotonin evokes prolactin release and usually inhibits gonadotropin secretion. Inasmuch as aging male and female rats show a reduced capacity to secrete gonadotropins and an increased ability to secrete prolactin (9,13), we postulated that old rats may exhibit a reduction in hypothalamic catecholamine activity, and perhaps an increase in serotonin activity, as compared with young sexually mature rats. This will be demonstrated here.

CATECHOLAMINES AND SEROTONIN IN HYPOTHALAMUS OF OLD AND YOUNG MALE RATS

In the present study, both concentration and turnover of dopamine(DA), norepinephrine(NE), and serotonin(5-HT) were measured in discrete areas of the hypothalamus and in other brain regions of male Wistar rats 21 or 3–4 months of age. Serum luteinizing hormone(LH), follicle-stimulating hormone(FSH), TSH, and prolactin also were assayed. Turnover or metabolism of DA and NE were determined by first injecting the rats i.p. with 250 mg α-methyl-paratyrosine (α-mpt)/kg body weight. This inhibits synthesis of DA and NE. One hour after α-mpt injection, the rats were decapitated and the brains were rapidly removed. The medial basal hypothalamus (MBH), the remaining hypothalamus, and the olfactory tubercles were dissected and immediately homogenized in 0.4 N perchloric acid. DA and NE were measured by the radioenzyme method of Coyle and Henry (6), and expressed as ng DA or NE per mg protein for MBH, and as ng DA or NE per gram wet weight for hypothalamus or olfactory tubercle.

* Present address: University of Florida, JHMHC, Departments of Physiology and Pharmacy, Gainesville, Florida 32610

TABLE 1. *Serum hormone levels in young and old male rats*[a]

	Study 1		Study 2	
	Young	Old	Young	Old
LH (ng/ml)	20 ± 3	6 ± 2	18 ± 4	9 ± 2
FSH (ng/ml)	220 ± 10	166 ± 14	330 ± 26	186 ± 48
PRL (ng/ml)	10 ± 1	29 ± 8	15 ± 2	37 ± 8
TSH (ng/ml)	650 ± 112	445 ± 132	576 ± 92	485 ± 134

[a]Serum LH, FSH, PRL, and TSH in young and old male rats. From Simpkins, et al. (19).

To measure 5-HT turnover, rats were first injected with 75 mg pargyline HCl/kg body weight to inhibit metabolism of 5-HT, and at the end of 30 min the rats were killed and brain tissues were removed as before. The brain tissues were assayed for 5-HT and 5-hydroxyindoleacetic acid (5-HIAA). Serotonin and 5-HIAA were measured by the method of Curzon and Green (7), as modified by Hyppä, et al. (11), and the results are expressed as ng 5-HT or 5-HIAA per gram of fresh tissue. Serum LH, FSH, prolactin, and TSH were measured by standard radioimmunoassays (RIAs) used in this laboratory.

The serum hormone concentrations (Table 1) show that LH and FSH were significantly lower and prolactin was significantly higher in the old than in the young rats. There were no significant differences in serum TSH levels, although they tended to be somewhat lower in the old rats.

DA and NE concentrations in the MBH and remaining hypothalamus (Figures 1, 2) were significantly lower in old than in young rats. One hour after α-mpt treatment, DA in the MBH decreased significantly more in the young rats,

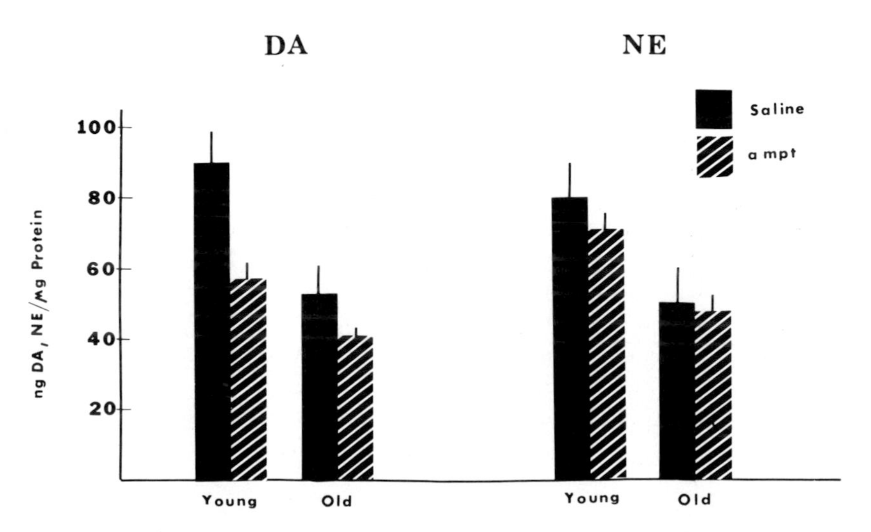

FIG. 1. DA and NE concentrations in medial basal hypothalamus before and after α-mpt treatment. Vertical lines indicate 1 standard error of mean. Each bar represents means of 6–8 rat determinations (after Simpkins et al., 19).

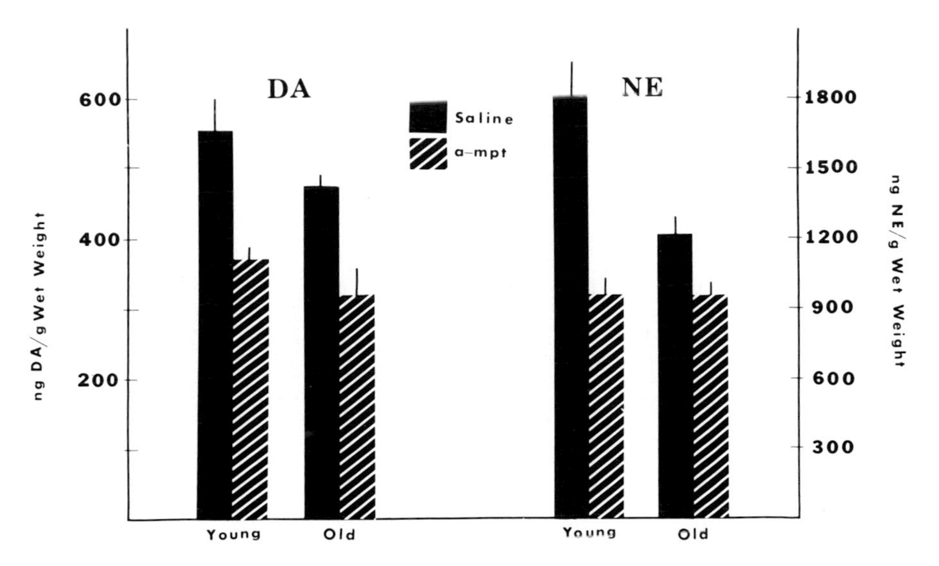

FIG. 2. DA and NE concentrations in remainder of hypothalamus before and after α-mpt treatment in same rats as in Fig. 1 (after Simpkins et al., 19).

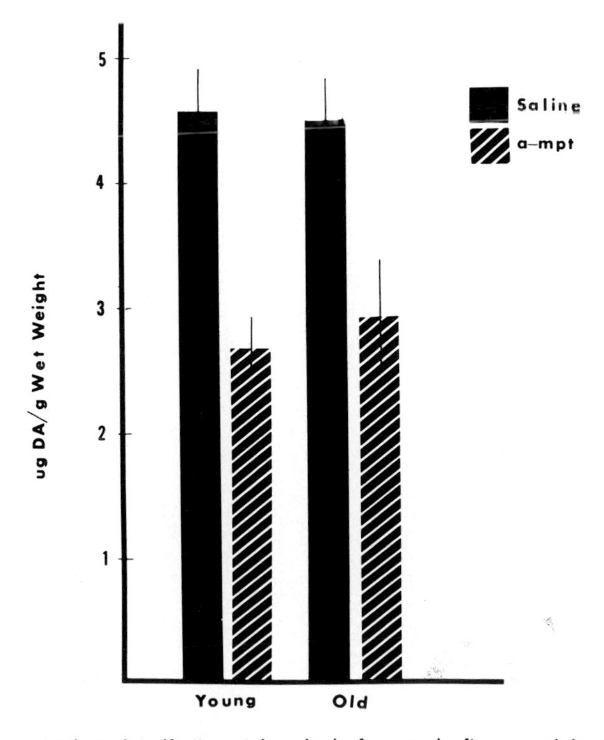

FIG. 3. DA concentrations in olfactory tubercle before and after α-mpt treatment in same rats as in Fig. 1 (after Simpkins et al., 19).

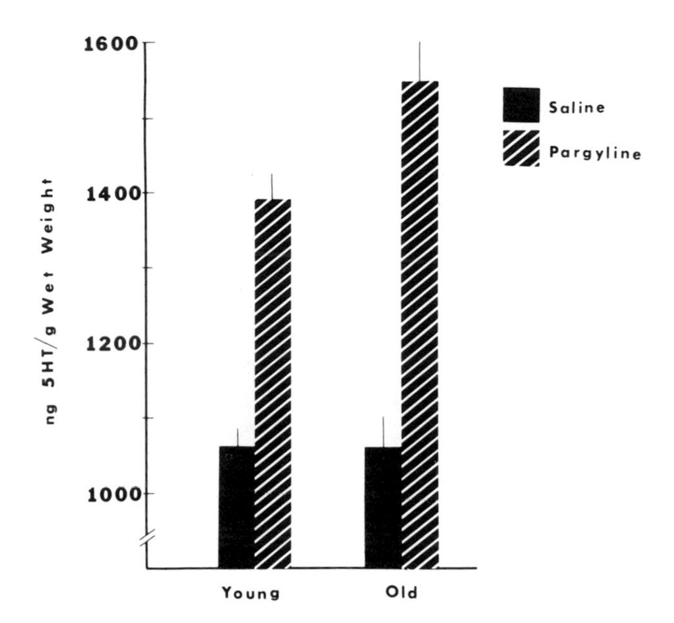

FIG. 4. 5-HT concentrations in hypothalamus before and 30 min after pargyline treatment in same rats as in Fig. 1 (after Simpkins et al., 19).

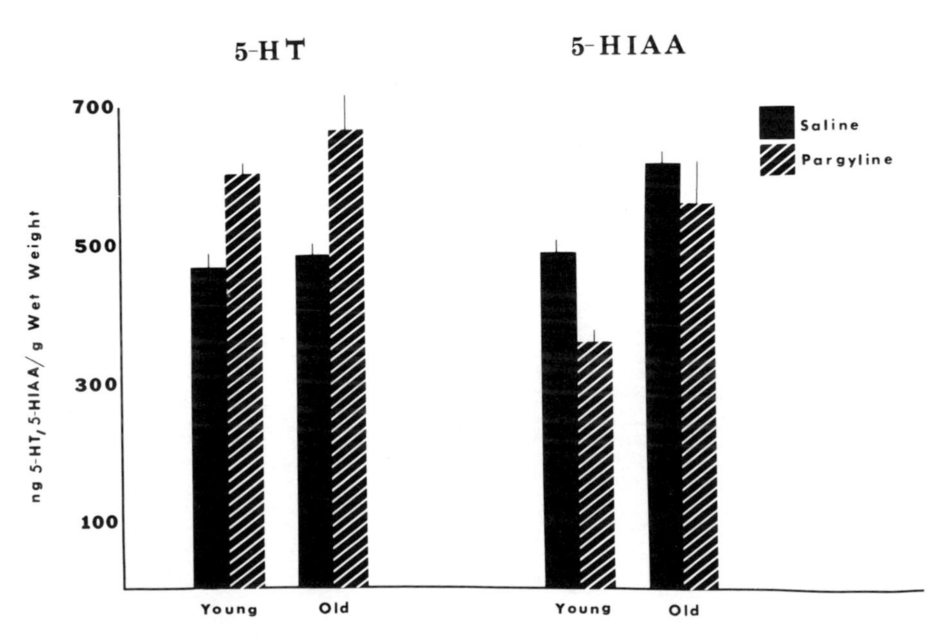

FIG. 5. 5-HT and 5-HIAA concentrations in brain before and 30 min after pargyline treatment in same rats as in Fig. 1 (after Simpkins et al., 19).

suggesting a greater turnover of this amine in the young rats. No differences were found in turnover of DA in the remaining hypothalamus between old and young rats, suggesting that age-related reductions in DA activity are confined to the MBH.

NE concentrations in the MBH and hypothalamus of old rats were significantly lower than in young rats (Figures 1, 2), but α-mpt treatment did not result in any differences in NE turnover between the two age groups in the MBH. However, NE turnover in the remainder of the hypothalamus was significantly lower in the old rats. No differences were found between the old and young rats in either concentration or turnover of DA in the olfactory tubercles (Figure 3).

Serotonin concentration in the hypothalamus and remainder of the brain did not differ between old and young rats (Figure 4). But by 30 min after pargyline treatment, hypothalamic 5-HT rose by 34 $\pm$ 5% in old rats and only by 20 $\pm$ 3% in young rats ($p < 0.05$). The increase in brain 5-HT after treatment with pargyline was the same in old and young rats. Brain 5-HIAA was significantly higher in old rats, and after treatment with pargyline, the decrease in 5-HIAA was significantly less in old rats than in young ones (Figure 5). These results indicate that 5-HT turnover in the hypothalamus and brain of old rats was greater than in young rats.

DISCUSSION

In general, the above results indicate that in old male rats, as compared with young male rats, catecholamine activity in the hypothalamus is depressed and serotonin activity is enhanced. These changes in turnover of the hypothalamic biogenic amines of the old rats appear to be related to the lower serum LH and FSH and higher serum prolactin levels found in the old rats. As already indicated, NE is stimulatory to gonadotropin release, whereas DA is inhibitory to prolactin release (14). Hence a reduction in both NE and DA would be expected to reduce gonadotropin and increase prolactin release. Serotonin has usually been found to be inhibitory to gonadotropin and stimulatory to prolactin release. Hence an increase in hypothalamic 5-HT activity in old rats would be expected to depress gonadotropin and raise prolactin secretion. Thus both the decrease in hypothalamic catecholamines and the increase in hypothalamic serotonin in old rats contribute to inhibition of gonadotropin release and stimulation of prolactin release. Why these differences appear in the hypothalamus of aging male rats is unknown at present, but we have found similar differences in hypothalamic biogenic amines between old and young female rats (Huang, Simpkins, and Meites, *unpublished data).* There also are indications from the work of Finch (8) that the hypothalamus of old male mice has less capacity to convert precursors to DA and NE than the hypothalamus of young mature male mice.

The demonstration that aging male and female rats show lower catecholamine and higher serotonin activity than young mature cycling rats is believed to be related to the lesser ability of the hypothalamopituitary system of aging rats to respond to stimuli that normally increase gonadotropin release and greater capacity to respond to stimuli that normally release prolactin. Thus we have reported that after castration of old male and female rats, the rise in serum LH and FSH is much less than in mature young rats, and the fall in prolactin secretion is usually greater than in young rats (9,18). Castration of old rats by estrogen administration generally produced a lesser fall in LH and FSH release and a greater rise in prolactin than in the young rats. Aschheim (1) also has shown that estrogen administration can increase prolactin secretion more readily in old than in young mature female rats.

The hypothalamopituitary system of old female rats also is less responsive than that of young mature female rats to the positive feedback of ovarian steroids on gonadotropin release. Thus when old and young rats were ovariectomized and later given a single injection of estrogen followed three days later by a single injection of progesterone, the surge in serum LH that occurred several hours after progesterone administration was much less in old than in young females (12). This experimentally induced surge of LH is believed to be similar to the preovulatory surge of LH that normally occurs on the evening of proestrus in the mature cycling female rat. Insufficient hypothalamic norepinephrine activity and excess hypothalamic serotonin activity probably account to a large degree for reduced ability of the old rats to respond to the negative and positive feedback of gonadal hormones on gonadotropin release, and for their increased ability to secrete prolactin in response to estrogen stimulation.

Earlier we had reported that the hypothalamus of old constant estrous rats had lower LH-releasing activity than that of young female rats (4). This was recently confirmed (17), suggesting that less luteinizing-hormone-releasing hormone (LHRH) is available to be released by the hypothalamus of old rats to stimulate pituitary LH and FSH release than in young female rats. Prolactin release inhibiting activity (PIF) also was found to be lower in the hypothalamus of old rats, which could account for the greater release of prolactin in old rats. Presumably the lower NE and higher 5-HT activity in the hypothalamus of old rats is responsible for the lower LHRH and PIF activities, since biogenic amines in the hypothalamus are believed to act on pituitary hormone secretion primarily by regulating release of the hypophysiotropic factors in the hypothalamus (14). However, we also have reported that the pituitary of aging male and female rats is less responsive to direct stimulation by the decapeptide LHRH and releases less LH than the pituitary of young rats (2,20). Thus it appears that the hypothalamus of old rats not only releases smaller amounts of LHRH and PIF in the portal vessels to act on the anterior pituitary (although this is yet to be demonstrated), but that the pituitary itself is less responsive to the LHRH released. Whether the pituitary of old rats is also less responsive to PIF has not yet been demonstrated, but L-DOPA administration was observed to be less effective for inhibiting prolactin release in old than in young rats (16).

If a deficiency in hypothalamic catecholamines and an excess of hypothalamic

5-HT are mainly responsible for the lower capacity to secrete gonadotropins and the greater ability to secrete prolactin in old rats, can this be reversed by appropriate treatment with biogenic amines? In 1969 (5) we reported that injections of epinephrine-in-oil could induce ovulation and reinitiate estrous cycles in old constant estrous rats. This was confirmed in a subsequent publication (15). However, epinephrine presumably does not enter the brain because of the blood-brain barrier, and this could have been a nonspecific response. We later found that ether stress could similarly induce estrous cycles in old constant estrous rats (10), and suggested that such a stress might act by stimulating the ACTH-adrenal cortical system to release progesterone in addition to glucocorticoids. ACTH and progesterone injections were both shown to initiate normal estrous cycles in old constant estrous rats (10). Two drugs that do increase hypothalamic catecholamines—L-DOPA (precursor of catecholamines) and iproniazid (a monoamine oxidase inhibitor)—also were effective in producing cycling in old constant estrous rats (15). We have not yet established whether a treatment that reduces hypothalamic 5-HT levels can result in reinitiation of cycling in old female rats, or whether a combination of treatments that increase hypothalamic catecholamines and decrease 5-HT would be more effective than either treatment alone. Clemens (3) recently reported that lergotrile, an ergot drug, can induce estrous cycles in old pseudopregnant rats, presumably by inhibiting prolactin and promoting LH release.

The female rat appears to be remarkable in her inherent capacity to reproduce. Although there is a decline with age in the number of ova shed with each ovulation, and most of these animals stop cycling between 8 and 15 months of age, the hypothalamus, pituitary, and ovaries appear to retain their capacity to respond to adequate stimuli for most if not all of their life-span. Even the tiny, unstimulated ovaries of old anestrous female rats, 24 to 36 months of age, retain their capacity to respond to gonadotropins (Huang and Meites, *unpublished data),* although most of them have pituitary tumors. When bred with vigorous young male rats, some old constant estrous rats became pregnant but did not undergo parturition and the young died *in utero* (Huang and Meites, *unpublished data).* This may be due primarily to defects in the reproductive tract. Most of the old rats that were bred in this manner became pseudopregnant rather than pregnant. It is apparent that the reproductive decline in aging female rats (and probably also in males) is associated not only with fundamental changes in hypothalamic function, which we consider to be primary, but also with alterations in function of the pituitary, gonads, and reproductive tract. In addition, changes in secretion of hormones other than gonadotropins and prolactin, as well as the general decline in body functions with aging, may contribute to reproductive senescence in rats.

ACKNOWLEDGMENTS

This study was aided in part by NIH research grants AG00416 from the National Institute on Aging; AM04784 from the National Institute of Arthritis,

Metabolic and Digestive Diseases; and CA10771 from the National Cancer Institute.

REFERENCES

1. Aschheim, P. (1976): Aging in the hypothalamo-hypophyseal-ovarian axis in the rat. In: *Hypothalamus, Pituitary and Aging,* ed. A. E. Everitt and J. A. Burgess. Charles C Thomas, Springfield, Illinois, pp. 376–418.
2. Bruni, J. F., Huang, H. H., Marshall, S., and Meites, J. (1977): Effects of single and multiple injections of synthetic GnRH on serum LH, FSH and testosterone in young and old male rats. *Biol. Reprod.,* 17:309–312.
3. Clemens, J. A., and Bennett, D. R. (1977): Do aging changes in the preoptic area contribute to loss of cyclic endocrine function. *J. Gerontol.,* 32:19–24.
4. Clemens, J. A., and Meites, J. (1971): Neuroendocrine status of old constant estrous rats. *Neuroendocrinol.,* 7:249–256.
5. Clemens, J. A., Amenomori, Y., Jenkins, T., and Meites, J. (1969): Effects of hypothalamic stimulation, hormones and drugs on ovarian function in old female rats. *Proc. Soc. Exp. Biol. Med.,* 132:561–563.
6. Coyle, J. T., and Henry, J. (1973): Catecholamines in fetal and new-born rat brain. *J. Neurochem.,* 21:61–67.
7. Curzon, G., and Green, A. R. (1970): Rapid method for the determination of 5-hydroxytryptamine and 5-hydroxyindoleacetic acid in small regions of rat brain. *Br. J. Pharmacol.,* 39:653–655.
8. Finch, C. E. (1973): Catecholamine metabolism in the brains of male mice. *Brain Res.,* 52:261–276.
9. Huang, H. H., Marshall, S., and Meites, J. (1976): Capacity of old versus young female rats to secrete LH, FSH and prolactin. *Biol. Reprod.,* 14:538–543.
10. Huang, H. H., Marshall, S., and Meites, J. (1976): Induction of estrous cycles in old non-cyclic rats by progesterone, ACTH, ether stress or L-dopa. *Neuroendocrinol.,* 20:21–34.
11. Hyppä, M. T., Cardinali, D. P., Baumgarten, H. G., and Wurtman, R. J. (1973): Rapid accumulation of ^{3}H-serotonin in brains of rats receiving intraperitoneal ^{3}H-tryptophan: effect of 5,6-dehydroxytryptamine and female sex hormones. *J. Neural Transm.,* 34:111–124.
12. Lu, K. H., Huang, H. H., Chen, H. T., Kurcz, M., Mioduszewski, R., and Meites, J. (1977): Positive feedback by estrogen and progesterone on LH release in old and young rats. *Proc. Soc. Exp. Biol. Med.,* 154:82–85.
13. Meites, J., Huang, H. H., and Riegle, G. D. (1976): Relation of the hypothalamo-pituitary-gonadal system to decline of reproductive functions in aging female rats. In: *Hypothalamus and Endocrine Functions,* ed. F. Labrie, J. Meites, and G. Pelletier. Plenum Publishing Corp., New York, pp. 3–20.
14. Meites, J., Simpkins, J., Bruni, J., and Advis, J. (1977): Role of biogenic amines in control of anterior pituitary hormones. *IRCS J. Med. Sci.,* 5:1–7.
15. Quadri, S. K., Kledzik, G. S., and Meites, J. (1973): Reinitiation of estrous cycles in old constant estrous rats by central acting drugs. *Neuroendocrinol.,* 11:807–811.
16. Riegle, G. D., and Meites, J. (1976): Effects of aging on LH and prolactin after LHRH, L-dopa, methyl-dopa and stress in male rat. *Proc. Soc. Exp. Biol. Med.,* 151:507–511.
17. Riegle, G. D., Meites, J., Miller, A. E., and Wood, S. M. (1977): Effect of aging on hypothalamic and prolactin inhibiting activities and pituitary responsiveness to LHRH in the male laboratory rat. *J. Gerontol.,* 32:13–18.
18. Shaar, C. J., Euker, J. S., Riegle, G. D., and Meites, J. (1975): Effects of castration and gonadal steroids on serum luteinizing hormone and prolactin in old and young rats. *J. Endocrinol.,* 66:45–51.
19. Simpkins, J. W., Mueller, G. P., Huang, H. H., and Meites, J. (1977): Evidence for depressed catecholamine and enhanced serotonin metabolism in aging male rats; possible relation to gonadotropin secretion. *Endocrinol.,* 100:1672–1678.
20. Watkins, B. E., Meites, J., and Riegle, G. D. (1975): Age related changes in pituitary responsiveness to LHRH in the female rat. *Endocrinol.,* 97:543–548.

Physiology and Cell Biology of Aging
(Aging, Volume 8), edited by A. Cherkin, et al.
Raven Press, New York © 1979.

Relation of Diabetes to Aging and Atherosclerosis

Rachmiel Levine

Department of Metabolism and Endocrinology, City of Hope National Medical Center, Duarte, California 91010

The prevalence of diabetes increases with chronological age up to the eighth decade. Among the so-called complications of diabetes are structural and functional changes that are sometimes linked to the aging process—cataract; peripheral nerve degenerations; basement membrane changes in the capillary blood vessels; dystrophic changes in skin; but above all, an earlier and more progressive development of atherosclerosis of the larger arteries. Coronary, aortic, iliac, and femoral atheromata occur far more frequently in diabetics.

What are the metabolic and endocrine factors which might play a role in promoting arteriosclerosis in the diabetic population?

On the basis of many observations since 1950, in both experimental animals and in man, we would like to present the hypothesis that high levels of circulating insulin in the presence of high blood glucose values are capable of promoting the formation and deposition of fatty deposits in the arterial wall, and thus may be atherogenic.

In the following pages we shall examine some of the clinical and experimental observations that argue for such a view and for the therapeutic conclusions that flow from such a concept. However, it is first necessary to realize that the term diabetes refers not to a single etiologic and clinical entity but to at least two or more forms of the disease.

THE DIFFERENT FORMS OF DIABETES

Long before the present era, there were thoughtful and observant clinicians who classified human diabetics into two or more types, which behaved differently over the course of the disease. For example, in the 19th Century, Bouchardat (6) and Lancereaux (10) thought they could distinguish three distinct forms of diabetes. The most severe was labeled "lean" diabetes. Such patients were generally children or young adults who showed all the classical symptoms and signs, especially weight loss in the face of a good calorie intake. The disease (in those days) led inexorably and over a period of weeks or months to coma and death. Another group had a high blood sugar in association with established neurological disease or emotional stress. This was the "nervous" form. But by far the largest contingent consisted of those who had the disease in milder form and were overweight—"the diabetes of the fat." It is for this large group that much could be done by dietary weight reduction and severe muscular exercise, as was shown in the 1850s.

On the basis of measurements of circulating insulin in the 1950s (5) and 1960s (3,4), one can also divide diabetics into three groups.

> Group I: Lean, juvenile, severe type. Blood insulins are very low and do not rise after a meal. Require insulin daily.
> Group II: Moderate symptomatology, adult-onset. Fasting blood insulin levels are normal or low. Insulin rises slowly and inadequately after a meal. May not go into acidosis even without treatment. Require diet low in soluble carbohydrates, and an oral hypoglycemic agent or insulin.
> Group III: Adult, obese, few if any symptoms or signs. Fasting blood insulin levels are high. After a meal, insulin rises to abnormally high levels, but may do so sluggishly. Require strict calorie reduction and muscular exercise.

There are indications at present that the insulin-requiring group of severe cases may result from a viral or immunological attack on the insulin secretory cells (16). This form of diabetes is characterized by the statistical presence of certain HLA types of antigens. On the other hand, the obese diabetic does not belong to this category, and the etiology of his disorder is not on a presumed immunological basis.

We have also learned in the past 15 to 20 years that the upper intestinal tract produces and secretes a host of hormonal peptides, following the ingestion of food. Some of these enteric hormones stimulate insulin secretion by the B cell (13,22).

THE OBESE DIABETIC—ORIGIN AND TREATMENT

The adult obese form of diabetes may originate or be maintained by overfunctioning of one or more of the enteric peptides. It has been demonstrated that a reduction in food intake can lead to normalization of blood sugar even before a significant weight loss has had time to occur.

Muscular work is helpful in reducing the blood sugar level because contracting muscle produces a blood sugar lowering factor and/or sensitizes tissues to circulating insulin (9,11).

Hence, the regimes advocated in the 19th Century for the "fat" diabetic have found their modern endocrine explanation.

The difficulty all along has been the failure of inducing "fat" diabetics to eat less and work harder. When food is abundant and the need for physical labor has decreased, caloric reduction and rational, restricted food consumption seems almost impossible. This is especially true since these patients have no symptoms. It is very difficult to induce a change of "life style" in the face of good food and its pleasures.

Until quite recently, despite the problems created by dietary noncompliance, one could attempt to decrease the higher blood sugars by means of the orally active agents, such as the sulfonylureas and the phenformin group. Because of the fear of (a) inducing a greater number of coronary attacks, and (b) causing lactic acid acidosis, there is an understandable reluctance to use these agents (2,23,24). Hence, there is increased use of insulin as the blood sugar lowering agent. The obese, mild diabetic suffers from insulin resistance. Therefore, comparatively large doses of insulin are required.

Naturally, these will lower the blood sugar level. The glucose leaves the blood and is taken up by the tissues. Only some of it is oxidized; the bulk is transformed to fat in the adipose tissue and in the liver. The result may be a normal or near normal blood sugar, but also a buildup of the fat depots and a rise in blood lipoproteins (1,17). Insulin is lipogenic not only in the liver and in the fat cell; it also increases the synthesis of fat in the arterial wall itself (12,20,21). In the 1950s it was found that atherosclerosis on a high fat, high cholesterol diet was promoted by insulin (7,8,15). There is abundant evidence that insulin favors lipoprotein production, and we have recently found that it also favors lipoprotein transport in the endothelium of blood vessels (25).

Therefore, I would conclude that treatment of the obese, nonketotic diabetic who has high blood insulin and is resistant to the proper action of the hormone should *not* include the chronic administration of insulin. Blood sugar normalization can and should be achieved by proper diet and exercise. It means that we have to provide such diabetics with continuous education, orient them towards a new life style, and practice truly preventive medicine. In my opinion, the use of insulin as the blood sugar lowering agent in the obese diabetic treats a sign of the disease (hyperglycemia) only, but does nothing to correct the basic syndrome.

There exists a genetic strain of mice (ob/ob) who develop obesity, high blood sugars, high blood insulin levels, and insulin resistance. The syndrome can be almost completely normalized by the chronic, consistent reduction in food intake. This is an animal model for the type of diabetes discussed above (18,19).

There is another group of adult onset diabetics in whom the insulin levels are reduced but who are not obese or hormone-resistant. Such diabetics may well be treated with insulin injections.

CONCLUSION

Since about 70% of deaths in adult obese diabetes are due to vascular, atherosclerotic disease (coronaries, aorta, limb blood vessels), it behooves us to sort out the diabetic population and treat according to rational criteria.

Admittedly it is quite difficult to inculcate dietary compliance and regular exercise, but until the "magic cure" appears, this regimen remains the only proper form of therapy. Otherwise the fight against the vascular complications, and the associated morbidity and mortality, will not be advanced. In the obese, adult diabetic, additional insulin is not the proper weapon.

The regimen of caloric reduction reduces the traffic of food in the gastrointestinal tract, it reduces weight, and it reduces insulin secretion. In the overweight diabetic, and in experimental animals with spontaneous diabetes and obesity, this mode of therapy produces significant improvements in glucose tolerance, insulin resistance, and the high insulin secretion rates. This is especially interesting when we consider that for many years it has been known that "caloric undernutrition" increases the life-span of experimental animals (14).

REFERENCES

1. Ball, E. G. (1970): Some considerations of the multiplicity of insulin action on adipose tissue. In: *Adipose Tissue,* ed. B. Jeanrenaud and D. Hepp, p. 102. G. Thieme, Stuttgart.
2. Bengtsson, K., Karlsberg, B. and Lindgren, S. (1972): Lactic acidosis in phenformin-treated diabetes: a clinical and laboratory study. *Acta Med. Scand.,* 191:203.
3. Berson, S. A., and Yalow, R. S. (1961): Plasma insulin in health and disease. *Am. J. Med.,* 31:874.
4. Berson, S. A., and Yalow, R. S. (1970): Plasma insulin in diabetes mellitus. In: *Diabetes Mellitus,* (ed. Ellenberg and Rifkin), p. 308 ff. McGraw-Hill, New York.
5. Bornstein, J., and Lawrence, R. D. (1951): Plasma insulin in human diabetes mellitus. *Br. Med. J.,* 2:1541.
6. Bouchardat, A. (1875): *De La Glycosurie.* G. Bailiere, Paris.
7. Cruz, A. B., Amatuzio, D. S., Grande, F., and Hay, L. J. (1961): The effect of intra-arterial insulin on tissue cholesterol and fatty acids in alloxan-diabetic dogs. *Circ. Res.,* 9:39.
8. Duff, G. L., and Macmillan, G. C. (1949): The effect of alloxan diabetes on the retrogression of experimental cholesterol atherosclerosis. *J. Exp. Med.,* 89:611.
9. Goldstein, M. S., Mullick, V., Huddlestun, B., and Levine, R. (1953): Action of muscular work on transfer of sugars across cell barriers: comparison with action of insulin. *Am. J. Physiol.,* 173:212, 1953.
10. Lancereaux, E. (1877): Diabetic maigre. *Bull. Acad. Med., Paris,* 2:1215.
11. Levine, R., and Goldstein, M. S. (1955): On the mechanism of action of insulin. *Rec. Prog. Hormone Res.,* 11:343.
12. Mahler, R. (1971): The effect of diabetes and insulin on biochemical reactions of the arterial wall. *Acta Diabetol. Latina,* 8:Suppl 1:68.
13. Marks, V., and Samols, E. (1970): Intestional factors in the regulation of insulin secretion. *Adv. Metab. Disorders,* 4:1.
14. McCay, C. M. (1952): In: *Problems of Ageing,* ed. A. I. Lansing. Williams & Wilkins, Baltimore.
15. McGill, H. C., and Holman, R. L. (1949): Influence of alloxan diabetes on cholesterol atheromatosis in rabbit. *Proc. Soc. Exp. Biol. Med.,* 72:72.
16. Nerup, J., et al., (1974): HL-A antigens and diabetes mellitus. *Lancet,* 2:864.
17. Nikkila, E. A., Pyorala, K., and Taskinen, M. R. (1971): Role of insulinemia in arterial disease. *Acta Diabetol. Latina,* 8:Suppl 1:56.
18. Renold, A. E. (1968): Spontaneous diabetes and/or obesity in laboratory rodents. *Adv. Metab. Disorders,* 3:49.
19. Stauffacher, W., Lambert, A. E., and Renold, A. E. (1967): Effect of insulin of diaphragm and adipose tissue of obese mice. *Diabetologia,* 3:320.
20. Stout, R. W. (1968): Insulin-stimulated lipogenesis in arterial tissue in relation to diabetes and atheroma. *Lancet,* 2:702.
21. Stout, R. W. (1970): Development of vascular lesions in insulin-treated animals fed a normal diet. *Br. Med. J.,* 3:685.
22. Turner, D. S. (1969): Effects of amino acids and intestinal hormones on insulin release *in vitro. Diabetologia,* 5:57.
23. University Group Diabetes Program Report I (1971): A study of the effects of hypoglycemia on vascular complication in patients with adult-onset diabetes. *Diabetes,* 19:Suppl. 2, pp. 747 ff.
24. University Group Diabetes Program Report II (1974): Evaluation of phenformin therapy. *Diabetes,* 24:Suppl. 1, pp. 65 ff.
25. Wayland, H., and Levine, R.: Transport of fluorescent-tagged lipoproteins across capillary wall. *Unpublished observations.*

Physiology and Cell Biology of Aging
(Aging, Volume 8), edited by A. Cherkin, et al.
Raven Press, New York © 1979.

In Pursuit of Molecular Mechanisms of Aging

Richard C. Adelman

Temple University Institute on Aging, York and Tabor Roads,
Philadelphia, Pennsylvania 19141

Current efforts in research on the biology of aging are based on two fundamental observations: (a) the vast differences in apparent maximal life-spans among the various animal species; and (b) the progressive diminution in the capability for physiological performance as a particular population ages beyond sexual maturity. It remains to be seen whether or not these phenomena represent two facets of the same problem.

This chapter is based on the latter approach. The rationale involves several stages. The initial approach is to document readily measurable manifestations of aging in a rigorously defined population of experimental animals. Then it is necessary to localize such phenomena in a specific cell population of a particular tissue. Subsequently, when a specific biochemical event is identified as the culprit—whose modification is expressed or at least contributes to the documented manifestation of aging in the intact animal—it becomes time to pursue the most exciting questions. There are: (a) At what age is the modification first expressed? (b) What is the nature and origin of those factors responsible for initial development of the modification in question? (c) How general is this relative to other types of manifestations of aging? (d) How does it relate to the potential health and longevity of the species under investigation, as well as other species?

One feature that probably characterizes all aging populations is the progressively modified ability to adapt to changes in the surrounding environment. At the gross physiological level, this may refer to the impaired ability of the elderly pedestrian to avoid an oncoming car. To the biochemist, it may refer to the capability for production of a set of gene products when a specific cell population within an organism responds to a particular hormonal signal. The potential in this type of approach is limited only by the ability to trace back the sequence of responsible events relevant to the initially selected adaptation. One such phenomenon that is now well documented is the capacity for enzyme adaptation.

ENZYME ADAPTATION

In response to a broad spectrum of environmental challenges, the ability to initiate adaptive fluctuations in the activities of several dozen enzymes is impaired during aging in a variety of tissues from several different species (4). As illustrated in Figure 1, enzyme adaptation during aging can be categorized into four types of response. Some enzyme adaptations are altered in time course and/or magnitude of response during aging, whereas others, for reasons not yet understood, apparently are not altered at all. Specific examples of enzyme adaptations within each of these categories were reviewed previously (4).

These patterns of enzyme adaptation during aging are susceptible to considerable variation related to differences in sex, strain, species, and conditions of environmental maintenance of the employed experimental animal model (4). This addresses a major issue of gerontological research: the need for rigorous definition of the genetic and environmental factors applied to the development and maintenance of colonies of experimental animals into old age. Except when indicated otherwise, all data reported in this article were obtained from a colony of male Sprague-Dawley rats maintained into old age for this investigator at the Charles River Breeding Laboratories. The rats are maintained behind a pathogen-defined barrier under rigorously controlled environmental and genetic conditions. For example, they are provided throughout their lifetime with a pasteurized, sterilized diet which is kept constant with regard to both percent

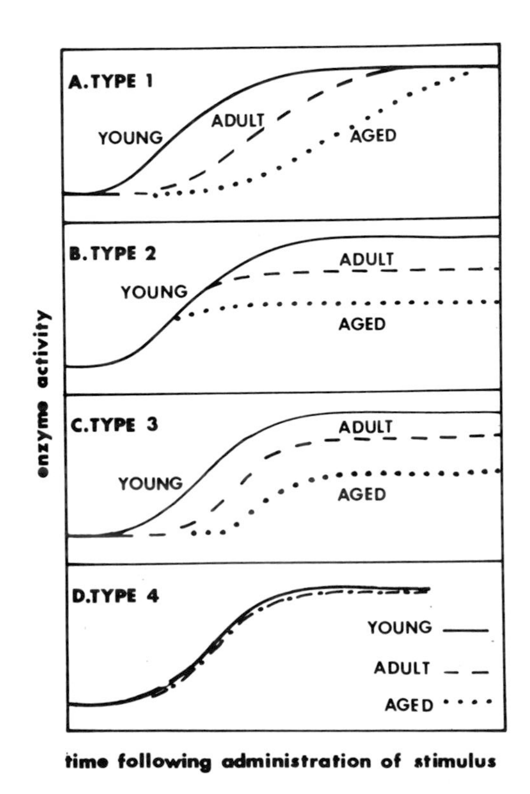

FIG. 1. Categories of enzyme adaptation during aging (from ref. 4).

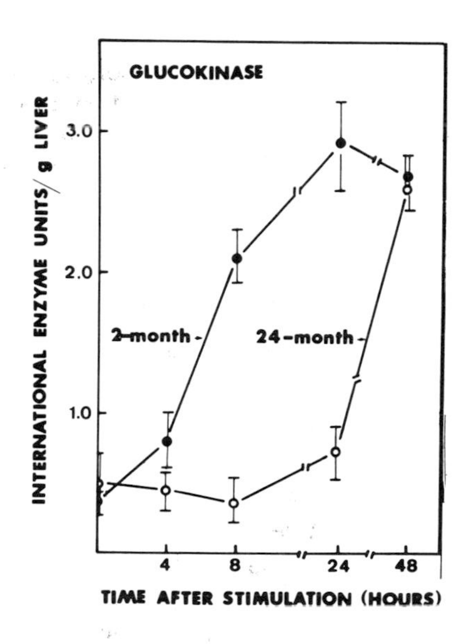

FIG. 2. Glucose-stimulated adaptation in hepatic glucokinase activity of rats aged 2 and 24 months (from ref. 1).

composition and component source. The mean life-span of these rats is approximately 30 months, and the maximal life-span is approximately 40 months. Until nearly 24 months of age they are remarkably free of detectable gross pathology, as will be described in a separate publication (9).

Figures 2 and 3 show two specific examples of age-dependent enzyme adaptation that occupied the attention of this laboratory. The increase in hepatic gluco-

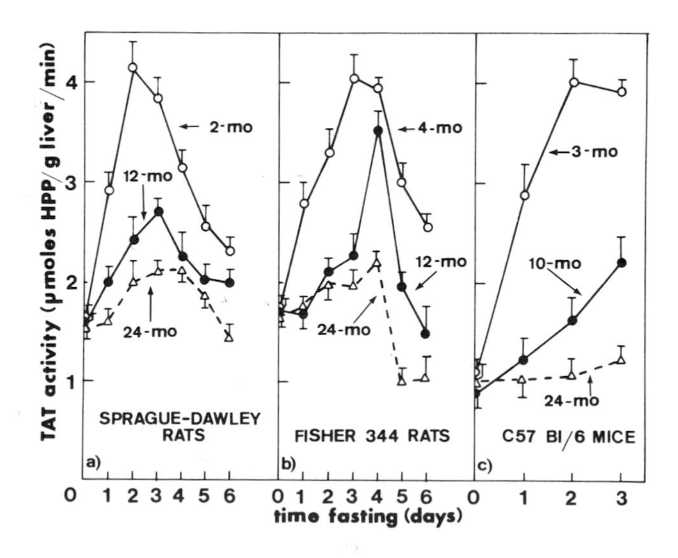

FIG. 3. Starvation-stimulated adaptation in hepatic tyrosine aminotransferase activity of aging rats and mice (from ref. 6).

kinase activity is provoked by intragastric injection of an aqueous solution of glucose to fasted rats (1). As male or female rats age from 2 to 24 months, the time required to initiate this enzyme adaptation lengthens from 3–4 to 10–12 hr, whereas magnitude of response is not altered. The increase in hepatic tyrosine aminotransferase activity is provoked by starvation (6). As male rats age from 2 to 24 months, the magnitude of response is reduced, whereas very little effect on time course is evident. The significance of the comparative evaluation of the tyrosine aminotransferase adaptation in several animal models is discussed below in the section on hormonal regulatory mechanisms.

There are several implications to these results. First, they indicate that manifestations of aging can be monitored readily in experimental animals. Second, the likelihood for successful elucidation of the sequence of responsible events seems considerable because so much information is already available at the levels of intracellular and extracellular control of these two enzymes. Third, and undoubtedly most intriguing of all, although these manifestations of aging are expressed most blatantly in the oldest rats examined, they are progressive in nature and begin very early in life. This is illustrated most dramatically in Fig. 4 for the adaptation in hepatic glucokinase (GK) activity. The adaptive latent period—the time that elapses between administration of glucose and the observed initiation of increased enzyme activity—lengthens progressively and is directly proportional to chronological age between at least 2 and 24 months (2,3). Therefore, whatever is ultimately responsible for the alterations in hepatic enzyme adaptation (as well as other manifestations of aging) should be regarded as one more facet of developmental biology rather than as something directed exclusively to the oldest members of a population.

In an effort to localize the origin of the age-dependent alterations in the capability for hepatic enzyme adaptation, three different experimental approaches were pursued. Although each approach suffered from its own limitations, the

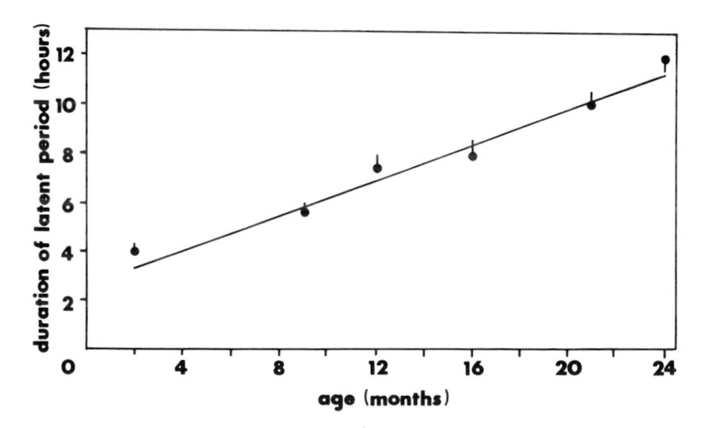

FIG. 4. Relationship between duration of glucokinase adaptation latent period and chronological age (from ref. 4a).

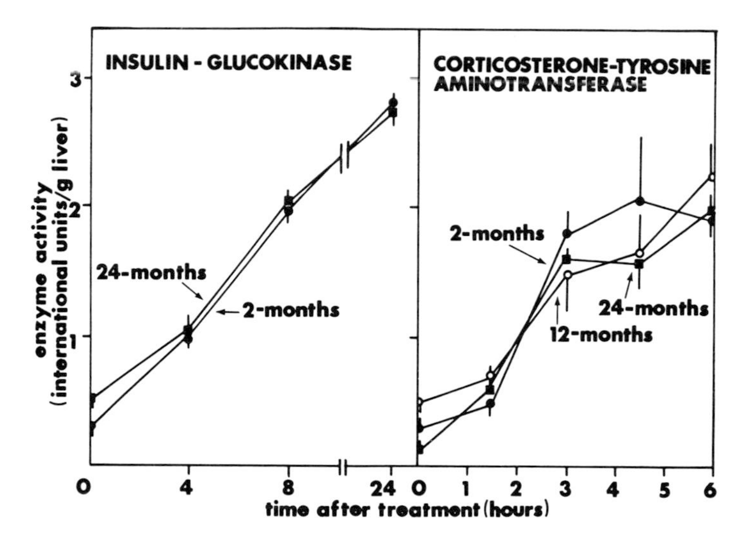

FIG. 5. Insulin-stimulated adaptation in glucokinase activity and corticosterone-stimulated adaptation in tyrosine aminotransferase activity during aging (from ref. 4a).

combination of results indicate that the alterations in enzyme adaptation probably reflect age-dependent modifications in extrahepatic regulatory mechanisms. (a) Adaptive increases in the activities of hepatic glucokinase and tyrosine aminotransferase are not modified during aging in response to injection of hormones that are known to interact directly with liver to enhance the rate of *de novo* enzyme synthesis *in vivo* (5,10). This is illustrated in Figure 5 for responses to administration of insulin and corticosterone, respectively. The difficulty in this experimental approach is that the minimal amount of injected hormone required for a detectable increase in enzyme activity generates massive nonphysiological amounts of hormone in the circulation (8). (b) Binding of at least certain hormones to their hepatic receptors is not altered in rats beyond 12 months of age (11). This is indicated in Table 1 for the binding of porcine insulin to purified fractions of hepatic plasma membranes isolated from rats of the ages indicated. One difficulty in this experimental approach concerns the absence of information relevant to the array of cellular events that occur subsequent

TABLE 1. *Effects of aging on kinetic constants of insulin binding to purified hepatic plasma membranes*

Constant	Age		
	2 months	12 months	24 months
K dissociation (molar)	1.1×10^{-9}	9.0×10^{-10}	1.3×10^{-9}
Binding capacity (ng/mg membrane protein)	1.17	0.57	0.50

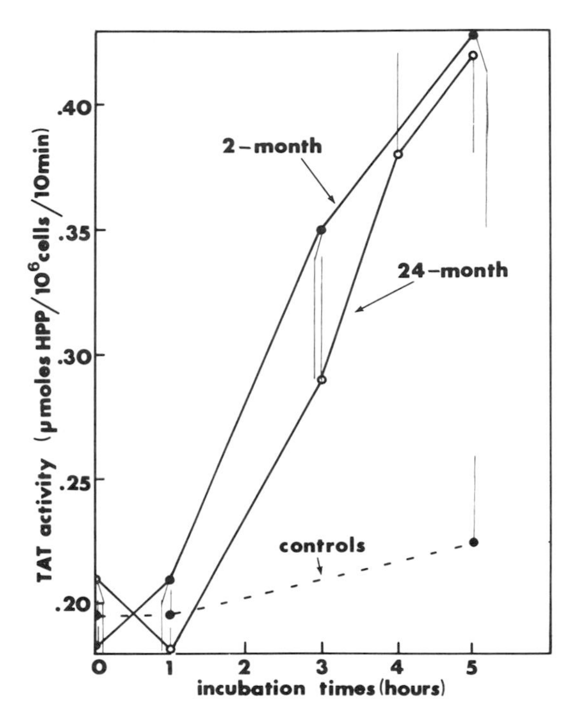

FIG. 6. Hydrocortisone-stimulated tyrosine aminotransferase activity in isolated liver cells from rats of different ages (from ref. 7).

to hormone-receptor interaction. A second difficulty concerns the controversial nature of reported effects of aging on the binding of another hormone, glucocorticoids, to its receptor molecules (13–16). (c) When suspensions of hepatocytes are prepared from rats of different ages and incubated with hormones *in vitro* (7), both time course and magnitude of adaptive increases in enzyme activity are identical. This is illustrated in Figure 6 for the hydrocortisone-stimulated increase in tyrosine aminotransferase activity. The difficulty in this experimental approach is that no information is obtainable relevant to the total number of cells which respond to hormonal stimulation in the intact liver of rats of different ages.

HORMONAL REGULATORY MECHANISMS

The regulation of insulin and corticosterone levels in blood was examined in rats of different ages. In each case disturbances that are capable of contributing to altered patterns of hepatic enzyme adaptation are evident during aging.

Figure 7 illustrates glucose-stimulated fluctuations in the concentration of immunoreactive insulin in serum collected from portal vein blood in rats of the indicated ages. Under identical experimental conditions, there is a remarkable analogy between the age-dependent delay in time of onset for the glucose-pro-

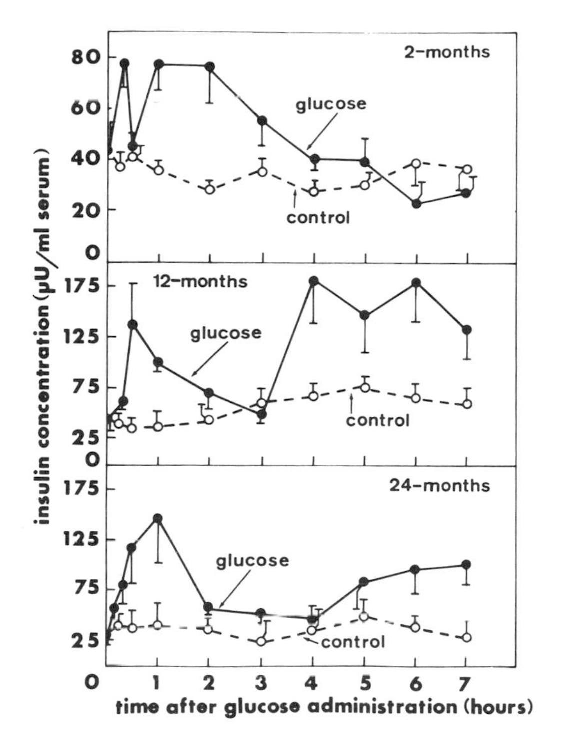

FIG. 7. Glucose-stimulated levels of immunoreactive insulin in portal vein blood from rats of different ages (from ref. 12).

voked adaptation in hepatic glucokinase activity and the delayed onset of the second phase of the insulin response from 0.5 hr to 3–4 hr to 6–7 hr at 2, 12, and 24 months of age, respectively (1,12). That such a phenomenon probably reflects age-dependent changes in the pancreatic regulation of insulin secretion was indicated by the detection of similar results when islets of Langerhans are isolated from rats aged 2 to 24 months and are perifused with glucose *in vitro.* Current efforts are focusing on the heterogeneous nature of the immunoreactive insulin species that are secreted, as well as on distinctive differences in the control of insulin secretion by islet populations that are extremely heterogeneous in size.

Figure 8 shows starvation-enhanced fluctuations in the concentration of corticosterone in serum collected from aortal blood in rats and mice of the indicated ages. Under identical experimental conditions, there is a remarkable resemblance between the age-dependent reduction in magnitude of response for the starvation-provoked adaptation in hepatic tyrosine aminotransferase activity and the reduced magnitude of the corticosterone response as the rats and mice age from approximately 2 to 24 months (6,8). Initial measurements were undertaken in Sprague-Dawley rats, in which body weight increased from approximately 150

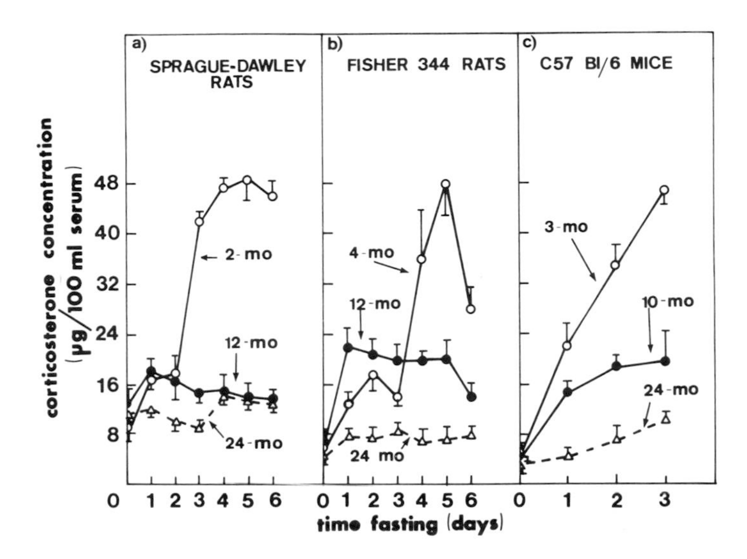

FIG. 8. Starvation-stimulated levels of corticosterone in aortal blood from rats of different ages (from ref. 6).

to 750 g between 2 and 24 months of age. Thus, it was not possible to distinguish between a manifestation of aging and the equally likely possibility that old large rats did not consider the employed extent of starvation employed in the experiment to be a particularly severe stress, i.e., there was no need to elicit secretion of corticosterone. However, identical results in three animal models that differ enormously in lifetime patterns of growth indicate that the impaired corticosterone response is a true expression of aging rather than a consequence of species- or strain-specific control of body weight (8). Additional results further indicate that the impaired capability for adrenal steroidogenesis *in vivo* probably reflects a deficiency in specific neuroendocrine control mechanisms.

CONCLUSIONS

One well-documented manifestation of biological aging is modification in the capability for adaptive regulation of enzyme activity. Progressive alterations in hepatic enzyme adaptation during aging probably reflect a variety of disturbances in endocrine regulatory mechanisms. Such endocrine changes undoubtedly contribute to the expression of important age-associated disease, such as maturity-onset diabetes. However, this knowledge creates little need to evoke an endocrinological theory of aging, just as very similar information provides little need to evoke a neurological theory of aging or an even larger body of similar knowledge provides little need to evoke an immunological theory of aging.

What is apparent is that beginning relatively early in life, each of these major regulatory systems of intact animals deteriorates in predictable fashion. Each of them contributes to the expression of age-associated diseases. Each of them interacts with one another in only partially understood ways. The deterioration of each will ultimately be understood in concrete cellular and molecular terms. When such biochemical events are localized and identified, then it will become possible to explore those fundamental processes as they develop and manifest themselves in different manners in various tissues, as well as how they relate to potential health and longevity.

ACKNOWLEDGEMENTS

Research described in this article was supported in part by grants AG-00368, AG-00431, and CA-12227 from the NIH and an Established Investigatorship from the American Heart Association.

REFERENCES

1. Adelman, R. C. (1970): *J. Biol. Chem.,* 245:1032.
2. Adelman, R. C. (1970): *Nature,* 228:1095.
3. Adelman, R. C. (1971): *Exp. Gerontol.,* 6:75.
4. Adelman, R. C. (1975): In: *Enzyme Induction,* edited by D. V. Parke, p. 303. Plenum Press, London.

4a. Adelman, R. C., and Britton, G. W. (1975): *Biosci.,* 25:639.

5. Adelman, R. C., and Freeman, C. (1972): *Endocrinol.,* 90:1551.
6. Adelman, R. C., Britton, G. W., Rotenberg, S., Ceci, L., and Karoly, K. (1978): In: *Effects of Genetics on Aging,* ed. D. Harrison. National Foundation March of Dimes Original Article Series, XIV: 355–364.
7. Britton, G. W., Britton, V. J., Gold, G., and Adelman, R. C. (1976): *Exp. Gerontol.,* 11:1.
8. Britton, G. W., Rotenberg, S., Freeman, C., Britton, V. J., Karoly, K., Ceci, L., Klug, T. L., Lacko, A. G., and Adelman, R. C. (1975): In: *Explorations in Aging,* edited by V. J. Cristofalo, J. Roberts, and R. C. Adelman, p. 209. Plenum Press, New York.
9. Cohen, B. J., Anver, M. R., Ringler, D. H., and Adelman, R. C. (1978): *Fed. Proc. (in press).*
10. Finch, C. E., Foster, J. R., and Mirsky, A. E. (1969): *J. Gen. Physiol.,* 54:690.
11. Freeman, C., Karoly, K., and Adelman, R. C. (1973): *Biochem. Biophys. Res. Commun.,* 54:1573.
12. Gold, G., Karoly, K., Freeman, C., and Adelman, R. C. (1976): *Biochem. Biophys. Res. Commun.,* 73:1003.
13. Parchman, L. G., Cake, M. H., and Litwack, G. (1978): *Mech. Age. Devel.,* 7:227.
14. Petrovic, J. S., and Markovic, R. Z. (1975): *Dev. Biol.,* 45:176.
15. Roth, G. S. (1974): *Endocrinol.,* 94:82.
16. Singer, S., Ito, H., and Litwack, G. (1973): *Intl. J. Biochem.,* 4:569.

Physiology and Cell Biology of Aging
(Aging, Volume 8), edited by A. Cherkin, et al.
Raven Press, New York © 1979.

Synaptic Growth in Aged Animals

Carl W. Cotman and Stephen W. Scheff

Department of Psychobiology, University of California, Irvine, California 92717

At the turn of the century, it was suggested that the connections between nerve cells might change and that these changes might modulate different behaviors. Recent evidence is in fact consistent with this idea. Neuronal circuitry is adjustable at the level of synaptic growth. In response to partial denervation, undamaged fibers may sprout and form new functional connections [see (3)]. Moreover, different environmental influences may induce synapses to grow/or reorganize, or both (8; see also Bennett, *this volume*.) Such results indicate that the brain has the potential to alter its circuitry.

The plasticity of central neurons in aged animals is a particularly critical issue because the aged brain is more susceptible to neuronal loss than the young brain. Neuronal loss is a normal consequence of the aging process as well as a result of such common disorders of the aging nervous system as stroke, tumors, and senile dementia. Since the formation of new synapses in the mature brain (reactive synaptogenesis) may ameliorate the functional effects of neuronal loss, we have compared the ability of fibers in aged and young adult rat brains to grow in response to lesions.

In this chapter, we will describe our studies on the changes in synaptic growth which occur in the hippocampus as a result of partial denervation in aged animals, and we will compare and contrast the changes with those occurring in young adult animals.

BACKGROUND

The hippocampus has a variety of involvements: it is an associative area, it serves in some aspects of short-term memory, and it serves in changing animals' behavioral responses. Most recently it has been proposed to be a modulator of spatial behaviors. It is a well-defined system—probably the only simpler area in the mammalian brain is the cerebellum.

The hippocampus is a horn-shaped structure underlying the cerebral cortex. The main input into the hippocampus comes from the cerebral cortex (area entorhinal). Projections from the cortex synapse in a portion of the hippocampus called the dentate gyrus (Figure 1). The dentate gyrus occupies a key position in hippocampal-cortical transactions. The dentate is the first integrative area

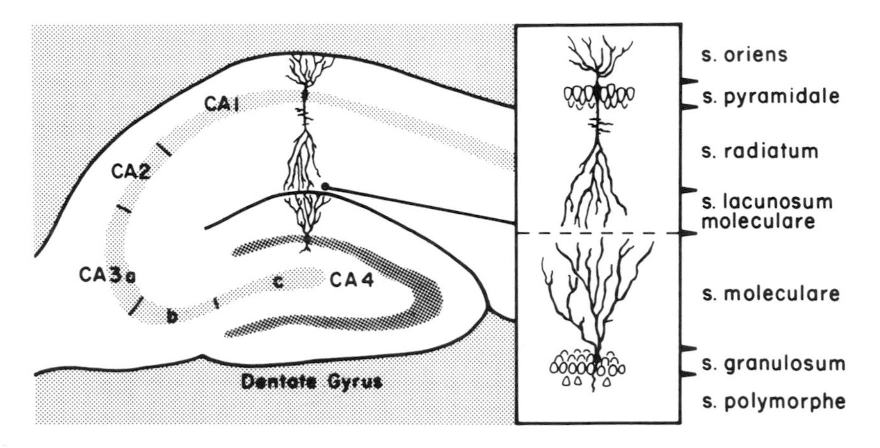

FIG. 1. Schematic diagram of the hippocampus showing the major subfields. (From ref. 5.)

for signals coming from the cerebral cortex. Decisions are made and sent to other hippocampal fields, where they are integrated further, and finally disseminated to other areas of the brain, as well as fed back to the cortex. As the initial control point, the capacity of the dentate to modify its circuitry is of particular relevance.

The primary cells of the dentate gyrus are granule cells, which are arranged in a neat layer (see Figure 1). Their dendrites arborize in a zone referred to as the molecular layer. The ordering of inputs on the granule cells is exquisite. The inputs are strictly ordered so that the inner ¼ of the dendritic field receives a projection from the contralateral and ipsilateral hippocampus (the commissural and associational projections, respectively), whereas the outer ¾ of the dendritic field receives a massive ipsilateral projection from the entorhinal cortex (see Figure 2). A sparse input from the contralateral entorhinal cortex and the septum is also present along with the ipsilateral entorhinal input. The lamination of synaptic fields on the granule cell dendrites is one of the most precise in the CNS. There is virtually no overlap between the commissural-associational and entorhinal inputs.

EFFECT OF PARTIAL DENERVATION IN YOUNG ADULT RATS

We have studied the response of the dentate gyrus following denervation of different portions of the input to the granule cells (4,5). Here we shall describe the responses following a unilateral entorhinal lesion.

After an entorhinal lesion, there is an immediate and precipitous loss of synapses in the dentate molecular layer, which continues for nearly 240 days (Figure 3, open circles). The entorhinal projection accounts for 85% of the input to the outer dendritic field, or nearly 60% of the total input to these cells (11).

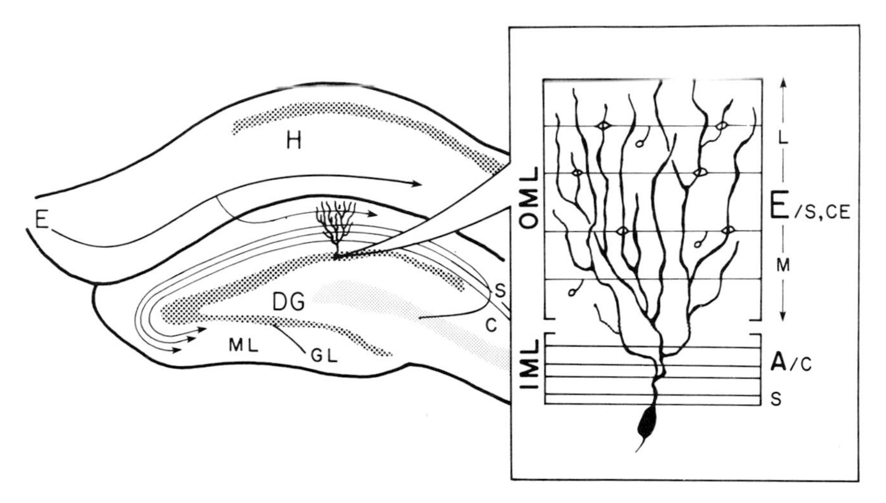

FIG. 2. Organization of inputs on the granule cells of the dentate gyrus. E, entorhinal; C/A, commissural and associational; S, septal; ML, molecular layer. (From ref. 2.)

Such a loss might be traumatic indeed. What happens to the granule cells and the residual inputs following an entorhinal lesion?

Initially the number of normal synapses is very low but then, over time, it recovers to approximately the number present prior to denervation. Growth starts 4–10 days after the operation, rapidly increases for the first 30 days, and continues at a slower rate thereafter. In the electron microscope the recovered area looks surprisingly normal. The dendrites are slightly distorted, but the synapses are for the most part normal in appearance (12).

A critical question is what are the sources of input that grow to repopulate

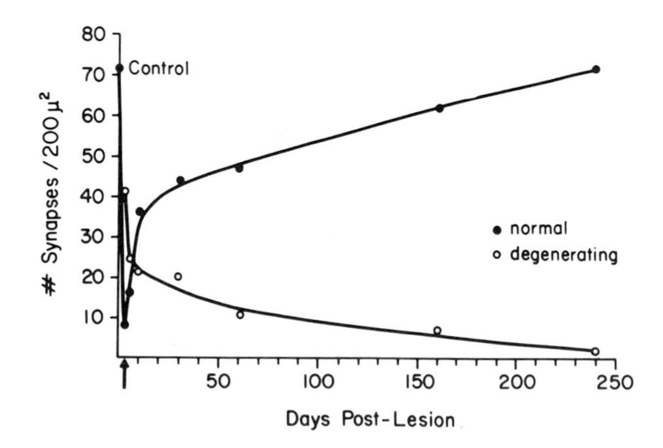

FIG. 3. Changes in the number of degenerating *(open circles)* and normal synapses *(closed circles)* following a unilateral entorhinal lesion. (From ref. 6.)

this area? Are they similar to the type that is lost? The most specific replacement would be to recreate an entorhinal field by fibers arising from the remaining entorhinal cortex. However, that is definitely not the case. Fibers from the remaining contralateral entorhinal cortex proliferate, but these account for only a few of the new synapses (7). In addition, septal, commissural, and associational fibers grow (see Figure 4). Catecholaminergic fibers do not show axon sprouting. Where it has been possible to test the new synapses by electrophysiological measures, they have been found to be functional.

Thus, commissural and associational fibers grow outside their normal zone into the denervated zone; septal inputs appear to retract. The consequence of a unilateral lesion is to selectively reorder granule cell inputs. A new pattern of afferent lamination is created.

It appears that the denervation results not only in a morphological reorganization, but also in a chemical one. The neurotransmitter of the entorhinal pathway appears to be glutamate (15,18), whereas the reactive septal fibers and commissural fibers that replace them appear to be cholinergic and aspartergic, respectively. GABAnergic interneurons present in the molecular layer also appear reactive (15). Thus reinnervation is associated with a change of neurotransmitter types. The specificity of the response is clearly independent of the nature of the transmitter.

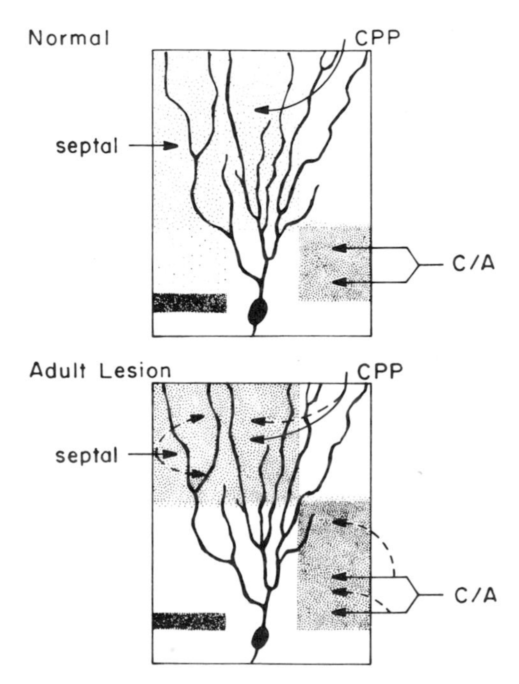

FIG. 4. Reorganization of synaptic inputs on dentate granule cells. The dashed lines indicate growth responses following an entorhinal lesion. (From ref. 5.)

We can summarize the properties of reactive synaptogenesis in the dentate gyrus of the young adult. Following partial denervation, new synapses begin to form in a few days, and by 240 days appear to replace nearly all those that were lost. Septal, commissural, associational fibers, and perhaps interneurons respond, whereas catecholaminergic ones do not. Thus there is a hierarchy of responsive growth. Commissural and associational fibers expand beyond their normal lamina, creating a new synaptic order in the dentate gyrus.

EFFECT OF PARTIAL DENERVATION IN AGED RATS

Is there regrowth of connections following cell loss in the aged rats? If so, does it take place at the same rate and extent as in young adult rats? Before describing the responses to lesions, we need to examine properties of the hippocampus in aged, unoperated animals to provide a basis for comparison.

In a detailed examination of the biochemical properties of the cholinergic septal projection to the hippocampus (Table 1), we found that the quantity of total protein in the hippocampus in a 26-month-old animal is larger than that in a 3-month-old. The activity of acetylcholinesterase does not change. Choline acetyltransferase activity increases significantly, but the change is not large, at best 15–20%. Na-dependent, high affinity choline uptake decreases slightly. Cholinergic receptors assayed by examining the binding of the antagonist quinuclidyl benzilate (QNB)(19) show little evidence of a change. Overall, although there are changes, this cholinergic system appears relatively stable. Certainly it is more stable than some of the catecholaminergic systems in the basal ganglia.

In general, we find that the status of synapses in the hippocampus of aged rats is similar to that of the younger ones. As reported elsewhere (6), the appearance and abundance of fibers appear unchanged, as does the average density of synapses per unit area.

At various times following a unilateral entorhinal lesion, the molecular layer in aged rats was analyzed by electron microscopy. Animals operated on at 3 months of age were compared to those operated on at 24 months of age. As shown in Figure 5, at two days postlesion, both groups show an approximately equal density of intact synapses per 100 μ^2. At four days postlesion, the younger animals begin to show a slight increase in the number of synapses and, by ten days postlesion, the younger animals have more synapses than the older animals. By 60 days postlesion, the synaptic density is normal in the three-month-old animals, whereas the aged animals have significantly fewer synapses. Since normal synaptic densities are similar in both age groups, this failure to repopulate the terminal field cannot be explained by presurgical differences in synaptic density between groups. Also the differences could not be explained by shrinkage, since measurements indicate equal changes in both age groups.

The overall appearance of synapses in the outer molecular layer of the aged animals 60 days after the lesion is generally similar to that of synapses in the younger animals, given equivalent recovery times (Figure 6). However, there

TABLE 1. *Cholinergic properties in the hippocampus of mature and aged rats*

	Total			Specific activity		
	3 month	26 month	% difference	3 month	26 month	% difference
Protein (mg)	14.55 ± 0.61	18.25 ± 0.19	125 ($p < 0.01$)			
Acetylcholinesterase (μmol/30 min) (μmol/g/30 min)	11.8 ± 0.4	12.4 ± 0.4	105	808 ± 7	680 ± 19	84 ($p < 0.0025$)
Choline acetyltransferase (nmol/30 min) (nmol/g/30 min)	238.4 ± 8.4	280.0 ± 4.8	117 ($p < 0.01$)	16.4 ± 0.4	15.4 ± 0.2	94 ($p < 0.05$)
Uptake (pmol/4 min) (pmol/4 min/mg)	55.7 ± 7.7	48.3 ± 1.9	88	9.29 ± 1.14	6.96 ± 0.54	75
QNB (pmol) (fmol/mg)	10.7 ± 0.9	12.1 ± 0.7	113	735 ± 34	648 ± 35	88

Acetylcholinesterase and choline acetyltransferase were assayed as previously described (14). The measurement of high affinity choline uptake was carried out as described by Shelton, et al. (16) and QNB binding as described by Yamamura and Synder (19). (Nadler, J. V., Shelton, D. L. and Cotman, C. W., unpublished observations).

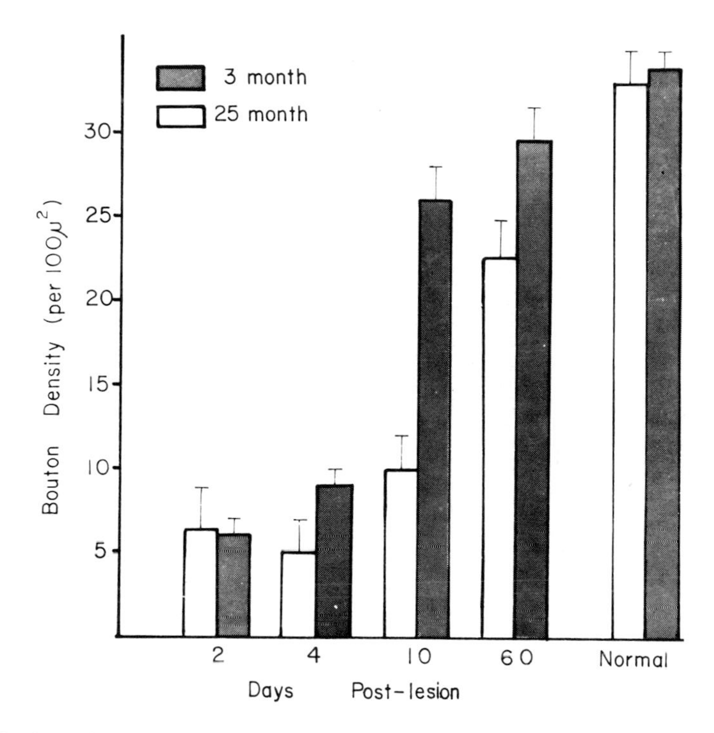

FIG. 5. Number of synapses per unit area of young adult and aged rats at various times after a unilateral entorhinal lesion (From ref. 0.)

are some differences. There is, for example, a greater number of U- and W-shaped synapses and a greater number of multiple synaptic contacts on a single axon terminal in the aged animals. Astrocytic processes also appear more prominent. The general shape and overall length of synaptic junctions is similar.

The response of individual fiber systems to denervation was also examined. The expansion of the commissural-associational systems can be readily and accurately followed by staining the fibers with a silver stain, the Holmes' stain. Figure 7a shows the appearance of the commissural-associational zone 2 days after an entorhinal lesion. Over time this plexus expands (Figure 7b). At 12 days postlesion, the younger animals demonstrated approximately a 22% increase over normal, whereas the aged animals demonstrated only a 10–11% increase (see Figure 8). Three months after the operation, the commissural-associational zone still has not expanded to the same extent as in the younger animals. In the group of aged animals there is much more variation between animals than in the younger group. In some animals the response is nearly identical to that in the younger group, whereas in others there is practically no outgrowth of the fiber plexus. Experiments conducted with AChE histochemistry on the response of the septal system indicate that, like the commissural-associational system, it is reactive, but the reaction is less pronounced.

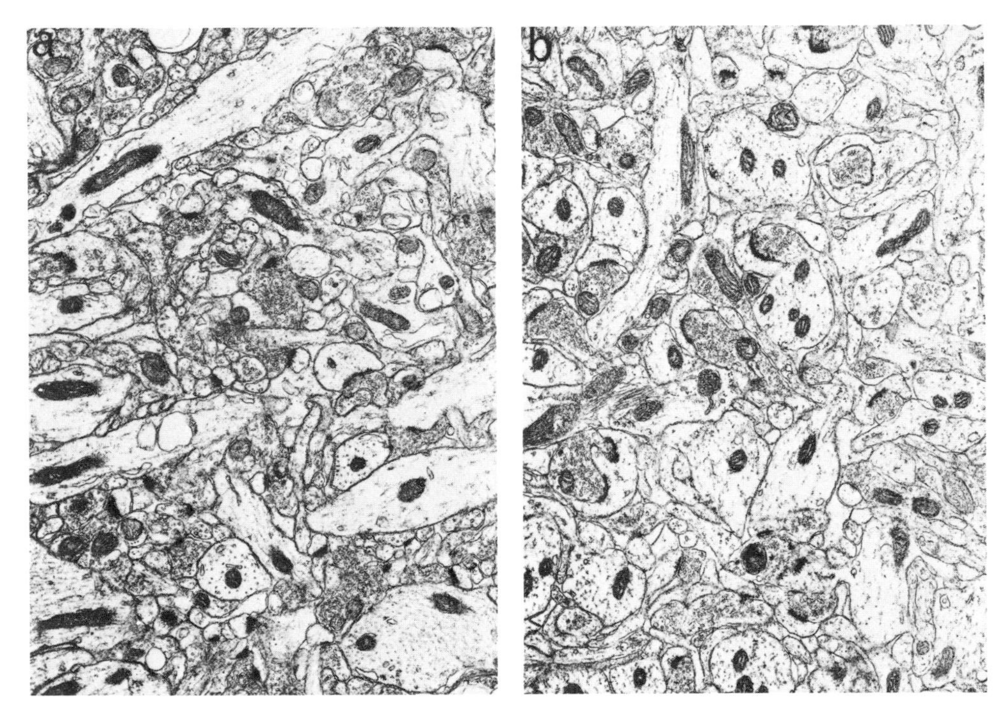

FIG. 6. At 60 days postlesion the denervated side regained some of its synaptic density and began to show a normal appearance. **a)** A reinnervated 25-month-old animal. **b)** A reinnervated 3-month-old animal. (From ref. 6.)

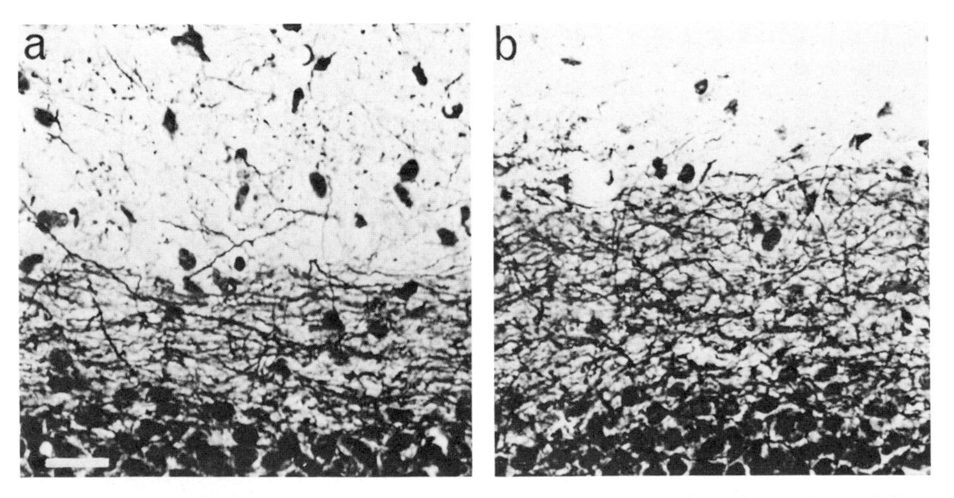

FIG. 7. Holmes' fiber stain of the dentate gyrus showing the commissural-associational fiber plexus. **a.** Pattern 2 days after a unilateral lesion. **b.** Pattern 15 days after the lesion. Note the expansion at 15 days compared to 2. Calibration bar = 25 μm. (From ref. 6.)

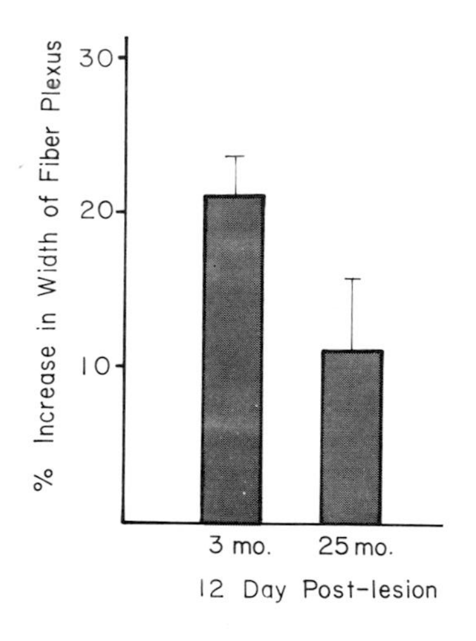

FIG. 8. Outgrowth of the commissural-associational fiber plexus in 3-month-old and 25-month-old rats at 12 days postlesion. (From ref. 6.)

We have also examined the reactions of central catecholaminergic fibers in the septum and of central sympathetic fibers in the hippocampus following a fimbrial transection. In both cases the sprouting response is less pronounced in the aged animals (6). The paucity or, in some cases, near-absence of anomalous growth by sympathetic fibers in older animals is particularly important because it indicates that the deficiency in growth is not restricted to central neurons or to any particular brain region. As a general rule, reactive growth declines with age.

In response to denervation, many of the remaining undamaged neurons grow and form new synapses to replace those lost in either young or old animals. However, this regenerative response is slower and apparently not as extensive in old rats, at least over the time period examined.

POSSIBLE MECHANISMS FOR DIMINISHED PLASTICITY IN AGED ANIMALS

The reason for this diminished plasticity is largely unknown, but some clues are emerging. It may reflect a reduction in the ability of neurons to synthesize or assemble materials necessary for growth. Alternatively, growth-inducing substances may not be so readily elaborated, or growth-suppressing substances may be more abundant.

It has been demonstrated that circulating corticosterone levels are elevated in aged animals and are almost three times as high as in younger animals (1). Since the hippocampus possesses specific receptor sites for corticosterone (9), it is important to examine whether high corticosterone levels could be responsible for certain age-related changes observed in the hippocampus, such as the decreased plasticity.

If elevated corticosterone levels depress plasticity in aged rats, then young rats with elevated levels should show a similar depression. Accordingly, hydrocortisone—an analog of corticosterone—was administered subcutaneously to young adult subjects. The young adult animals were pretreated six days before the lesion and for fifteen days after surgery. Control animals received the vehicle alone; other animals received no treatment other than the lesion. Yet another group of animals was treated with the hormone, but was given no lesion.

The spread of the commissural-associational fiber plexus was examined by the Holmes' stain. In young adult animals, as was previously described, the fiber plexus expands outwardly approximately 22% by 15 days after a unilateral entorhinal lesion. Animals treated with corticosterone, on the other hand, show a marked reduction in this outgrowth (Figure 9). The fiber plexus spreads only 13.5% at 15 days after operation. This percentage of spread corresponds very closely to what we found in the aged animals given 15 days' survival. The aged animals were observed to exhibit a spread of only 12.5% (Figure 9).

These same brains were stained for AChE activity with similar results. Animals treated with corticosterone generally demonstrate a lighter staining pattern than nontreated controls. Thus, the evidence suggests that a decrease in the growth response may in some way be connected to an increase in circulating levels of corticosterone. We do not know the exact mechanism by which the change in hormone levels might alter the induced growth response following injury. However, we speculate that an increased involvement of the astrocytes might somehow hinder the outgrowth of the fibers by physically or chemically suppressing growth.

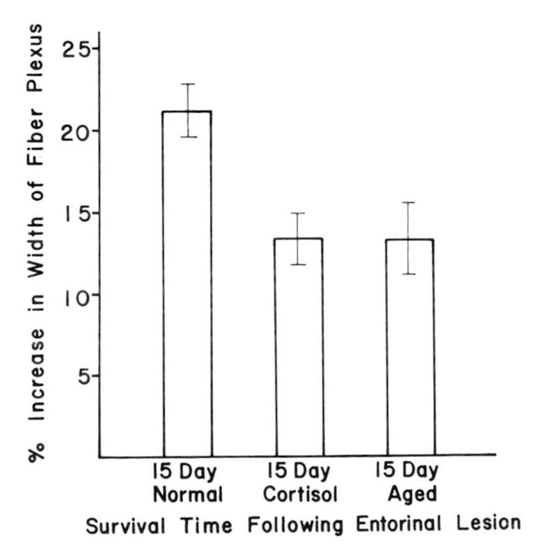

FIG. 9. Outgrowth of the commissural-associational fiber plexus following a unilateral entorhinal lesion in 3-month-old rats with and without cortisol treatment compared to 25-month-old untreated rats.

Examination of coronal sections of the hippocampus from animals treated with hydrocortisone revealed quite different astrocytic effects. Astrocytes in treated animals appear hypertrophied despite the fact that no lesion was made. Their processes are thickened, and cell bodies are irregular in shape. The distribution of the astrocytes does not change. This effect is observed throughout the hippocampus—not only in the molecular layer of the dentate gyrus, but also in the various hippocampal fields. Astrocytes are known to exhibit such hypertrophy normally in response to brain injury in adjacent regions and during Wallerian degeneration (13,17). However, electron microscopic analysis failed to reveal any massive degenerative debris, so the astrocytes must be serving some other role. Of particular significance is that the astrocytes in the normal animals treated with hydrocortisone appear identical to those described in aged animals (6,10). Perhaps the astrocytes indirectly regulate growth, or they may be merely a responsive element in a suppressed neuropil.

CONCLUSION

Age-related differences in reactive fiber growth may have functional significance. In some instances reactive synaptogenesis appears to underlie recovery or retention of normal function after damage to the CNS, but in other instances it seems to cause abnormal behavior, or at least contribute to it. In the aged brain we would expect this process to operate not only in cases of severe damage, such as that induced in the present study, but also in the replacement of connections that are lost as a result of the natural aging process. If the new connections can replace the old functionally, then reactive synaptogenesis may be regarded as a compensatory mechanism that counteracts the negative effects of aging. A reduction in growth capacity with age would therefore be detrimental. On the other hand, if the new connections interfere with normal function, reactive synaptogenesis would be harmful and the aged brain would benefit from a diminished growth capacity. Such issues need direct evaluation and must await a more complete understanding of the significance of reactive synaptogenesis. In any case, our finding that neuronal circuitry appears more rigid in old animals needs to be considered.

ACKNOWLEDGMENTS

This work was supported by a Research Grant from the National Institute on Aging, AG 00538. We wish to thank Julene Mueller for her secretarial assistance.

REFERENCES

1. Britton, G. W., Rotenberg, S., Freeman, C., Britton, V. J., Karoly, K., Ceci, L., Klug, T. L., Lack, A. G., and Adeleman, R. C. (1975): Regulation of corticosterone levels and liver enzyme activity in aging rats. In: *Explorations in Aging,* edited by V. J. Cristofalo, J. Roberts, and R. C. Adelman, p. 209. Plenum Press, New York.

2. Cotman, C. W. (1978): Synapse formation and plasticity in the developing dentate gyrus. In: *Limbic Mechanisms, The Continuing Evolution of the Limbic System Concept,* edited by K. Livingstone and O. Hornykeiwicz (Eds.), pp. 47–65. Plenum Press, New York.
3. Cotman, C. W., editor (1978): *Neuronal Plasticity.* Raven Press, New York.
4. Cotman, C. W., and Lynch, G. S. (1976): Reactive synaptogenesis in the adult nervous system: the effects of partial deafferentation on new synapse formation. In: *Neuronal Recognition,* edited by S. Barondes, pp. 69–108. Plenum Press, New York.
5. Cotman, C. W., and Nadler, J. V. (1978): Reactive synaptogenesis in the hippocampus. In: *Neuronal Plasticity,* edited by C. W. Cotman, pp. 227–271. Raven Press, New York.
6. Cotman, C. W., and Scheff, S. W. (1978): Compensatory synapse growth in aged animals after neuronal death. In: *Mechanisms of Ageing and Development,* edited by B. L. Strehler. Elsevier Sequoia, SA Lausanne *(in press).*
7. Cotman, C. W., Gentry, C., and Steward, O. (1977): Synaptic replacement in the dentate gyrus after unilateral entorhinal lesion: electron microscopic analysis of the extent of replacement of synapses by the remaining entorhinal cortex. *J. Neurocytol.,* 6:455–464.
8. Diamond, M. C. (1978): The aging brain: some enlightening and optimistic results. *Am. Scientist,* 66:66–71.
9. Gerlach, J. L., and McEwen, B. S. (1972): Rat brain finds adrenal steroid hormone: radioautography of hippocampus with corticosterone. *Science,* 175:1133–1136.
10. Landfield, P. W., Rose, G., Sandles, L., Wohlstadter, T. C., and Lynch, G. (1977): Patterns of astroglial hypertrophy and neuronal degeneration in the hippocampus of aged memory-deficient rats, *J. Gerontol.,* 32:3–12.
11. Matthews, D. A., Cotman, C., and Lynch, G. (1976): An electron microscopic study of lesion-induced synaptogenesis in the dentate gyrus of the adult rat. I. Magnitude and time course of degeneration. *Brain Res.,* 115:1–21.
12. Matthews, D. A., Cotman, C., and Lynch, G. (1976): An electron microscopic study of lesion-induced synaptogenesis in the dentate gyrus of the adult rat. II. Reappearance of morphologically normal synaptic contacts. *Brain Res.,* 115:22–41.
13. Murray, H. M., and Walker, B. E. (1973): Comparative study of astrocytes and mononuclear leukocytes reacting to brain trauma in mice. *Exp. Neurol.,* 41:290–302.
14. Nadler, J. V., Cotman, C. W., and Lynch, G. S. (1973): Altered distribution of choline acetyltransferase and acetylcholinesterase activities in the developing rat dentate gyrus following entorhinal lesion. *Brain Res.,* 63:215–230.
15. Nadler, J. V., White, W. F., Vaca, K. W., Redburn, D. A., and Cotman, C. W. (1977): Characterization of putative amino acid transmitter release from slices of rat dentate gyrus. *J. Neurochem.,* 29:279–290.
16. Shelton, D. L., Nadler, J. V., and Cotman, C. W. (1978): Development of high affinity choline uptake and associated acetylcholine synthesis in the rat fascia dentata. *Brain Res. (in press).*
17. Vaughn, J. E., and Pease, D. C. (1970): Electron microscopic studies of Wallerian degeneration in rat optic nerves. *J. Comp. Neurol.,* 140:207–225.
18. White, W. F., Nadler, J. V., Hamberger, A., Cotman, C. W., and Cummins, J. T. (1977): Pre- and postsynaptic evidence favoring glutamate as transmitter of the hippocampal perforant path. *Nature,* 270:356–357.
19. Yamamura, H. I., and Snyder, S. H. (1974): Muscarinic cholinergic binding in rat brain. *Proc. Natl. Acad. Sci. USA,* 71:1725–1729.

Physiology and Cell Biology of Aging
(Aging, Volume 8), edited by A. Cherkin, et al.
Raven Press, New York © 1979.

Pathology and Biochemistry of the Neurofilament in Experimental and Human Neurofibrillary Degeneration

Dennis J. Selkoe, Alice M. Magner, and Michael L. Shelanski

Departments of Neurology and Neuroscience, Harvard Medical School, Children's Hospital Medical Center, Boston, Massachusetts 02115

The remarkable maturing of our population that has occurred in recent decades has particular consequence for those working in the neurological sciences, at both the experimental and the clinical level. In the past 100 years, the rise in life expectancy throughout the world has resulted primarily from a reduction in mortality at the beginning of the chronological scale, because of the control or eradication of many bacterial infectious diseases. On the other hand, the life expectancy of someone who achieved the age of 65 in the United States has risen by only three years between 1900 and 1970. The decline in the birth rate and the potential amelioration of the major fatal diseases of late life, which has so far remained elusive, signify that a much greater absolute and relative number of individuals in society will be reaching an age at which they will be subject to the neuronal abiotrophies that occur with senescence. This is borne out by recent epidemiological data on what is by far the most common neuronal degeneration of late life, senile dementia of the Alzheimer type. Roth and colleagues[1] have reported that the prevalence of this disorder has risen more than 20% during the past decade in the particular community which they surveyed, Newcastle-Upon-Tyne in England. They state that marked dementia now affects 6.2% of the population over 65 in that city, and accounts for 65% of all nursing home admissions. Other preliminary and incomplete attempts to determine a prevalence of senile degenerative dementia have resulted in estimates of over 1.2 million cases in the United States, or 6 individuals per 1,000 in the entire population (22). Terry (38) has pointed out that such data, despite their limitations, clearly suggest that senile dementia is approximately 14 times as common as multiple sclerosis, even in high-risk northern latitudes, and about 100 times as common as amyotrophic lateral sclerosis. Yet to date, the major tools of neurobiology and biochemistry, which are now being intensively used

[1] M. Roth (1977): At Workshop Conference on Alzheimer's Disease-Senile Dementia and Related Disorders. Bethesda, MD. National Institutes of Health, Bethesda, Maryland, June 6–9.

to elucidate other less common neurological disorders, have scarcely been applied to a search for etiologic clues in presenile and senile dementia.

We would like to review some very formative attempts, in our laboratory and others, to learn something about the molecular pathology of such neuronal degenerations, particularly as they relate to a particular group of neuronal proteins, the fibrous proteins. Past attempts to obtain biochemical information about neuronal degenerative diseases like Alzeheimer's disease have tended to employ techniques of quantitative compositional analysis, i.e., small or large areas of autopsied cerebral hemisphere have been sampled and homogenized, and a variety of structural constituents and enzymes have been assayed. The results of such studies have shown considerable interlaboratory differences in both the degree and direction of reported compositional changes. Those specific changes in structural components (such as total proteins, total lipids, cerebrosides, and gangliosides) that are generally agreed upon appear to be the result of neuronal loss and cortical atrophy that has become profound by the terminal stage of the disease. Since no clear picture of the mechanism—or even the results—of age-related neuronal degeneration has emerged from this quantitative approach, we are now beginning to see the application of techniques of contemporary neuronal cell biology, which have been used for some time in work on other human neurological disorders such as myasthenia gravis and multiple sclerosis, to study the process of neurofibrillary degeneration, which is the pathological hallmark of Alzheimer's disease. This new cell-by-cell type of analysis relies strongly on simultaneous biochemical and ultrastructural study of the diseased neuron and employs techniques such as neuronal/glial isolation, immunohistochemistry at the light and electron microscopic levels, analytical protein separation, and neuronal tissue culture.

NEUROPATHOLOGICAL BACKGROUND

Before describing our recent work on the chemistry of the neuronal filament in the normal and diseased nervous system, we believe it would be appropriate to provide some background on why someone interested in the neurochemistry of memory in general and in age-related intellectual deterioration in particular would wish to study the neuronal fibrous proteins. As is well known, the classical morphological alterations of human brain found in Alzheimer's disease and in normal individuals of great age are the neuritic, or senile, plaque and the neurofibrillary tangle. (When we use the term Alzheimer's disease, we will be referring to both presenile and senile dementia due to neuronal degeneration.) There is now considerable evidence that common degenerative senile dementia and the classical presenile dementia that Alzheimer described in 1907 are one and the same disease, indistinguishable on clinical, neuropathological, or ultrastructural grounds. Most investigators in the field now use the term Alzheimer's disease for both forms. The epidemiological question of whether the incidence of the disorder rises gradually with age, suggesting a single pathogenetic process, or whether it has a bimodal age distribution before and after age 65, has not yet been settled.

The neuritic plaque is a cortical lesion consisting of many abnormal neuritic processes lying in an extracellular matrix that is largely composed of amyloid. The enlarged processes in this structure have been shown to be mostly presynaptic elements, namely, axonal boutons that contain abnormal fibrous organelles identical to those of the neurofibrillary tangle, as well as degenerating lysosomes and mitochondria (17). The postsynaptic dendrites and the synaptic cleft appear normal. The neurofibrillary tangle, on the other hand, is a wholly intracellular lesion which consists of a tangled mass of abnormal cytoplasmic fibrils that are highly argyrophilic when stained with silver preparations, and then resemble a Brillo pad. The ultrastructure of the abnormal fibers found in these two classical lesions of Alzheimer's disease has been the subject of some debate (23,38,43). There is now growing agreement that the abnormal organelles are made up of a pair of helically wound 9 to 10 nm filaments, rather than a pinched or "twisted" microtubule-like structure, as was previously believed. This morphological issue has considerable importance for the interpretation of the protein chemical data we will present.

The third classical lesion of Alzheimer's disease, granulovacuolar degeneration of cortical neurons, is only inconstantly seen in such brains and even less frequently in the normal aged brain. The granulovacuolar change does not involve abnormalities of fibrous proteins. We will also not discuss here the controversial question of the degree of cortical neuronal loss in Alzheimer's disease versus normal aging.

Although neurofibrillary degeneration of the perikarya and the neuropil is most marked and specific in Alzheimer's disease, it is also found in several other neurological disorders that have dementia as a prominent feature. These include postencephalitic Parkinson's disease, the parkinsonism-dementia complex found in remarkably high incidence in the Chamorro Indians on Guam, and the slow measles virus infection, subacute sclerosing panencephalitis. Of particular interest is the development of a neurofibrillary degeneration that is ultrastructurally identical to Alzheimer's disease in a high proportion of patients with Down's syndrome (trisomy 21) after the third decade (13). Careful psychometric testing can reveal the development of this superimposed "presenile" dementia in these already retarded adults (9). It has been postulated that the neurofibrillary degeneration may be a manifestation of accelerated cellular aging, as may occur in other organs of the progeric-like mongols.

It is noteworthy that the unusual paired helical filaments of the Alzheimer neurofibrillary tangle and the neuritic plaque occur only in the human and only in brain. Lesions that resemble primitive senile plaques have been reported in the cortex of very old monkeys and dogs. However, the fine structure of such plaques is clearly distinguishable from the human change by the absence of the twisted fibrous organelles. Recently, there has been some experimental interest in a model of the neuritic plaque developed in the mouse (42). It seems that certain strains of inbred mice will develop cortical plaques as a response to the intracerebral inoculation of the scrapies virus. At the light microscopic level, these plaques are identical to those in Alzheimer's disease. Although this work raises the spectre of a slow virus etiology for Alzheimer's disease, it must be pointed out that no abnormal fibrous organelles are found in the neurites of the scrapie-induced plaque. Thus, the paired helical filaments remain uniquely human.

One may readily question the importance of neurofibrillary degeneration in the pathogenesis of memory loss and dementia when these lesions are also the most important structural change found in the brains of intellectually normal old people. However, recent quantitative histopathological studies (2,3,40,41) have demonstrated very significant differences in the density and distribution of plaques and tangles in cortex between the normal aged and those with the clinical syndrome of Alzheimer's disease. The clinical syndrome of Alzheimer's disease is a progressive, generalized loss of intellectual function, particularly of recent memory (although there is preservation of alertness), which appears as the *initial* neurological disturbance, clearly preceding the development of positive physical findings on the neurological exam. Among the so-called irreversible dementias, Alzheimer's disease stands out from the other neuronal degenerations, such as Huntington's chorea, and from multi-infarct dementia by the virtual absence of any important motor disturbance or other focal neurological signs during the early development of the syndrome.

The quantitative analyses of Tomlinson and others have documented the steady accumulation of neuritic plaques and neurofibrillary tangles with advancing age. Microscopic sections of cortex reveal small numbers of neuritic plaques in approximately 15% of normals by age 50, in roughly 50% by age 70, and in 75% by age 90 (41). Although the percentages may seem impressive, the actual number of plaques per microscopic field in individuals who have been carefully tested and found to have normal intellect within 6 months of death is very low—less than 5 plaques per field in 70% of such subjects and less than 10 plaques per field in 90%. The mean plaque count in nondemented subjects over age 65 is only 3 per field, compared to 15 per field in demented patients of similar age (difference significant at $p < 0.001$) (41). Moreover, within the nondemented group, those patients who made minimal errors on intellectual testing had mean plaque counts of 5.6 per field compared to only 1.5 in those with full preservation of mental status. The authors of these studies have further pointed out that in mentally normal subjects showing plaque formation, the lesions are discrete and widely separated, with only middle cortical layers involved, whereas in the demented, plaques tend to cluster and overlap, with involvement of all layers and the appearance of abnormal neuropil between plaques.

A similar situation obtains for the neurofibrillary tangle (NFT). These intraneuronal lesions may be found in small numbers in the brains of a few normals in middle age and in a majority after the age of 65. By age 90, few patients are devoid of neurofibrillary tangles. However, the lesions in controls are few in number and largely confined to hippocampus and amygdala. Even there, scanning light microscopy has documented that the number of tangle-bearing hippocampal neurons is 6 to 40 times greater in Alzheimer's disease than in age-matched controls (2). A pattern of moderate or large numbers of tangles widely scattered throughout neocortex is totally restricted to demented patients (41).

Taken together, the available clinical, psychometric, and histopathological data suggest that the dementia that occurs in approximately 10% of those over 65 can be clearly distinguished from the pattern of brain aging in the vast majority of the elderly population. It is possible, then, that Alzheimer's disease represents a pathological entity, not the mere acceleration of normal aging. One may conclude that neurofibrillary change either is pathogenetically associated with, or is at least a clear marker for, a gross abnormality of neuronal function, which results in memory impairment.

THE NEURONAL FIBROUS PROTEINS

Let us now consider the three major classes of neuronal fibrillar organelles and focus on recent and ongoing attempts to understand the biochemistry of the normal neuronal filament.

Microtubules

These structures have received the greatest attention to date, and their biochemistry has been elucidated in considerable detail. Microtubules are long cylindrical organelles of 24 nm diameter with an apparently hollow lumen and a 6 nm wall made up of globular subunits arranged as 13 longitudinal protofilaments (29). They can be up to several microns in length. They are ubiquitous in eukaryotic cells but are found in particularly high concentration in neurons, where they are often referred to as neurotubules. These are seen both in the perikaryon and in the processes. In addition, microtubules make up the spindle fibers of the mitotic apparatus and are the major structural elements of cilia and flagella. In these motile processes, small "side arms" can be seen on one of each of the tubule pairs. These "arms" are composed of an ATPase called dynein which is critical for flagellar motion.

Biochemically, microtubules are long polymers of the subunit protein, tubulin. Tubulin is a dimeric molecule with a molecular weight of 110,000 daltons and a sedimentation coefficient of 6S. The tubulin dimer is in turn composed of two nonidentical monomers with approximate molecular weights of 55,000 (α-tubulin) and 53,000 (β-tubulin). Isolated tubulin invariably is found to have guanine nucleotides bound to it. This binding of guanosine 5'diphosphate (GDP) and -triphosphate (GTP) appears to have great importance in the polymerization of tubulin subunits into microtubules (15). It is now apparent that the formation of microtubules from tubulin dimers is remarkably sensitive to a variety of physicochemical variables, including calcium concentration, pH, ionic strength, and temperature. For example, the warming of a cold solution of tubulin subunits to room temperature or above in the presence of GTP will allow their polymerization into ultrastructurally "normal"-appearing microtubules (15). Cooling the sample to 4° C will result in an equally rapid disassembly into tubulin dimers. The mechanism of such assembly is still a matter of debate, but it appears to involve a self-nucleation mechanism in which disc- or ring-shaped structures composed of tubulin serve as nucleation centers for further assembly. GTP binding and hydrolysis also occur during the polymerization of tubulin. The details of this process have recently been reviewed by several authors (16,29).

An important factor in the biochemical study of microtubules has been their affinity for a number of widely used pharmacological agents (29). Foremost among these is colchicine, which binds to tubulin at a ratio of one mole of colchicine per mole of tubulin dimer. The specific affinity of colchicine for microtubule subunits allowed the original isolation and subsequent characterization of this protein: hence, its descriptive name "colchicine-binding protein." Another antimitotic agent, vinblastine, also induces the breakdown of microtubules into subunits by binding to the tubulin molecule at a different site than colchicine does. Additional binding sites in tubulin for podophyllotoxin and other mitotic spindle inhibitors, and for cations like calcium, have either been demonstrated or postulated. Such sites may be important for both the proper assembly and the function of microtubules.

Although the precise intracellular functioning of the microtubules remains unclear, these organelles have been implicated in the following processes: cellular motility; transport; axoplasmic flow; neurotransmitter release; chromosome movements in cell division; and determination and maintenance of cell shape.

Microfilaments

The second class of neuronal fibrillar organelles, microfilaments, are the thinnest of the fibrous structures under consideration, averaging 5 to 6 nm in diameter. They are found in highest concentration in the growing tip, or growth cone, of the axon in the embryonic neuroblast or in cultured nerve cells. Because of their prominence in the growth cone, a motile role for the microfilaments was proposed. It has been shown biochemically that these thin filaments are in fact composed of actin (14,32). As would be expected, the microfilaments can combine with heavy meromyosin or with subfragment 1 of the myosin molecule (21). It seems, therefore, that these thin filaments subserve motile functions in axon outgrowth and synaptic renewal. No role for these filaments in cellular aging or any disease state has yet been proposed, and we will not consider them further.

Intermediate Filaments (IF)

Intracellular filaments of intermediate diameter, ranging from 8 to 11 nm in width, have long been recognized by electronmicroscopists but have received considerably less attention from biochemists, who have concentrated their efforts on the thin filaments (5 to 6 nm) and the microtubules (24 nm). Morphologically, intermediate filaments are defined as long cytoplasmic rods of approximately 10 nm diameter with no apparent lumen and often small projections from their sides ("arms"). When filaments of this type are found in neurons, they are called neurofilaments; when seen in astroglia, they have been called glial filaments. Neuropathologist and neuroanatomists have contended that these two filaments are of different morphology and have differential histochemical staining. Workers in other fields have referred to similar structures as tonofilaments or 10 nm filaments. Very recently, evidence has accumulated that intermediate filaments may, like the thin filaments and microtubules, be the same or similar from species to species. It is still far from certain that intermediate filaments are as highly conserved in evolution as are microtubules, nor is it clear that within a single mammalian species all the intermediate filaments are biochemically identical.

Early ultrastructural studies of brain demonstrated large numbers of intraneuronal 10 nm filaments usually distributed randomly throughout the cytoplasm, as well as tightly packed bundles of filaments of similar diameter in astrocytes. The presence of highly similar but presumably nonidentical filaments in those two cell types posed serious problems for biochemists who wished to isolate and characterize these organelles. It was necessary to develop a method that would give filaments exclusively from one source or the other. A solution to this problem has been to use the NF-rich myelinated axons of cerebral white matter as a starting material. Purification of such axons has been accomplished by gentle homogenization followed by flotation of the axons, whose myelin acts as a "life-vest" in aqueous media, away from cell bodies and organelles (36). The axons can then be denuded of myelin and purified by gradient centrifugation.

Biochemical studies of this presumably neuronal filamentous material have shown that the intermediate filament isolated by this method is markedly different from the microtubule. To begin with, compared to the readily disassembled microtubules, these filaments are highly insoluble in aqueous media of various ionic strengths. Solubilization can be achieved in urea, guanidine, or detergents such as sodium dodecyl sulfate (SDS). When the SDS-solubilized filaments are run on SDS-urea acrylamide gels, the molecular weight of the major protein subunit is 51,000 from bovine white matter, while those from rat are closer to 53,000, with the latter migrating as a closely spaced doublet. Two-dimensional tryptic peptide maps of the bovine major filament band show some similarities to those of β-tubulin (25), but the differences are sufficient and the molecular weights, although close, differ enough to suggest that tubulin and the 51,000 MW filament subunit are distinct proteins. These biochemical differences are borne out by immunological studies using modified Ouchterlony immunodiffusion, which reveals no cross-reaction between antibody prepared against this presumptive NF band and tubulin antigen (46). The reverse experiment, antitubulin antisera reacted with the filament antigen, is likewise negative. These results have recently been confirmed by Liem, et al. (25), using a radioimmunoassay. Anti-brain filament antiserum shows no reaction against either tubulin or its separated monomers. Likewise, antitubulin shows only a very weak cross-reaction with bovine filaments. The results do not support the hypothesis that these two proteins interconvert or that one is the precursor of the other.

Although this major brain filament protein is thus clearly distinguishable from tubulin, its relationship to glial filaments is less clear. Initial work showed that a presumably astroglial filament preparation (GFA) isolated by the method of Dahl (10) has a major band that migrates in proximity to the major 51,000 MW brain filament band on acrylamide gels (46). By immunodiffusion, isolated brain filaments and the glial filament fractions both reacted with antibodies to the glial filament protein, showing a line of identity (46). Conversely, antisera raised to the brain filament major band (BF) cross-reacted with the GFA protein with no visible spurring at the intersection of the precipitant lines. Neither the anti-BF nor the anti-GFA reacted with actin, total tubulin, α-tubulin, or β-tubulin over the wide range of antigen and antibody concentrations assayed (46).

These results were initially interpreted as suggesting great similarity, if not identity, of the subunit proteins of glial and neuronal filaments, both of which were distinct from tubulin. However, the interpretation became more difficult when immunohistological studies using an indirect fluorescent technique were carried out. Antibodies raised against the major protein band (51,000 MW) of the bovine brain axonal filament preparation show very strong staining of fibrous astroglial processes in brain, particularly in cerebellum and hippocampus (47). This pattern is identical to that seen with an antibody to the GFA protein. No clear neuronal staining in the central nervous system is seen. On the other hand, peripheral nerve, where no astroglial cells are present, does show staining with this antibody both in the axon per se and in the nerve sheath (47). If the serum is preabsorbed with the major filament band eluted from the gels, then both the astroglial and the peripheral nerve staining are completely blocked.

The results presented thus far would suggest that either (a) the intermediate filaments of both neurons and glia are closely related, or (b) the bovine axonal filament preparation, to which our antisera were raised, contains significant numbers of glial filaments, which then co-migrate with neuronal filaments on gel electrophoresis; as a result, the antiserum contains activities against both antigens. The latter interpretation is supported by the presence of some tightly packed bundles of filaments resembling glial filaments in electron micrographs of the bovine filament preparations. It is also supported by the very strong immunofluorescent staining of the astroglia with the anti-brain filament antisera (47). On the other hand, this prominent astroglial staining may be due to the much greater concentration of a common filament antigen in glial cells compared to neurons.

An additional methodological explanation of the immunohistological results may be offered. During the isolation of bovine filament, there could be a selective loss via solubilization or degradation of neuronal filaments. Schlaepfer (33) has reported that neurofilaments from rat sciatic nerve are soluble at low ionic strength. Such an occurrence could leave the preparation relatively enriched in glial filaments, which then migrate at the 51,000 MW position on gels and are the basis of the strong astroglial staining by the raised antisera.

It can be seen that a definite identification of the neuronal filament is not yet at hand. To further complicate matters, several higher molecular weight bands are found on the gels of bovine brain filament preparations. These migrate at approximately molecular weights of 68,000, 160,000, and 210,000. Although in the bovine CNS axonal filaments, these minor proteins together comprise between 15 and 60% of total protein, a preparation prepared from rat peripheral nerve shows 68,000 and 160,000 MW bands as prominent, with little if any 51,000 dalton component. Antibody against the 160,000 dalton component gives strong neuronal staining and no glial staining (26). Of interest in this regard are the recent findings of Schlaepfer (33) that low ionic strength extraction of rat peripheral nerve results in the isolation of neurofilaments as monitored electron microscopically. Purification of these peripheral nerve filaments followed by gel electrophoresis results in a predominant 68,000 dalton band which comigrates with serum albumin. Immunofluorescent studies with antisera to this protein give apparent staining of neuronal filaments (34).

Hoffman and Lasek (18) studied the electrophoretic mobility of labeled polypeptides that had been transported down the peripheral axon from the perikaryon of the rat anterior horn cell via the slow component of axonal transport. They found three bands, which they referred to as the slow component triplet, and these had molecular weights of 68,000, 160,000, and 212,000. On the basis of several assumptions, they postulated that these bands represented primary structural polypeptide components of the neurofilament.

It can be concluded that there is considerable controversy over the biochemical identity of the proteins comprising the mammalian neuronal filament. We have purposely not reviewed several biochemical studies of axonal filaments isolated from various invertebrate species. Most of this work indicates protein subunits of 68,000 and 160,000 daltons for the neurofilament.

The analysis of the proteins comprising the paired helical filaments of Alzheimer neurofibrillary degeneration has been retarded by this difficulty in characterizing the normal neurofilament. In the remainder of this discussion, we relate our most recent efforts to resolve some of these problems through a biochemical and immunohistological analysis of an important experimental model of neurofibrillary degeneration. Finally, we will stress the implications of these results for our present understanding of the biochemistry of human neurofibrillary degeneration.

BIOCHEMISTRY OF EXPERIMENTAL NEUROFIBRILLARY DEGENERATION

In recent years, a growing number of chemical toxins have been identified that cause a striking increase in cytoplasmic filaments within central or peripheral neurons. The neurotoxic effect of several of these agents was first identified in humans as the result of industrial or accidental exposure. These agents include a heterogeneous group of compounds such as the organic solvents *n*-hexane (45) and methyl-*n*-butyl ketone (27); the lathyrogenic agent, iminodipropionitrite (6); as well as acrylamide (31) itself, all of which can produce a profound sensorimotor peripheral neuropathy in man and laboratory animals. If injected intracisternally, colchicine, vinblastine, and podophyllotoxin—all of which were mentioned earlier as being capable of binding to and precipitating microtubules *in vitro*—can induce striking perikaryal bundles of NF in spinal neurons (44). The mechanism by which any of these agents might induce the aggregation of cytoplasmic filaments remains obscure.

Among the neurotoxins that can cause neurofibrillary pathology, aluminum has aroused the greatest interest, in both experimental and clinical circles. Remarkably, the neurotoxicity of aluminum was first established in 1897, when a German worker, Döllken (12) injected aluminum tartrate into rabbits and found that they came down with a neuronal degeneration in different parts of the brain. During the 1930s and 1940s, aluminum paste began to be injected intracortically to produce experimental epileptogenic foci. In 1965, Klatzo and colleagues (24) first studied the ultrastructure of aluminum lesions in cortex and discovered that they contained intraneuronal bundles of 10 nm neurofilaments that showed agryrophyllic staining and bore a striking light microscopic resemblance to neurofibrillary tangles. A major ultrastructural difference was that the aluminum-induced filaments did not contain the periodic twists that are the hallmark of the paired helical filaments (39).

When the aluminum salts are injected into brain or subarachnoid space, the animals maintain an asymptomatic interval of 10 to 15 days. After that, a rapidly progressive encephalomyelopathy ensues, marked by seizures and quadriparesis and leading to death in a few days from status epilepticus or inanition. Crapper and associates (7) studied the early course of the intoxication, when the animals were overtly asymptomatic, and found that (a) their performance on a task requiring retention of new information (short-term memory) progressively declined, and (b) the level of performance was linearly correlated with cortical aluminum levels.

Moving from experimental to human pathology, Crapper, et al. (7) decided to assay aluminum in the biopsied and autopsied cortex of four patients, with verified Alzheimer's disease, and found levels 3 to 4 times that of control cortex in some areas. The highest levels were noted in mesial frontal and temporal cortex. A later study by the same group extended these observations to 12 demented patients, including 2 with Down's syndrome and Alzheimer's disease, and again significantly elevated aluminum levels were found in an average of 28% of the cortical samples from each of the 12 brains (8). The distribution of cortical aluminum levels closely matched the topographic distribution of neurofibrillary tangles, although it did not seem to correlate with the presence of senile plaques. The source of the aluminum and its possible etiologic relevance to the dementia remain entirely obscure. Although these results are clearly preliminary, they are given additional credence by the recent report of elevated cortical aluminum levels in 12 patients with chronic renal failure who developed the syndrome of dialysis dementia (1). These patients had been taking high doses of aluminum-containing antacids daily for longer than 3 years and may also have been exposed to high concentrations of aluminum in the water used to prepare the renal dialysate. Interestingly, the brains of these patients did not show neurofibrillary tangles. Moreover, there is virtually no evidence to date that aluminum mill workers or bauxite miners are unusually prone to develop encephalopathy or presenile dementia.

Nonetheless, aluminum-induced encephalomyelopathy is perhaps the best available experimental model of neurofibrillary degeneration because of its correlation with impairment of short-term retention in animals as well as the restriction of pathological changes to a marked neurofilamentous degeneration of neurons. Furthermore, aluminum may conceivably have importance as a human neurotoxin associated with dementia in the elderly. For these reasons, and in order to learn more about the intermediate filament, we have recently undertaken the following studies.

Adult rabbits were given 0.25 ml of 1% aluminum chloride in an artificial CSF solution via a slow intracisternal injection using a stereotaxic device. Following an asymptomatic interval of 9 to 16 days, the animals began to develop the characteristic encephalomyelopathy, which progressed to a moribund state over a period of 2 to 5 days. Following sacrifice by intracardiac perfusion, histological examination of Bodian and Nissl preparation revealed the widespread presence of cytoplasmic neurofibrillary bundles in medium-sized neurons in the central gray matter and anterior horns of the spinal cord at all levels, as well as in certain nuclei of the basis pontis (Fig. 1). There was essentially no involvement of hemispheral structures.

Ultrastructurally the neuronal lesions consisted of large single or multiple perinuclear sworls of 10 nm filaments of variable packing density showing multiple tiny irregularly spaced "side arms." These filaments conformed in appearance to the normal intermediate filaments of the neuron.

Initially, we attempted to react these filament bundles with the available antiserum prepared against the major 51,000 dalton band of bovine brain filament mentioned earlier (46). No fluorescent staining of these lesions in 10 micron frozen sections of spinal cord could be observed. Indeed, the swollen neurons of the anterior horn containing the filament bundles appeared as intensely black areas or "holes," showing not even the normal background fluorescence of the rest of the gray matter. Similar immunofluorescent experiments using an antitu-

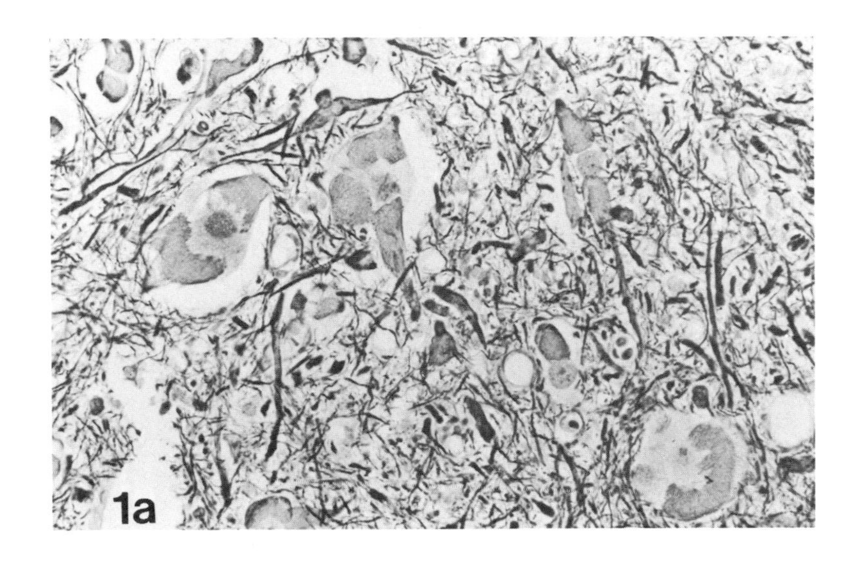

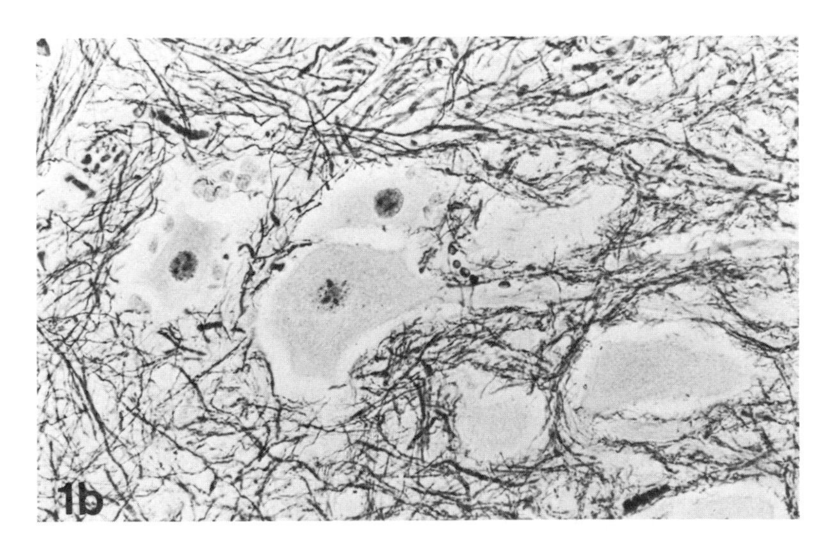

FIG. 1. Light microscopy of Bodian-stained sections of rabbit spinal cord: **a)** anterior horn cells in experimental aluminum myelopathy show multiple, large, cytoplasmic filament bundles which stain darkly (×240); **b)** control cord shows normal cytoplasmic staining (×240).

bulin antiserum also showed lack of staining of the filamentous lesions. On the other hand, antiserum against the 160,000 dalton component of the bovine filament preparation did give intense, specific fluorescent staining of the filament bundles (35).

To further clarify the result, we attempted to isolate and biochemically characterize the central nervous system neurons undergoing neurofibrillary degeneration induced by aluminum (35). Using whole fresh cord from aluminum-treated and control rabbits as a starting material, we modified available neuronal/glial separation techniques (5,19,28), employing mechanical sieving and sucrose density gradient centrifugation in the absence of proteolytic enzymes. The neuronal perikaryal fraction obtained contained approximately 20 to 25% of the total large anterior horn cells that have been estimated by morphological counts to exist in rabbit spinal cord. The cells appeared well preserved (Fig. 2), with clearly visible nuclei and nucleoli and neurites as long as 2 to 4 times the

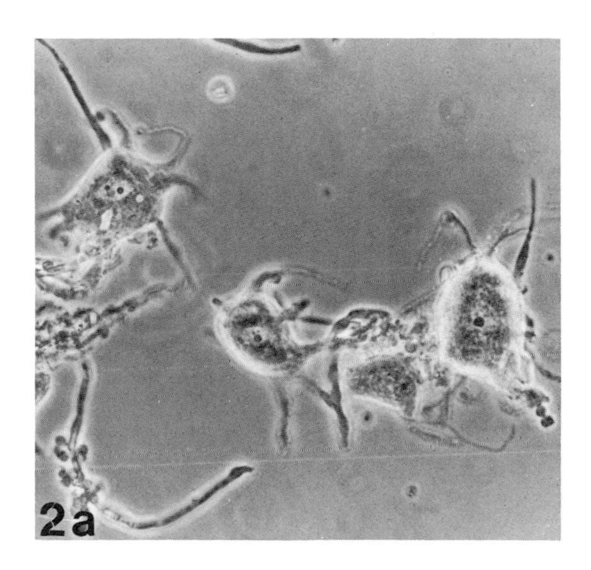

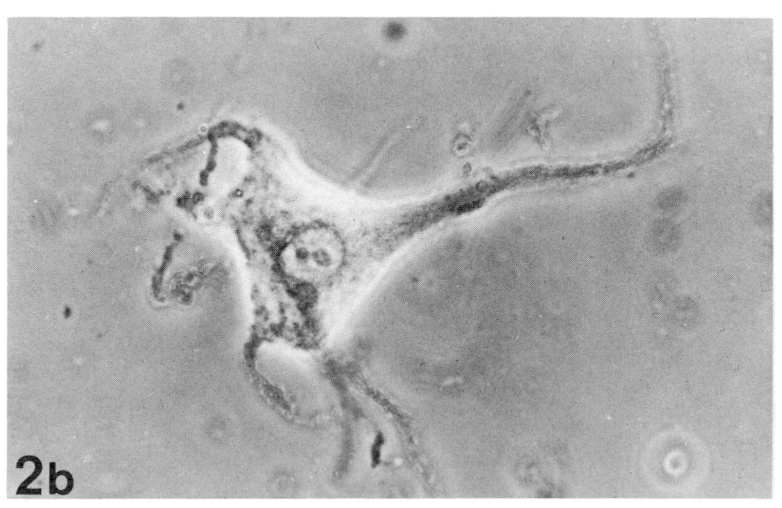

FIG. 2. Isolated spinal neurons from rabbit with aluminum-induced neurofibrillary degeneration: **a.** several perikarya with clear nucleoli and intact short processes (×180); **b.** higher magnification of isolated anterior horn cell (×400).

perikaryal diameter, despite the fact that electron microscopy demonstrated an almost total lack of preservation of the plasma membrane. The neuronal fraction was contaminated principally by small capillary fragments and cell-free nuclei, which could then be removed by a reverse sieving step. What appeared to be well-preserved fibrillary astrocytes were found in small numbers in a separate fraction from the sucrose gradient.

Electron microscopy of the isolated neuronal fractions from the aluminum model revealed the preservation of the large perinuclear neurofilament bundles, which, of course, were not found in identical isolates from control cords (Fig. 3). Estimation of protein per cell was approximately 6.4 ng/cell in the experimental fractions and 5.3 ng/cell in the controls. These values are in rough agreement with reported estimates of protein content of isolated large bovine anterior horn cells (5).

When the neuronal fractions were taken up in a 1% SDS-8M urea sample buffer and electrophoresed on 5–15% SDS-polyacrylamide continuous gradient gels, distinct differences were found. The experimental fraction contained 2 bands that were less visible in the electrophoretic pattern of control neurons: an approximately 68,000 dalton band that migrated with bovine serum albumin; and a band of approximately 160,000 daltons migrating well behind the β-galactosidase standard. In addition, a third band—actually a doublet of roughly 210,000 daltons, migrating just behind myosin—was seen in the aluminum-treated neurons but not in control cells.

Densitometric scans of the control and experimental gels confirmed these differences (Fig. 4). The second arrow in Fig. 4 points to the 68,000 MW band,

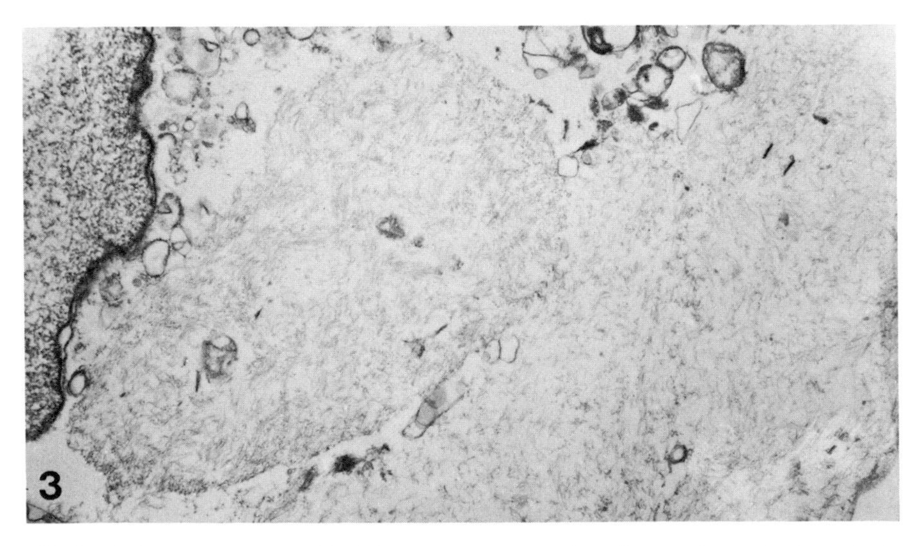

FIG. 3. Electron microscopy of isolated anterior horn cell perikaryon from aluminum-treated rabbit reveals preservation of large perinuclear neurofilament bundle (×18,700).

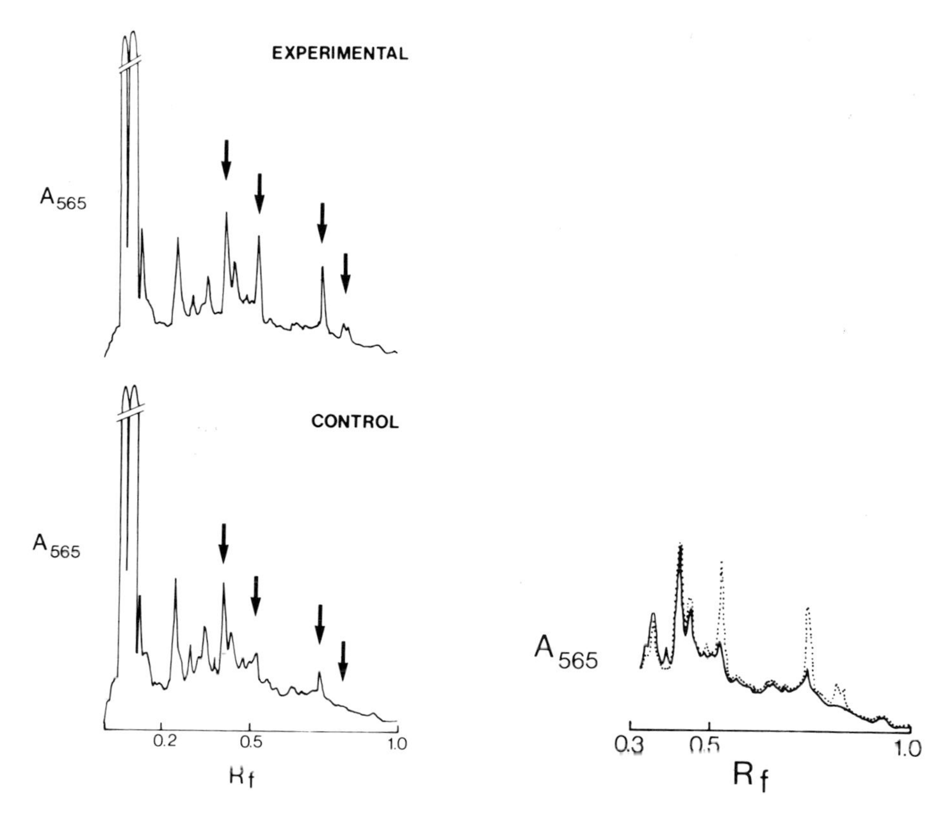

FIG. 4. Densitometric scans of SDS-polyacrylamide electrophoretic gels of isolated neuronal fractions from experimental and control rabbit cord. **Left:** arrow furthest to left indicates the approximately 51,000 dalton protein band; second arrow—68,000 dalton band; third arrow—160,000 dalton band; fourth arrow—210,000 dalton band. **Right:** superimposition of experimental and control profiles shows the major differences at the 68,000 and 160,000 dalton peaks.

which is approximately four times greater in quantity in experimental cells. The third arrow indicates the 160,000 dalton band, which is similarly increased in the experimental fraction. The fourth arrow shows the doublet at slightly over 200,000 daltons, which is present in experimental but not control neurons. Superimposing the densimetric scans shows the virtual identity of the protein patterns, except at the three peaks just described. Of particular interest is the fact that the 51,000 dalton peak is not accentuated.

The only reproducible differences in the protein electrophoretic patterns of control and experimental cells are at the three indicated bands. Coupled with the finding that the only quantitative differences by EM are in the much enriched number of 10 nm filaments in the experimental fractions, this suggests that these bands may represent forms of the subunit protein of the aluminum-induced neuronal filaments. This is reinforced by the immunohistological finding of strong fluorescent staining of the aluminum-induced filament bundles with antisera against the 160,000 dalton protein.

The results to date suggest the following conclusions.

1. Spinal neurons undergoing experimental neurofibrillary degeneration contain augmented amounts of proteins that correspond to a high degree in electrophoretic mobility with axonal polypeptides from mammalian ventral motor neurons tentatively identified by Hoffman and Lasek (18) as subunits of the 10 nm neurofilament.
2. The filaments in aluminum neurofibrillary degeneration are immunologically recognized by antibodies raised to the 160,000 dalton protein of normal bovine brain filament preparations, further supporting the notion—based on ultrastructural appearance—that these filaments are indeed normal neuronal filaments.
3. The proteins of the pathologically induced filaments are distinct from tubulin on both biochemical and immunological grounds.

Of great urgency now is the comparison of the aluminum-induced filament with that accumulating in the neurons of Alzheimer's disease. We are presently preparing to react the anti-160 antibody with *in situ* neurofibrillary tangles and neuritic plaques from hippocampal cortex. Many other human disorders characterized by neurofibrillary degeneration may be investigated with these techniques to determine similarities and differences as far as their protein chemistry is concerned. A single previous study (20) of enriched fractions from hippocampal cortex in Alzheimer's disease showed an accentuated band at 50,000 MW. In the light of the findings on the experimental model and of the data on the normal neuronal and glial filaments, these results may now be reinterpreted. It seems clear from work in our laboratory that antibodies to a 51,000 dalton brain filament protein react strongly with astroglial filaments (47). Moreover, the neuronal fractions used in the Alzheimer experiments (20) were taken from severely gliotic areas, i.e., hippocampus, and the filaments were not purified to morphological homogeneity. Thus the possibility exists that the protein identified as the presumptive subunit of the abnormal paired filaments in Alzheimer cortex may actually be a GFA contaminant.

All of the work on the neuronal fibrous proteins must be placed in perspective with the recent very exciting data on the synaptic biochemistry in Alzheimer's disease. Three independent groups in Great Britain have all documented a marked and selective abnormality of the presynaptic cholinergic system in Alzheimer cortex (4,11,30). Levels of choline acetyltransferase (CAT), the principal synthetic enzyme for acetylcholine, are reduced by 50 to 90% in hippocampus, amygdala and neocortex, with lesser reductions in deep gray nuclei. Although initial determinations were made in autopsied cortex, agonal or postmortem autolysis was eliminated by the finding of normal activity of glutamic acid decarboxylase (GAD), an enzyme particularly sensitive to antemortem hypoxia (11). Subsequently, biopsied cortex in Alzheimer's disease showed a similar marked decrease in CAT activity with preservation of GAD (37). Levels of acetylcholinesterase (AChE) are also dramatically reduced in limbic and neocortex. Regions of cerebral cortex showing the greatest reductions in CAT and AChE activity correlate roughly with areas of maximum neurofibrillary tangles (11). Preliminary studies show no decrease in activity of the catecholamine transmitter enzymes, tyrosine hydroxylase, aromatic amino acid decarboxylase, dopamine-β-hydroxylase, and monoamine oxidase, thus emphasizing the selective vulnerability of cholinergic neurons in this disease.

It would be of considerable interest to know what relationship, if any, exists between the widespread neurofibrillary degeneration found in Alzheimer's cortex and these abnormalities of the presynaptic cholinergic system. Since this will be difficult to approach directly in the human disease, the aluminum-induced changes in the spinal neurons with which we have been working may represent a suitable model to explore which alteration (if either) is primary. We plan to determine levels of CAT and AChE in the spinal cord of rabbits undergoing neurofibrillary degeneration.

CONCLUSIONS

The application of newer techniques in neuronal cell biology has recently provided some intriguing clues to the pathogenesis of Alzheimer's disease. Yet even a tentative understanding of the etiology of this common and devastating disorder is not at hand. A major obstacle to progress in discerning the cause of Alzheimer's disease has been our previous failure to separate this entity from the cerebrovascular disorders that so commonly accompany senescence. Recent work has gone a long way toward resolving this difficulty and has also brought into much sharper focus the distinction between cerebral aging as a normal biological process and the specific and malignant neuronal degeneration that affects a minority of our elderly population. Our principal challenge may now rest in the proliferation of seemingly unrelated leads which make a unitary hypothesis for the etiology of Alzheimer's disease inadmissible and require us to direct our resources to various lines of evidence simultaneously. It is tantalizing to speculate that endogenous (genetic) or exogenous (toxic) factors may act on the nerve cell's nuclear apparatus and thereby cause accumulation of altered proteins, expressed morphologically as paired helical filaments. These may in turn interfere with neuronal metabolism and transport, thus affecting maintenance of the cell's axons and dendrites and leading ultimately to a deficiency of presynaptic transmitters, cholinergic or otherwise. Since such a relationship remains entirely unproved, we must at present be satisfied with the inevitable multifactorial hypothesis of Alzheimer's disease.

ACKNOWLEDGEMENTS

The authors gratefully acknowledge the assistance of Shu-Hui Yen and Ron Liem in carrying out the immunohistological studies. Carol Van Horn provided expert assistance in electron microscopy.

This work was supported in part by NIH postdoctoral fellowship NS01152 (D.J.S.) and NIH Grants NS11504 and AG00704.

REFERENCES

1. Alfrey, A. C., LeGendre, G. R., and Kachy, W. D. (1976): The dialysis encephalopathy syndrome: possible aluminum intoxication. *N. Engl. J. Med.*, 294:184–189.
2. Ball, M. J. (1976): Neurofibrillary tangles and the pathogenesis of dementia: a quantitative time study. *Neuropathol. Appl. Neurobiol.*, 2:394–410.
3. Blessed, G., Tomlinson, B. E., and Roth, M. (1968): The association between quantitative measures of dementia and of senile change in the cerebral gray matter of elderly subjects. *Br. J. Psychiatr.*, 114:797–811.
4. Bowen, D. M., Smith, C. B., White, P., and Davison, A. N. (1976): Neurotransmitter related enzymes and indices of hypoxia in senile dementia and other abiotrophies. *Brain*, 99:459–496.
5. Capps-Covey, P., and McIlwaine, D. L. (1975): Bulk isolation of large ventral spinal neurons. *J. Neurochem.*, 25:517.
6. Chou, S. M., and Hartmann, H. A. (1964): Axonal lesions and waltzing syndrome after IDPN administration in rats, with a concept-axostasis. *ACTA Neuropathol.*, 3:428–450.
7. Crapper, D. R., Krishnan, S. S., and Dalton, A. J. (1973): Brain aluminum distribution in Alzheimer's disease and experimental neurofibrillary degeneration. *Science*, 180:511–513.
8. Crapper, D. R., Krishnan, S. S., and Quittkat, S. (1976): Aluminum, neurofibrillary degeneration and Alzheimer's disease. *Brain*, 99:67–80.
9. Crapper, D. R., Skopitz, M., Scott, J. W., Haschinski, V. C. (1975): Alzheimer degeneration in Down's syndrome: electrophysiological alterations and histopathology. *Arch. Neurol.*, 32:618–623.
10. Dahl, D. (1976): Glial fibrillary acid protein from bovine and rat brain. Degradation in tissue and homogenates. *Biochim. Biophys, Acta*, 420:142.
11. Davies, P., and Maloney, A. J. F.: (1976): Selective loss of central cholinergic neurons in Alzheimer's disease. *Lancet*, 2:1403.
12. Döllken, V. (1897): Uber die wirkung des aluminums mit Berucksichtingung der aluminum verursachten Lasionen im CNS. *Arch. Exp. Pathol. Pharm.*, 40:48.
13. Ellis, W. G., McCulloch, J. R., and Corley, C. L. (1974): Presenile dementia in Down's syndrome. *Neurology*, 24:101–106.
14. Fine, A., and Blitz, A. L. (1974): Muscle-like contractile proteins and tubulin in synaptosomes. *Proc. Natl. Acad. Sci.*, 71:4472.
15. Gaskin, F., Cantor, C. R., and Shelanski, M. L. (1974): Turbidimetric studies of the *in vitro* assembly and disassembly of porcine neurotubles. *J. Mol. Biol.*, 89:737.
16. Gaskin, F., and Shelanski, M. L. (1977): Microtubules and intermediate filaments. *Essays Biochem.*, 12:115. 1977.
17. Gonatas, N. K., Anderson, W., and Evangelista, I. (1967): The contribution of altered synapses in the senile plaque: An electron microscopic study in Alzheimer's dementia. *J. Neuropath. Exp. Neurol.*, 26:25–39.
18. Hoffman, P. N., and Lasek, R. J. A. (1975): The slow component of axonal transport. *J. Cell Biol.*, 66:351–366, 1975.
19. Iqbal, K., and Tellez-Nagel, I. (1972): Isolation of neurons and glial cells from normal and pathological human brain. *Brain Res.*, 45:296–301.
20. Iqbal, K., Wisniewski, H. M., Shelanski, M. L., Brostoft, S. (1974): Protein changes in senile dementia. *Brain Res.*, 77:337–341.
21. Ishikawa, H., Bischoff, R., and Holtzer, H. A. (1969): Formation of arrowhead complexes with heavy meromyosin in a variety of cell types. *J. Cell Biol.*, 43:312–328.
22. Katzman, R. (1976): The prevalence and malignancy of Alzheimer disease. *Arch. Neurol.*, 33:217–218.
23. Kidd, M. (1963): Paired helical filaments in electron microscopy of Alzheimer's disease. *Nature*, 197:192–193.
24. Klatzo, I., Wisniewski, H. M., and Streicher, E. (1965): Experimental production of neurofibrillary degeneration. I. Light microscopic observations. *J. Neuropathol. Exp. Neurol.*, 24:187–199.

25. Liem, R. K. H., Yen, S.-H., Loria, J. C., and Shelanski, M. L. (1977): Immunological and biochemical comparison of tubulin and intermediate brain filament protein. *Brain Res.*, 132:167–171.
26. Liem, R. K. H., Yen, S.-H., Salomon, G., and Shelanski, M. L. (1978): Intermediate filaments in nervous tissue. *J. Cell Biol. (in press).*
27. Mendell, J. R., Saida, K., Ganansia M. F., Jackson, D. B., Weiss, H., Gardier, R. W., Chrisman, C., Allen, N., Couri, D., O'Neill, J., Marks, B., and Hetland, L. (1974): Toxic polyneuropathy produced by methyl-n-butyl ketone. *Science,* 185:787–789.
28. Norton, W. T., and Poduslo, S. E. (1970): Neuronal soma and whole neuroglia of rat brain: a new isolation technique. *Science,* 167:1144–1146.
29. Olmsted, J. B., and Borisy, G. G. (1973): Microtubules. *Ann. Rev. Biochem.,* 42:507.
30. Perry, E. K., Perry, R. H., Blessed, G., and Tomlinson, B. E. (1977): Necropsy evidence of central nervous system cholinergic deficits in senile dementia. *Lancet,* 1:189–190.
31. Prineas, J. A. (1969): The pathogenesis of dying-back polyneuropathies. II. An ultrastructural study of experimental acrylamide intoxication in the cat. *J. Neuropathol. Exp. Neurol.,* 28:589–621.
32. Puszkin, S., Berl, S., Puszkin, E., and Clarke, D. D. (1968): Actomyosin-like protein isolated from mammalian brain. *Science,* 161:170–172.
33. Schlaepfer, W. W. (1977): Immunological and ultrastructural studies of neurofilaments isolated from rat peripheral nerve. *J. Cell Biol.,* 74:226–240.
34. Schlaepfer, W. W., and Lynch, R. G. (1977): Immunofluorescence studies of neurofilaments in the rat and human peripheral and central nervous system. *J. Cell Biol.,* 74:241–250.
35. Selkoe, D. J., Liem, R. K. H., Yen, S.-H., Shelanski, M. L. (1978): Biochemical and immunological characterization of neurofilaments in experimental neurofibrillary degeneration induced by aluminum. *Brain Res. (in press).*
36. Shelanski, M. L., Albert, S., DeVries, G. H., and Norton, W. T. (1971): Isolation of filaments from brain. *Science,* 174:1242.
37. Spillane, J. A., White, P., Goodhart, M. J., Flack, R. H. S., Bowen, D. M., and Davison, A. N. (1977): Selective vulnerability of neurons in organic dementia. *Nature,* 266:558–559.
38. Terry, R. D. (1976): Dementia: a brief and selective review. *Arch. Neurol.,* 33:1–4.
39. Terry, R. D., and Pena, C. (1965): Experimental production of neurofibrillary degeneration. II Electron microscopy, phosphate histochemistry and electron probe analysis. *J. Neuropathol. Exp. Neurol.,* 24:200–210.
40. Tomlinson, B. E., Blessed, G., and Roth, M. (1970): Observation of the brains of demented old people. *J. Neurol. Sci.,* 11:205–243.
41. Tomlinson, B. E., and Henderson, G. (1976): Some quantitative cerebral findings in normal and demented old people. In: *Neurobiology of Aging,* ed. R. D. Terry and S. Garshon., pp. 183–204. Raven Press, New York.
42. Wisniewski, H. M., Bruce, M. E., and Fraser, H. (1975): Infectious etiology of neuritic (senile) plaques in mice. *Science,* 190:1108–1110.
43. Wisniewski, H. M., Narong, H. K., and Terry, R. D. (1976): Neurofibrillary tangles of paired helical filaments. *J. Neurol. Sci.,* 27:173–181.
44. Wisniewski, H. M., Shelanski, M. L., and Terry, R. D. (1968): Effects of mitotic spindle inhibitors on neurotubules and neurofilaments in anterior horn cells. *J. Cell Biol.,* 38:224–229. 1968.
45. Yamamura, Y. (1969): *n*-Hexane polyneuropathy. *Folia Psychiatr. Neurol. Jap.,* 23:45–47.
46. Yen, S., Dahl, D., Schachner, M., and Shelanski, M. L. (1976): Biochemistry of the filaments of brain. *Proc. Natl. Acad. Sci.,* 73:529–533.
47. Yen, S.-H., Van Horn, C., and Shelanski, M. L. (1976): Immunohistologic localization of the neurofilament protein in the mouse (Abstract). *J. Neuropathol. Exp. Neurol.,* 35:346.

Physiology and Cell Biology of Aging (Aging, Volume 8), edited by A. Cherkin, et al.
Raven Press, New York © 1979.

Brain Plasticity, Memory, and Aging

Edward L. Bennett* and Mark R. Rosenzweig†

**Laboratory of Chemical Biodynamics, Lawrence Berkeley Laboratory, Berkeley, California 94720; and † Department of Psychology, University of California, Berkeley, California 94720*

It is generally assumed that memory faculties decline with age. A discussion of the relationship of memory and aging and the possibility of retarding the potential decline is hampered by the fact that no satisfactory explanation of memory is available in either molecular or anatomical terms. However, this lack of explanation of memory does not mean that there is a lack of suggested mechanisms for long-term memory storage. Present theories of memory usually include, first, neurophysiological or electrical events, then a series of chemical events that ultimately leads to long-lasting anatomical changes in the brain. Evidence is increasing for the biochemical and anatomical plasticity of the nervous system and its importance in the normal functioning of the brain. Modification of this plasticity may be an important factor in senescence.

This discussion reports experiments that indicate that protein synthesis and anatomical changes may be involved in long-term memory storage. Environmental influences can produce quantitative differences in brain anatomy and in behavior. In experimental animals, enriched environments lead to more complex anatomical patterns than do colony or impoverished environments. This raises fundamental questions about the adequacy of the isolated animal, which is frequently being used as a model for aging research. A more important applied question is the role of social and intellectual stimulation in influencing aging of the human brain.

It is not necessary to emphasize the importance of neural function and the importance of preserving that function for as long as possible. Unfortunately, we know little about the mechanism(s) of formation of long-term memory, and this makes the problem of understanding the effects of age on neural function a difficult one. In this chapter, we shall attempt to outline a model for the formation of long-term memory. The observations of Cotman's group on brain plasticity may have intriguing implications for memory, and I will mention some related findings of our group at Berkeley which has focused on brain plasticity for a number of years.

MODEL OF MEMORY FORMATION

To guide research in this area, it is helpful to have a model, and we like to use the model published by Shashoua (Fig. 1) (14). In this model he suggests that sensory stimuli perceived by the organism are subsequently transduced through short-term electrical and chemical events, and that some of this information is eventually entered into a long-term form of storage in the neuronal membrane.

Our focus is on the mechanisms involved in the long-term storage. It is now generally accepted that a number of chemical steps are involved. Many investiga-

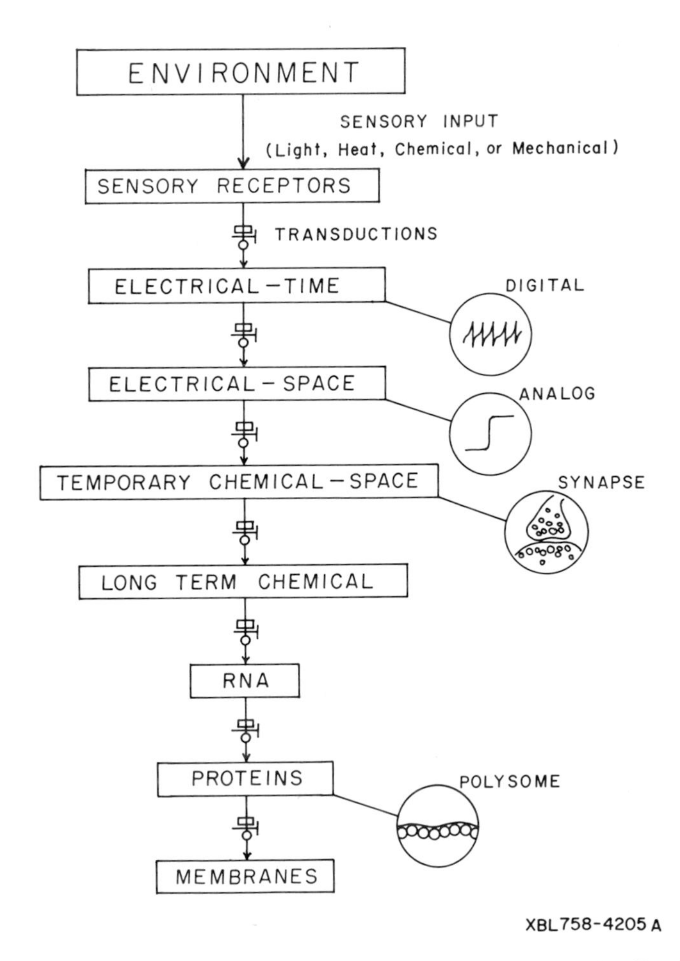

FIG. 1. A theoretical model for long-term memory storage proposed by Shashoua (14). In this model for information storage, sensory input from the environment is perceived by sensory receptors (eyes, ears, nose, skin, etc.) and transduced through several electrical transformations into short-term chemical changes. These are subsequently elaborated into long lasting membrane changes (from ref. 14).

tors have concentrated on the formation of new RNA as a key step in long-term memory storage (10,12). Our research has led us to emphasize the importance of the formation of protein and, perhaps, its transport down the axon to the synapse with subsequent modification of the membrane, or even the formation of new synapses; that is, the last steps depicted in Fig. 1 are key factors. We tend to place less emphasis on speculations that the production of new and unique molecules is the critical element in the formation of long-term memory. Our biases, and we admit that these are biases, would hold that in forming memories, we are making more of the same kinds of neuronal molecules that we would be normally synthesizing.

ANISOMYCIN, A PROTEIN SYNTHESIS INHIBITOR, AND MEMORY

We shall now describe a method we have used quite successfully to study the involvement of protein synthesis in the formation of long-term memory. In addition, we have developed a paradigm to study the effects of other drugs that may modify the formation of memory.

The drug used is anisomycin (ANI) (Fig. 2), an inhibitor of protein synthesis. For studies of memory formation using puromycin or cyloheximide, ANI is a superior drug, at least in the mouse. By preventing transpeptidation, ANI effectively inhibits protein synthesis. In the mouse, a far from lethal dose of ANI produces inhibition of protein synthesis for several hours, whereas near-lethal doses of cycloheximide are required to produce equivalent inhibition of protein synthesis. As a result of the low toxicity of ANI, one can control the time course and duration of inhibition of cerebral protein synthesis by administering successive doses of ANI at 2-hr intervals (Fig. 3).

During the past several years, we have investigated the effectiveness of ANI as an amnestic agent in a variety of passive and active-avoidance behavioral tasks schematically depicted in Fig. 4. Using the passive avoidance step-through test, we found some years ago that as one increases the number of doses of ANI, amnesia increases, but that the particular shape of the curve relating amnesia and inhibition of protein synthesis depended greatly on factors such as training strength (which is influenced by many parameters, including the behavioral test employed, and intensity and duration of footshock), duration of the interval between the training and testing, and strain of the mouse. If the training is marginal and the test is difficult, then one dose of ANI may pro-

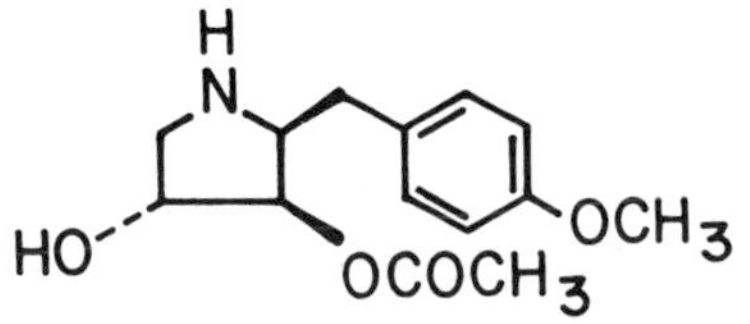

FIG. 2. Anisomycin (ANI), 2-p-methoxyphenylmethyl-3-acetoxy-4-hydroxypyrollidine ($C_{14}H_{19}NO_4$) is an effective inhibitor of cerebral protein synthesis and also a powerful amnesic agent in mice.

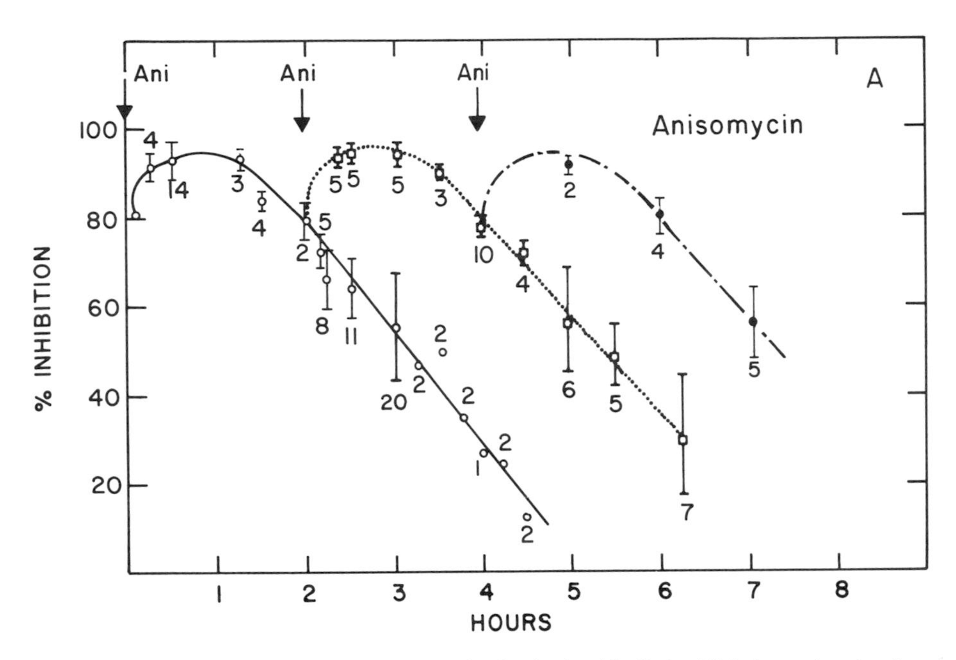

FIG. 3. The inhibition of cerebral protein synthesis obtained in Swiss-Webster male mice from one *(open circles),* two *(open squares),* and three *(closed circles)* successive subcutaneous injections of 0.5 mg ANI. The number of animals used and the standard deviation of each data point are shown (from ref. 6).

duce amnesia. With stronger training, several successive doses may be required to cause amnesia. By varying these factors, one can get a dose-response curve of the relationship of long-term memory formation to training strength and to the duration of protein synthesis inhibition. Furthermore, it has been found that protein synthesis inhibition by ANI will cause amnesia for a variety of training tasks, including tasks learned for positive as well as negative reinforcement (7).

DRUG INFLUENCES ON MEMORY FORMATION

In our studies, ANI has been used in combination with posttrial injections of a number of pharmacological agents in order to evaluate the effects of the latter drugs on memory formation and, in the process, to learn more concerning the mechanisms of memory formation. Typically, in such investigations, the drug has been given prior to the training task, and one has the problem of trying to dissociate the drug's effects on training from its effects on memory formation. We have developed a paradigm that allows us to circumvent this difficulty. Normally, ANI is given 15 min prior to training, then one or more additional doses are given at 2-hr intervals. At some point after the training, the drug of interest may be administered, perhaps an excitant, such as meth-

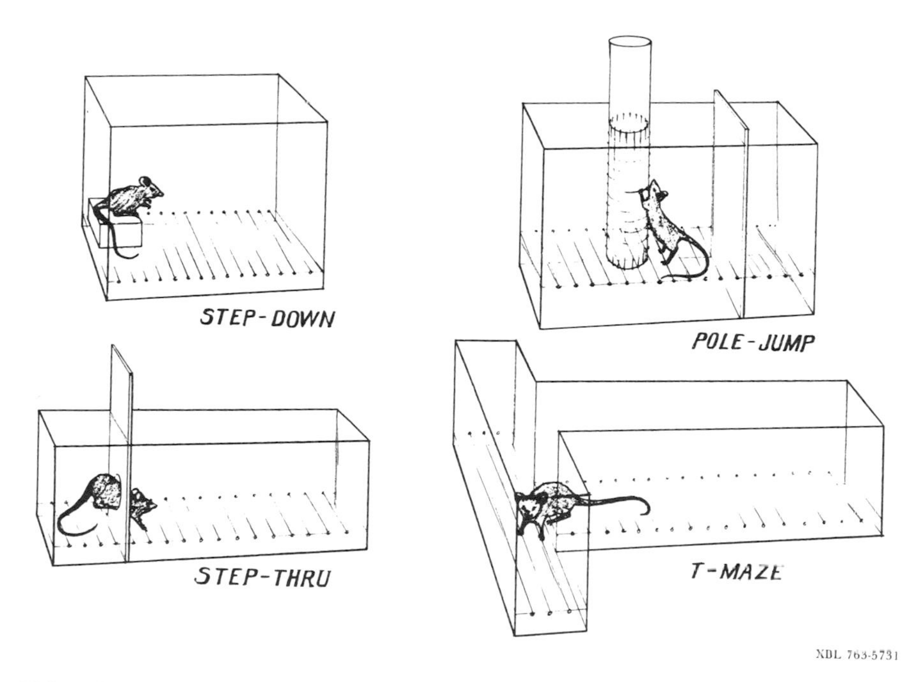

FIG. 4. Schematic representation of four types of behavioral apparatuses useful for studying memory. The step-down task is normally used as a passive-avoidance task; the step-through test can be used as a passive or active avoidance task. In the passive avoidance task, the animal must remain either on the platform or in the small compartment to avoid shock; in the active avoidance test, the animal must move to the larger compartment on the right, frequently after a cue such as a bell or light, to avoid shock. For the pole-jump, the animal must jump onto the pole to avoid shock. The T-maze may be used as either a spatial (right-left) or a visual (light-dark) discrimination task. The relative difficulty of these tasks is approximately in the order named, but can vary with the exact details of the training procedure.

amphetamine or strychnine, a depressant such as chloral hydrate or phenobarbital, ACTH derivatives, etc. (6,8,9). The effect of depressants on ANI-induced amnesia is shown in Fig. 5. Under the training and testing conditions used in this experiment, two doses of ANI did not produce amnesia, but three successive doses did. If a depressant such as chloral hydrate or phenobarbital was administered after training instead of a dose of ANI, the mice also became amnesic. The opposite effect can be seen if one uses a stimulant. In this case, a sufficient number of doses of ANI are administered to produce amnesia, and then a stimulant such as strychnine, picrotoxin, or methamphetamine is given post-training. These stimulants counteract the ANI-induced amnesia.

At this time, it might be useful to think of some sort of construct by which these effects may be explained. Experiments showed that the effects of the stimulant and depressant drugs on memory could not be explained in terms of direct effects on protein synthesis: neither type of drug had sufficient effect on synthesis to alter memory. Rather, we interpret their effects on memory in terms of

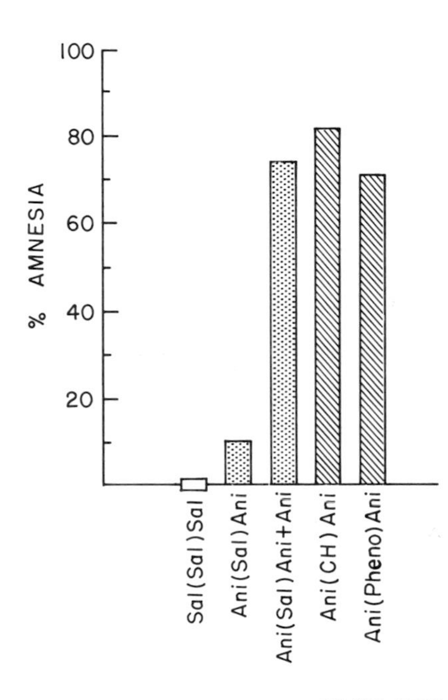

FIG. 5. The effect of chloral hydrate (CH) and phenobarbital (Pheno) on ANI-induced amnesia. The depressants were administered ip 30 min after training in the step-through passive avoidance task. ANI was administered 15 min prior to training, and 1-¾ hr or 1-¾ and 3-¾ hr after training. The amnesia produced by two doses of ANI and the depressant was significantly greater than that produced by only two doses of ANI and approximately equal to that produced by three doses of ANI (from ref. 6).

level of arousal following acquisition; this plays an important role in determining the length of time over which the biosynthetic phase of memory will last (8). Figure 6 (4) shows simplified curves not too different from those presented by Cherkin (3). Units on the axes, "days after training," and "strengths of memory" are arbitrary. The several curves represent the strengths of memory formation

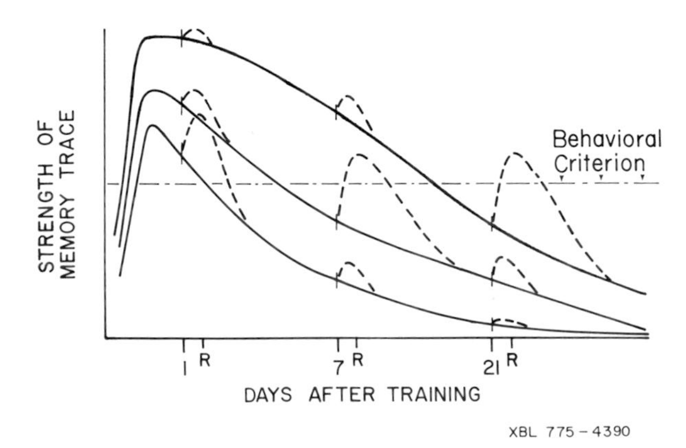

FIG. 6. Schematic representations of the strength of memory traces as a function of time after training. Memory for a given training test will depend upon a number of factors including number of trails, and length and intensity of shock. Events occurring after training can either increase or decrease the nature of the "strength of memory trace" curve and will determine if the behavioral criterion used to measure the memory will be reached and thus, if the subject is "amnesic" or "nonamnesic" (from ref. 4).

under several conditions. If the training strength is marginal, the memory strength may not even reach a level at which we would say that memory formation had occurred. We suggest that while long-term memory formation is taking place, factors such as a stimulant can in some manner increase the rate of long-term memory formation.

We believe this paradigm in which ANI-treated mice are given posttrial injections of the drug of interest will be useful to screen the potential of drugs to facilitate or accelerate the formation of long-term memory. It would be of interest to use this system to test some of the drugs that Scott discusses (this volume).

In a recent experiment, we have shown that colchicine given shortly after training is also an effective amnesic agent. Colchicine blocks transport of materials down the neural axons but does not affect neural impulses. We interpret the results as a demonstration that the protein that is synthesized in the cell body must be transported down to the synaptic endings; we do not yet know how it is deposited there, but ultimately it produces a change in synaptic function, and therefore a long-lasting effect which we refer to as "memory" (2).

EFFECTS OF ENVIRONMENT ON BRAIN FUNCTION

Some investigators refer to environment as a hostile influence. On the contrary, we believe that environment and external stimuli may be necessary for the normal function of the animal. We are concerned about some experiments now being reported with older animals because we fear that the environments employed may have limited the value of the results. The pathogen-free animals that investigators are frequently using in aging studies probably have had reduced environmental stimulation throughout their lives; that is, they have been raised in what we would characterize as an "impoverished condition." Experiments that we have carried out over many years that we will describe briefly here demonstrate that there are measurable effects in the brain depending upon the environment in which the animal has been raised. We would also like to suggest that the environment may be useful as therapy to promote the recovery of the animal from brain injury.

Three main environmental conditions have been used to provide differential experience to animals, typically from 30 to 60 days of age. We refer to these conditions as enriched, standard colony, or impoverished, but it should be clear that "enriched" and "impoverished" are used only in a relative sense. In the enriched condition (EC), a dozen rats are maintained in a cage about 75 cm square containing rat "toys," i.e., numerous objects with which the rat can interact. These objects are changed daily. A more typical sort of environment for the laboratory rat is the standard colony (SC), in which three rats live in a cage 20 × 32 cm with no added inanimate objects. The third environment is the "impoverished condition" (IC), and this may be the one which most closely approximates that of the pathogen-free animal. In IC, a single animal lives in a cage the same size as that used for SC. Among the differences we have found between animals raised in EC and IC, and to a lesser extent between

EC and SC animals, are increased weight and thickness of cerebral cortex, and increased ratio of RNA to DNA. (The RNA/DNA ratio may be our best single measure for distinguishing between EC- and IC-raised rats). Greenough and Diamond in our group have found changes in parameters such as synaptic density, synaptic length, and dendritic branching. In other words, we have evidence that an enriched environment, when compared to an impoverished environment, leads to measurable morphological and biochemical changes in the brains of rats (1,11).

We have not done a great deal of work with older animals, but some years ago in our laboratory Riege compared the responsiveness of rats when they were placed into enriched or impoverished environments at ages as late as approximately 300 days. He found that animals at all ages were responsive to enriched environment with an increase in the ratio of cortex weight to subcortex weight (Fig. 7) (13). We would like to have data out to 700 days of age or so in

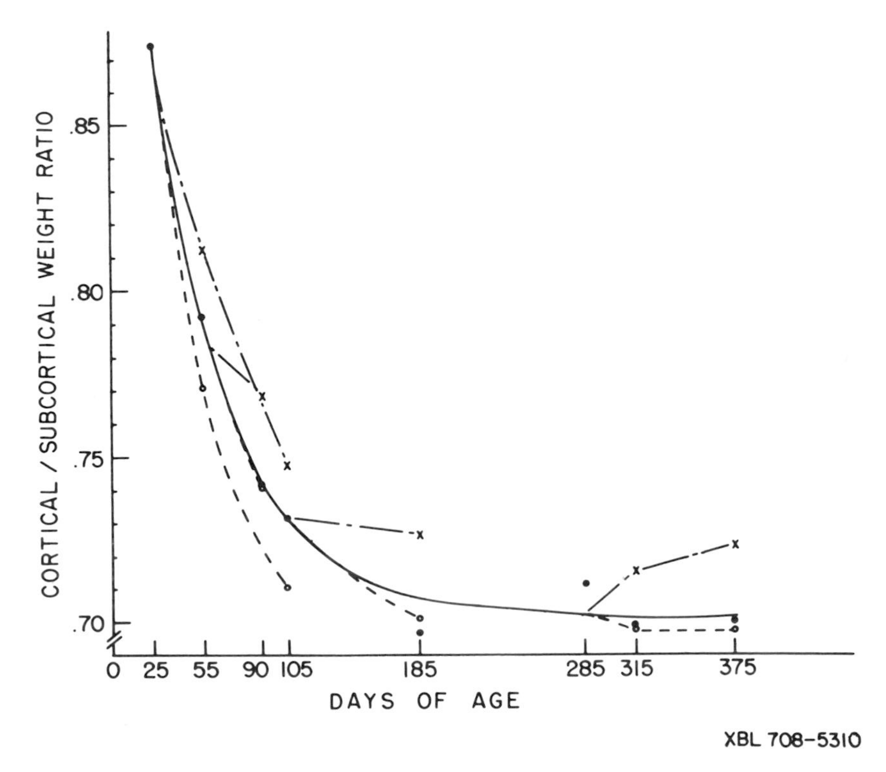

FIG. 7. Cortical/subcortical weight ratio of rats raised in standard colony conditions and then placed in enriched or impoverished environment for 30 or more days at 25, 60, 105, and 285 days of age. At each age, the ratio of cortex/subcortex weight was higher for rats placed in EC, and lower for rats placed in IC than for the rats maintained in standard colony conditions. Enriched (EC) (X — · — · — X); impoverished (IC) (0 ———— 0); or standard colony (SC) (· ———— ·).

order to provide a more definite answer for truly aged rats. Extrapolating from these results, we would like to suggest that we should be concerned with the environments and stimulation that animals (including human beings) receive throughout their lifetime.

Environmental factors may be particularly important during recovery from brain lesions. To test this, we have questioned the kinds of effects there are from differential environments in animals with lesions made in occipital cortex. In recent experiments, we compared the behavioral test scores in a Hebb-Williams maze test of both brain-lesioned and nonlesioned rats raised postoperatively in EC, SC, and IC. We have confirmed previous observations that intact rats raised in EC do better than SC- or IC-raised rats in the maze test. We have also found that the lesioned animals maintained in EC made fewer errors than the lesioned rats maintained in SC or IC; this is true whether the lesions were made neonatally (16), or shortly after weaning (17), or in young adult rats (15). Cotman is now investigating whether recovery measured in anatomical terms can be hastened by appropriate stimulation.

That an enriched environment helps to overcome the behavioral effects of surgical brain lesions and that even fully adult animals respond to environmental enrichment raises the question of whether enriched experience would also help to maintain intellectual function during aging (5). The loss of cells that occurs in advanced age, as well as the probability of at least small cerebrovascular accidents, indicates that the aging brain is likely to be suffering from naturally occurring lesions. In this case, ongoing enriched experience might provide concurrent therapy. Dr. Cherkin has already suggested that individuals who keep active intellectually may show less decline of the nervous system, although it may be difficult to separate cause and effect here.

In conclusion, we believe that experiments being carried out in many laboratories with young and young adult animals do raise questions and suggest directions for experiments being carried out with animal models of much greater ages. Although there is still much to find about neural mechanisms of learning and memory, we do not believe it is necessary to wait to begin to carry out related experiments with the aged.

ACKNOWLEDGMENTS

This research has received support from the National Institute of Mental Health (Grants R01-MH26704, R01-MH26327, and NH 26608–02) and from the Division of Biomedical and Environmental Research of the U.S. Department of Energy.

REFERENCES

1. Bennett, E. L. (1976): Cerebral effects of differential experience and training. In: *Neural Mechanisms of Learning and Memory,* edited by M. R. Rosenzweig and E. L. Bennett, pp. 279–287. MIT Press, Cambridge, Massachusetts.

2. Bennett, E. L., Flood, J. F., Landry, D. W., and Jarvik, M. E. (1978): Transport of protein and long-term memory trace formation. *Trans. Am. Soc. Neurochem.*, 9(Abstract).
3. Cherkin, A. (1966): Toward a quantitative view of the engram. *Proc. Natl. Acad. Sci. U.S.A.*, 55:88–91.
4. Davis, H. P., Rosenzweig, M. R., Bennett, E. L., and Orme, A. E. (1978): Recovery as a function of the degree of amnesia due to protein synthesis inhibition. *Pharmacol. Biochem. Behav.*, 8:701–710.
5. Diamond, M. C. (1978): The aging brain—some enlightening and optimistic results. *Am. Sci.*, 66:66–71.
6. Flood, J. F., Bennett, E. L., Orme, A. E., Rosenzweig, M. R., and Jarvik, M. E. (1978): Memory: modification of anisomycin-induced amnesia by stimulants and depressants. *Science*, 199:324–326.
7. Flood, J. F., and Jarvik, M. E. (1976): Drug influences on learning and memory. In: *Neural Mechanisms of Learning and Memory*, edited by M. R. Rosenzweig and E. L. Bennett, pp. 483–507. MIT Press, Cambridge, Massachusetts.
8. Flood, J. F., Jarvik, M. E., Bennett, E. L., and Orme, A. E. (1976): Effects of ACTH peptide fragments on memory formation. *Pharmacol. Biochem. Behav.*, Vol. 5, Suppl. 1, "The Neuropeptides," pp. 41–51.
9. Flood, J. F., Jarvik, M. E., Bennett, E. L., Orme, A. E., and Rosenzweig, M. R. (1977): The effect of stimulants, depressants, and protein synthesis inhibition on retention. *Behav. Biol.*, 20:168–183.
10. Glassman, E. (1969): The biochemistry of learning: an evaluation of the role of RNA and protein. *Ann. Rev. Biochem.*, 38:605–646.
11. Greenough, W. T. (1976): Enduring brain effects of differential experience and training. In: *Neural Mechanisms of Learning and Memory*, edited by M. R. Rosenzweig and E. L. Bennett, pp. 255–278. MIT Press, Cambridge, Massachusetts.
12. Hydén, H. (1973): RNA changes in brain cells during changes in behaviour and function. In: *Macromolecules and Behaviour*, ed. G. B. Ansell and P. B. Bradley, pp. 51–75. Macmillan, London.
13. Riege, W. H. (1971): Environmental influences on brain and behavior of year-old rats. *Develop. Psychobiol.*, 4:157–167.
14. Shashoua, V. E. (1974): RNA metabolism in the brain. In: *International Review of Neurobiology, Vol. 16*, edited by C. C. Pfeiffer and John R. Smythies, pp. 183–231. Academic Press, New York.
15. Will, B. E., and Rosenzweig, M. R. (1976): Effets de l'environnement sur la récuperation fonctionnelle après lèsions cérébrales chez des rats adultes. *Biol. Behav.*, 1:5–16.
16. Will, B. E., Rosenzweig, M. R., and Bennett, E. L. (1976): Effects of differential environments on recovery from neonatal brain lesions, measured by problem-solving scores and brain dimensions. *Physiol. Behav.*, 16:603–611.
17. Will, B. E., Rosenzweig, M. R., Bennett, E. L., Hebert, M., and Morimoto, H. (1977): Relatively brief environmental enrichment aids recovery of learning capacity and alters brain measures after postweaning brain lesions in rats. *J. Comp. Physiol. Psychol.*, 91:33–50.

Physiology and Cell Biology of Aging
(Aging, Volume 8), edited by A. Cherkin, et al.
Raven Press, New York © 1979.

A Review of Some Current Drugs Used in the Pharmacotherapy of Organic Brain Syndrome

Francis L. Scott

Pennwalt Corporation, Pharmaceutical Division, Rochester, New York 14603

In this chapter, I shall focus on one area of drug usage in geriatric medicine, omitting consideration of drugs used in many of the illnesses experienced by the aged population—cancer, Parkinson's disease, the major mental illnesses, etc. I shall also omit discussion of efforts in the area of pharmacolongevity (86) (i.e., of drugs targeted to prolong life-span), and shall instead concentrate on drugs that purport to improve the quality (rather than quantity) of that life-span. What I intend to cover, then, is the rather broadly defined category of illnesses entitled "organic brain syndrome" (OBS) (68).

Hollister, in a recent review (68), mentions that one-sixth of those over 65 years have some manifestations of this syndrome. One-third of the same group have some type of functional disorder such as depression or neurosis; many have both. Age per se does not necessarily indicate that an aged patient's emotional disorder must be due to OBS. Hollister also notes that the characteristic clinical hallmark of this condition is an insidious and progressive loss of memory; that the types of organic lesions formed in the aged brain are the well-known senile plaques and neurofibrillary tangles of Alzheimer; that arteriosclerotic changes are found in only about one-third of the patients displaying OBS in old age; that such changes only contribute substantially to the pathologic conditions in about 10% of the cases; and, finally, that a patient may have severe cerebral arteriosclerosis without much clinical evidence of coronary, renal or peripheral arteriosclerosis, and vice versa.

In the normal process of drug design, the following general steps are adopted. (a) We identify as clearly and as specifically as possible the disease state toward which our syntheses will be directed. The more knowledge we have of biochemical mechanisms, physiological pathways, etc., involved in that disease state, then the more precisely can we design our new drugs. (b) Clearly, from the very beginning, we need convenient animal models of the disease selected, the key point here being the essential relevance (such as can be determined) of those animal models in modality, mechanisms, etc., to the human condition they purport to copy. (c) We generally do not design in a historical vacuum, and so many of the medicinal chemical ideas begin with historically active compounds against that disease. These older drugs are what is called "lead compounds." (d) Once we have prepared our new group(s) of compounds, and found the desirable activity, the complex process of drug evolution begins: optimization of the desired activity by structural modifications of the early active materials, concurrent minimization of the undesirable side effects of these materials, large-scale synthesis (after suitable process development), extensive preclinical pharmacological and toxicological studies, clinical studies, and so on.

My purpose here is to outline the process by which such a drug-design group operates and to see how that drug-design process applies to the topic under consideration. First, we shall look somewhat at the aging process (the so-called disease state, in my outline) and determine if we can define its multivariant character. In other words, what shall be measured clinically to assess the effectiveness of our drugs? The question of pharmacological models in gerontological research shall be bypassed since it is covered elsewhere (96) and in this volume by Meier-Ruge et al. A review of a number of old and new "lead compounds" is presented here, a heterogenous group of structures from which we hope to gain insight and derive new guidelines for the design of the next generation of such drugs. The number of compounds discussed shall be limited and merely sample studies on each presented. Whenever possible, recent clinical data on the compounds selected will be presented uncritically, for the moment.

Organic Brain Syndrome: Definition and Assessment

A typical sequence in the development of organic brain syndrome within an individual may be as follows: there is a diminution in mental alertness and memory that may lead to bouts of disorientation and confusion with accompanying personality changes such as increasing irritability, hostility, and emotional lability (occasionally to the point of paranoia); finally, motivation and initiative may be completely lost, and an essentially vegetative existence for the individual concerned may be evident. Behavioral disturbances such as poor self-care, unsociability, lack of cooperation and just plain "obstreperousness," and such secondary symptoms as anxiety, depression, fatigue, dizziness, decreased appetite, and sleep loss may be part of this sequence.

One typical mode of assessment of these factors that appears reliable has been the Sandoz Assessment of Clinical States Rating Form—Geriatric, recently reviewed by Shader et al. (131). This measures 18 categories of cognitive, affective, interpersonal and psychomotor function (with a rating of 1 for not present and 7 for severe): 1) confusion, 2) mental alertness (or its absence), 3) impairment of recent memory, 4) disorientation, 5) mood depression, 6) emotional lability, 7) self-care (or lack of), 8) motivation and initiative (or lack of), 9) irritability, 10) hostility, 11) bothersomeness, 12) indifference to surroundings, 13) unsociability, 14) uncooperativeness, 15) fatigue, 16) appetite (or lack of), 17) dizziness, and 18) anxiety. Clearly, even though this enumerates many factors, the distinctions between some and the assessment of many is not a trivial task, and there is art as well as science in being an effective rater.

It is this singular lack of sharp focus, blurring of outlines, and overlapping of symptoms that makes our chosen "disease state" (OBS) difficult to define (80). Without precision in definition and without modern high-quality clinical trials, we are going to have difficulty deciding whether a given drug "works" or not. Therefore, we are faced with evaluating many drug actions, based on poorly designed clinical trials (though, this is an aspect rapidly improving), in an attempt to find compounds effective against OBS and suggestive in various ways of even more effective materials.

Before discussing specific compounds, we might consider some possible underlying mechanisms of OBS. As recently stated by Hoyer et al. (70), it has been established that abnormalities in psychoorganic dementia (their usage) are associated with both qualitative and quantitative changes in cerebral blood flow and cerebral oxidative metabolism. According to Hachinski, et al. (53), two types of idiopathic dementia can be distinguished—the "multi-infarct" type, accompanied by a reduced cerebral blood flow, and the "primary degenerative" type, in which cerebral blood flow is largely normal. Thus, we may anticipate that many of our historical and even newer compounds used in this condition will be involved in either increasing cerebral blood flow (by vasodilation) or enhancing cerebral metabolism, or both.

In a recent study (31), Cosnier et al. address this problem generally:

> Drugs prescribed in cerebral circulatory insufficiency are increasingly expected to have a predominant metabolic action. It should not, however, be forgotten that, where possible, the most logical therapy should be correction of the circulatory deficiency. Unfortunately, the precise circulatory effects of vasodilator substances, at the level of ischemic cerebral areas, are poorly known. Even for a substance which has been used a long time, like papaverine, opinions on this subject are conflicting. This is particularly due to the different methodologies used in pathopharmacological studies. In addition, it is worthless to extrapolate activities found during studies in healthy animals or humans to patients with cerebral circulatory insufficiency. The vessels of healthy cerebral areas are not comparable to those of ischemic regions where hypoxia causes metabolic disturbances with acidosis and vasodilation.

Regarding animal models of the aging process, hypoxic energy deprivation of the brain is a relevant model in experimental brain research, and I shall describe the data derived by Cosnier, et al. (31) as an indication of the types of study involved. [Meier-Ruge, et al. *(this volume)* discusses many other effective models of this condition.] They examined the effects of various drugs on local cerebral blood flow (LCBF) (measured by a thermoclearance technique) in the normocapnic and hypercapnic conscious rabbit. They induced hypercapnia, i.e., the presence of an abnormally large amount of CO_2 in the circulating blood, by making the animals breathe a mixture of air and 8% CO_2. In ischemic cerebral regions, there is a state of vasodilation and metabolic acidosis. Hypercapnia also causes vasodilation and lowers the pH. The hypercapnic brains of the animals concerned then seemed to show similarities with the ischemic brain in patients with cerebral circulatory insufficiency. Two types of pathophysiological response can be envisaged in the patient, and two groups of compounds were, in fact, distinguished in the study: (a) Products which increased LCBF *less* in hypercapnia than in normocapnia (or reduced blood flow in hypercapnia). It can be supposed that in patients treated with these products, the blood flow tends to increase in healthy brain areas and is hardly changed or is, in fact, reduced in diseased areas. In addition, if flow is preferentially increased in healthy areas, an intracerebral steal phenomenon may develop. In other words, blood supply to the healthy areas would be increased *at the expense* of the ischemic regions. (b) Products which increase LCBF *more* strongly in hypercapnia than in normocapnia. Under their action, flow would be increased in diseased areas and modified only slightly, if not at all, in healthy regions. This lack of activity in healthy regions would preclude any intracerebral steal. These two hypothetical mechanisms can only be considered valid when a supplementary vasodilation is possible in ischemic cerebral areas. This could be the case in cerebral circulatory insufficiency due to incipient atherosclerosis in elderly patients.[1]

Cosnier, et al. (31) found that in normocapnia an increase in LCBF was observed with naftidrofuryl (NAF), cinnarizine (CI), viquidil (VI), and heptaminol acefyllinate (HA). The LCBF was only slightly increased or unchanged after hydrogenated ergot alkaloids (HEA), cylandelate (CY), hexobendine (HE), ifenprodil (IF), piridoxilate (PI), vincamine (VC), and xanthinol niacate (XN). In hypercapnia, a more pronounced increase in LCBF than in normocapnia was seen with CY, HE, NAF, and VI, and a decrease or a lesser effect was observed with HA, IF, VC, and XN. The effects of CI, HEA, and PI were not modified.

Irrespective of the precise merits of that model, the study suggests that cerebral vasodilation may be desirable, in general, though it may not be sufficient in OBS—there are many who consider it irrelevant—and, more importantly perhaps, that metabolic factors may play a key role.

AN EXAMINATION OF SPECIFIC DRUGS

Drugs to Increase Cerebral Blood Flow

The first general group of drugs to be discussed have the common property of being vasodilators. The use of these drugs based on this property is very controversial, and involves a long history of negative comment. For example, Innes and Nickerson (75) point out that it is not likely that severely sclerotic

---

[1] While I shall consider some of the compounds these workers studied in some additional detail below, with most, because of space limitations, I shall not do so. I include them just to indicate a typical European sampling of such compounds.

cerebral vessels are capable of dilatation. In addition, the degree of dilatation of cerebral vessels depends largely on local factors that will already have induced the greatest dilatation of which these vessels are capable. Nickerson (105) makes the point even more forcibly:

> Reports of decreased functional impairment from the use of vasodilator drugs—assumed to increase cerebral blood flow—appear from time to time. However, subsequent work has consistently failed to confirm neurological improvement: this is not unexpected. Resistance in the cerebrovascular bed is very effectively controlled by local autoregulation. Control appears to be most directly related to extracellular pH, which, in turn, is very sensitive to changes in pO_2 and pCO_2. This autoregulation functions over a very wide range; indeed, under some circumstances, a cerebral blood flow adequate to prevent CNS damage can be maintained when the jugular bulb pressure is only a few millimeters of mercury less than the cerebral arterial pressure. Atherosclerosis reduces the transport of blood to the small vessels involved in autoregulation, and evidence of cerebral ischemia may then result from much smaller reductions in arterial pressure. Conversely, there may be areas of persistent hyperemia adjacent to a recent cerebral infarct in which autoregulation in response to an increase in blood pressure does not occur. As in skeletal muscle, there is a danger that induction of generalized cerebral vasodilatation will redistribute blood flow to the detriment of the most compromised areas. However, deterioration of function or other definite evidence of intracerebral 'steal' has not appeared, perhaps, because most vasodilators have only a limited effect in this vascular bed. No known drug can decrease cerebral vascular resistance as much as does an increase in arterial pCO_2, and local accumulation of carbon dioxide should have an equivalent effect. The effective autoregulation and the dismal history of previous attempts justify a skeptical response to reports of clinical benefits from the action of vasodilator drugs on the cerebral circulation.
>
> Many references to studies of the effects of drugs on the cerebral circulation are given in a review by McHenry (94).

In a more recent review, Caplan (27) concludes: " . . . there are insufficient data concerning the effectiveness of vasodilator drugs in stroke patients. There is little to recommend the use of vasodilator agents for nonspecific cerebral symptoms of the elderly."

Despite these caveats (and there are many others similar to these), let us examine those studies that support the other side in this class, and let us try to illustrate the problems and the efforts to solve them that are being made in this difficult clinical area.

Papaverine

Papaverine produces vasodilation by a direct action on arterial smooth muscle, and it causes a decrease in cerebral vascular resistance and an increase in cerebral vascular flow (78). Although it has been claimed (Marion Labs) to often relieve and reduce symptoms of cerebral ischemia associated with arterial spasm, and although a thermographic study (USV) has shown that it produces a very significant improvement in blood flow to the ophthalmic branch of the internal carotid

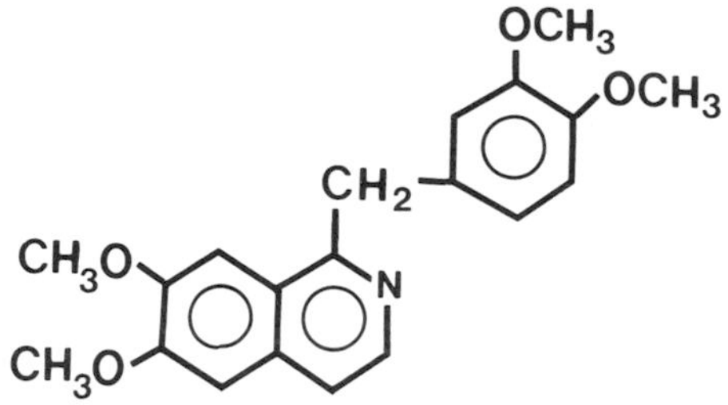

FIG. 1. Papaverine.

artery, a review by *Medical Letter* (95) states as follows: "No well-controlled trials have shown that any papaverine product will prevent or relieve ischemic cerebral vascular disease in man or improve the mental or physical states of elderly or senile patients." Perhaps the well-controlled trials are just emerging, however. In a recent double-blind study, Branconnier and Cole (22) treated two groups of geriatric outpatients (25 in each group, initially) with either 300 mg/day of papaverine hydrochloride or placebo over a period of two months. Neuropsychologic evaluations were made by means of the following tests: EEG, profile of mood states, subject-paced digit symbol substitution test, continuous performance test, Peterson and Peterson test for short-term memory, and clinical global impression. There were positive responses from baseline to active-drug treatment in four of the six evaluations. It must be remembered that statistically significant improvement with geropsychiatric drugs may not occur until the sixth week of treatment. Branconnier and Cole comment:

> It has long been believed (124) that papaverine HCl produces a therapeutic effect on the symptoms of OBS because of its vasodilatory effects. However, recent evidence by Obrist (109) suggests that reduced cerebral blood flow is the result of, and not the cause of, cerebral atrophy. Indeed, it has been independently demonstrated by Corsellis (30) and by Roth, et al. (122) that there is a strong positive correlation between the density and the location of neurofibrillary tangles and neuritic plaques and the severity of neuropsychologic impairment. Tomlinson et al. (138) believe arteriosclerosis to be the primary etiologic factor in only ten per cent of cases of dementia in the elderly and to be contributory in only another seven per cent.

To explain the positive results in their study, Branconnier and Cole suggest that papaverine may be producing its effect by a dopamine receptor blocking action. Ernst (37) and Gonzalez-Vegas (51) have shown that papaverine does display such a property. [Interestingly, despite its inhibition of striatal dopaminergic pathways, papaverine failed to demonstrate an inhibitory effect on dopaminergic hypothalamic-pituitary systems (29).] Branconnier and Cole further suggested that the issue of dopamine blockade or vasodilation might be resolved by a direct comparison of papaverine with a drug like cyclandelate (which has a potent vasodilating action but does not possess the structural activity requirements to block dopamine receptors) or with a low dose of a primary dopamine blocker like haloperidol.

In a double-blind placebo-controlled study, Culebras (32) recently examined the effects of papaverine on the electrical activity of the brain in 20 elderly patients with dementia associated with diffuse cerebrovascular disease. The changes in electroencephalographic activity observed in this study suggest that papaverine favorably affects neuronal metabolism, possibly through improved tissue perfusion, although the exact mechanism of action remains unknown.

The role of such other pharmacological properties of papaverine as its being a powerful inhibitor of phosphodiesterase, and thus an elevator of tissue cyclic-AMP (105,136), and its blockade of the intrarenal vascular receptor that blocks renin release (20) remains uncertain. Incidentally, because of the excessively rapid rate at which papaverine is metabolized and inactivated in the body, either small and frequent dosing or a sustained-release preparation is used (84).

Naftidrofuryl

Recent studies on naftidrofuryl raise several points. First, in a double-blind trial of naftidrofuryl (300 mg daily) versus placebo in two groups (30 in each) of "confused" elderly patients seen in general practice, Brodie (24) not only examined the intellectual capacity of the patients (by a very sparse evaluation involving ten simple questions) but also assessed their ability to "carry out daily living activities." Expressing his viewpoint as a general practitioner, he stated: "The value of any treatment for senile dementia will depend very largely on the therapeutic aim the physician has set himself. From a practical (and humane) point of view, the most desirable aim seemed to be some kind of recovery to enable the patient to live and function reasonably well in his normal environment." Brodie also noted that in view of recent findings by Hachinski (53), "the typical insidious slowly progressing dementia of old age (primary senile dementia) is not due to atherosclerosis." It is not surprising that "previously used drugs that act on the cerebral circulation have proved to be of little value" (4). However, as well as being a vasodilator, naftidrofuryl has a specific action on cerebral metabolism, especially as regards glucose catabolism, resulting in an overall general acceleration of brain metabolism (98). Hence, its clinical activity in OBS may be metabolically related.

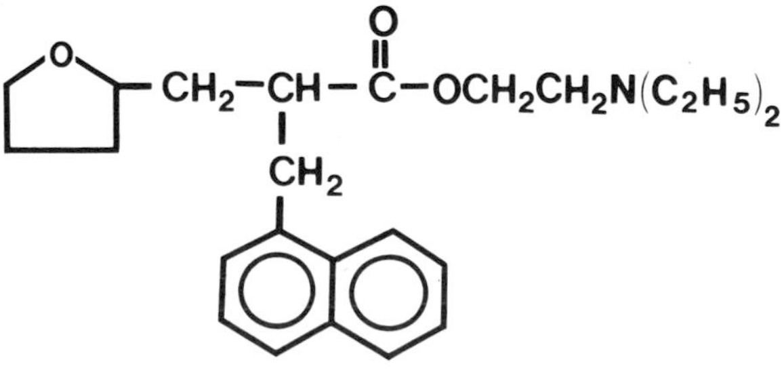

FIG. 2. Naftidrofuryl.

According to Brodie, a number of clinical studies with this drug during the period 1972–1975 described its effectiveness in patients with advanced senile dementia. In Brodie's own study, which was multicentered, the patient group receiving naftidrofuryl again showed significant improvement. Another recent study on this material conducted in an entirely different style was performed by a Boston group, Branconnier and Cole (23). In a double-blind study (60 patients, 90 days, 300 mg per person of drug daily), they evaluated the effects of drug on memory in a group of patients with OBS. They compared an automated test of paced stimulus material (Sperling's Perceptual Trace, SPT) with several other standard memory scales. The data indicated that the SPT indirectly assesses short-term memory, is resistant to practice effects (unlike the more commonly used Wechsler Memory Scale), and is drug-sensitive. They suggest that the SPT might in fact become the test of choice in the indirect assessment of short-term memory in the elderly.

Isoxsuprine

Isoxsuprine is a smooth muscle relaxant (89) with both β-adrenoceptor stimulant and α-adrenergic blocking properties. Using cerebral angiography, it has been found (46) to increase the rate of flow of contrast medium through the cerebral vessels. Using radiolabeling, it has been found to increase both cerebral blood volume and cerebral blood flow (69). In 1973, Elliott and his colleagues (36) reported the results of an open multicentered general-practitioner trial with 170 geriatric patients over a four-month period in which improvement in Crichton Score (120) was the main concern. Although these authors found that many patients improved during the treatment period, they concluded that only a double-blind controlled follow-up trial would resolve whether the improvement was due mainly to the closer attention to the patient that such a trial necessitates or to drug effects. Such a trial was subsequently performed by Hussain, et al. (73) in 1976. An automatically controlled learning task was used to objectively assess the mental performance of 17 patients with cerebrovascular disease. The effect on performance of the patients during treatment with isoxsuprine (in a sustained-release form) was measured during a double-blind placebo-controlled trial of 16 weeks' duration. There was a significant improvement in performance between the treated and untreated groups.

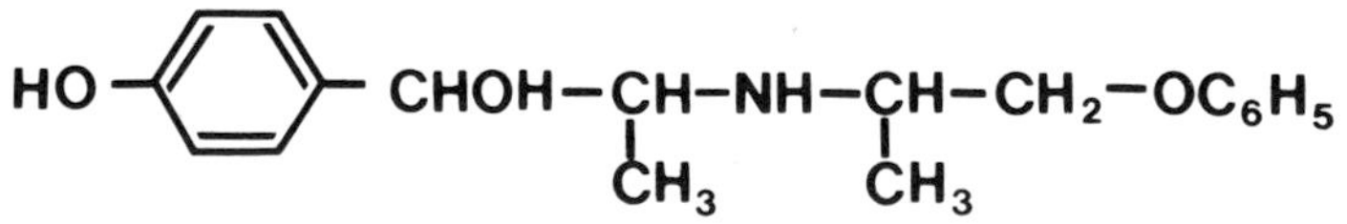

$$HO-C_6H_4-CHOH-\underset{CH_3}{\underset{|}{CH}}-NH-\underset{CH_3}{\underset{|}{CH}}-CH_2-OC_6H_5$$

FIG. 3. Isoxsuprine.

Cyclandelate[2]

Cyclandelate is a general smooth muscle relaxant that differs chemically from most of such compounds in that it contains no nitrogen. However, most of its pharmacological properties (18) are very similar to those of papaverine, although it is less toxic and has greater spasmolytic action. It can produce mild vasodilation in man, including increases in cerebral and muscle blood flow. In a recent double-blind crossover study (143), 24 men with senile dementia received careful physical and neurological examinations, blood chemistry analyses, and psychometric tests appraising visual spatial perception, memory, intelligence, and logical thinking. Cyclandelate was found to be no more effective than a placebo in

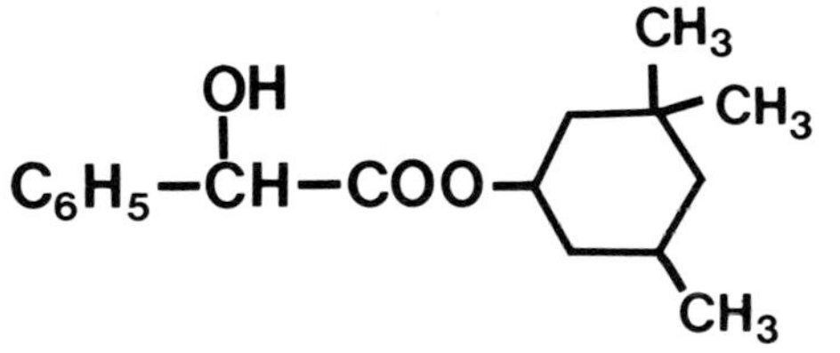

$$C_6H_5-\overset{\displaystyle OH}{\overset{|}{C}}H-COO-C_6H_8(CH_3)_3$$

FIG. 4. Cyclandelate.

improving higher cortical function in these patients. Many of the previous clinical trials have been briefly described elsewhere (4,11).

However, anecdotal information speaks otherwise. In 1975, Hall (55) described a study with the above drug in which 21 patients were subjected to a double-blind crossover study for a year, the treatment group receiving 400 mg of drug four times a day. Hall assessed all patients by means of gross behavioral tests that covered such items as the patients' social independence and their ability to feed and to dress themselves. The team also evaluated the patients' intelligence on a Wechsler adult IQ scale, and gave them psychiatric and neurologic examinations. Analyzing the data on the behavioral scales, Hall was disappointed to find no significant difference in his patients' behavior on the drug and off it. He was encouraged by the fact that during treatment, the patients showed some improvement in all the psychological factors tested, e.g., mental state, mood, orientation, memory abstraction, apraxia, and aphasia. The most impressive difference, however, was seen in the intelligence measurements. "In general, we found no measurable decline in IQ during cyclandelate therapy," he observed. "But, there was a statistically significant decline when the same patients were on placebo." The chief differences in comprehension and verbal ability were highly significant statistically. He believes that the drug acts prophylactically

---

[2] See also Addendum.

to arrest the downslide of mental performance rather than to reverse any earlier decline.

In the same article, Mayer (97) comments on his good results in vasodilator treatment of patients with arteriosclerotic dementia: "If you can diagnose these patients correctly, there's a great future for treating them with drugs that influence the circulation of the brain." But he stresses that arteriosclerosis is not the only factor in dementia and that other factors, such as neuronal atrophies, may be of significance. He points out other complexities in the vasodilator treatment: "It may not be cerebral vasodilation that is causing the improvement. These drugs that are supposed to be vasodilators also influence platelet aggregation—a side effect that may be important and beneficial. In addition, most of them influence cyclic AMP metabolism in the brain, and this itself may affect not only the blood vessels but the function of the brain."

Betahistine

On December 26, 1974, Unimed filed an NDA with the FDA seeking approval of SERC (betahistine HCl) for use in patients with cerebral vascular disease. The FDA held that the application is 'not approvable' and raised certain questions regarding the scientific data submitted. These scientific questions are being pursued for clarification and determination. Additional data are regarded by the FDA as necessary. Unimed is proceeding with additional clinical studies and will submit the results to the FDA to supplement the data already submitted. No rapid FDA decision can be expected ((142)).

Since that time in 1974, a number of papers have appeared which merit comment.

Seipel et al. (127) published a series of papers on the rheoencephalographic technique of evaluating betahistine. In part III of that series (128), they address themselves to careful patient selection in this area of clinical study. They conclude as follows:

> Thirty-six institutionalized geriatric patients with particularly severe and long-standing dementia were selected by chart review and were studied neurologically. Criteria for clinical diagnosis were derived, and it was shown that arteriosclerosis caused or contributed to their dementias in 33 patients and probably did so in two others. Arteriosclerotic dementia may not be rare and may be much more prevalent than may now be considered the case. As 23 patients showed carotid and/or vertebrobasilar involvement, large-vessel disease also may be

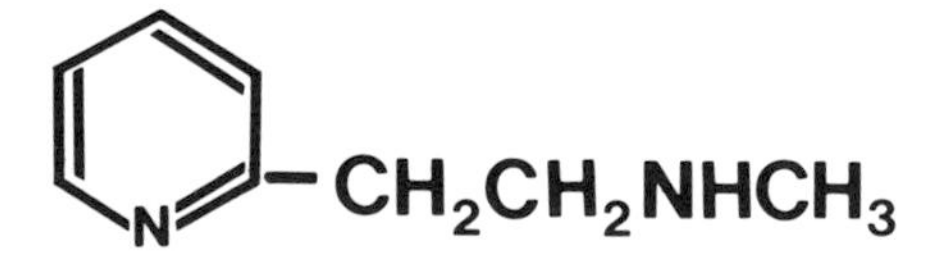

FIG. 5. Betahistine.

> more prevalent in geriatric dementia than is now thought and should be considered in all such patients. The results further confirm that careful review of ward records without reference to psychiatric diagnosis of etiology is a simple, rapid and reliable method of screening a large, institutionalized populations for possible arteriosclerotic dementia. The selectees should then be surveyed appropriately to confirm the presence of the disease. Since a careful review of the complete hospital records and a thorough neurologic examination proved sufficient, more definitive but more dangerous invasive diagnostic procedures may not be necessary for clinical pharmacologic and similar investigation.

From part IV of their study, several interesting conclusions were also drawn (129): "If one considers that arteriosclerotic arteries may be difficult, if not impossible, to dilate actively or passively, it is possible that prolonged microcirculatory vasodilatation may facilitate the establishment of enough collateral circulation that arterial vasodilatation could appear to have occurred." They conclude that study as follows:

> The effects of prolonged betahistine administration were studied in institutionalized geriatric patients with particularly severe and long-standing arteriosclerotic dementias. Thirty received drugs or placebos orally in fixed dosage for six months on a double-blind basis and were followed by ward behavioral and psychometric ratings. Six others received active medication and were additionally followed by intracranial rheoencephalography (IREG). The results show that betahistine caused definite, strong and highly significant cerebral and scalp arterial vasodilatation and circulatory improvement and that these caused equally definite, strong and highly significant global improvement in the patients' dementias. Betahistine thus acts in humans as a potent and efficacious cerebral and peripheral microcirculatory and arterial vasodilator which can significantly improve cerebrovasculary insufficiency and any associated dementia, no matter how severe either may be and despite the possible presence of large vessel disease.

In a final paragraph they state:

> Since the circulatory response detected by the IREG thus arose essentially completely from the effects of betahistine on the cerebral circulation alone, this appears to be the first direct demonstration that cerebral circulatory improvement can cause improvement in mental function in patients with even severe, long-standing and apparently clinically irreversible arteriosclerotic dementia and that such improvements can be effected pharmacologically.

In another recent study, betahistine, despite its being formally considered one of the histamine analogues, was not found to directly activate histamine receptors and by inference, its histamine-like effects in other tissue and whole animals result from the release of histamine (72).

Dihydrogenated Ergot Alkaloids

Hydergine, a mixture of three hydrogenated ergot-derived alkaloids, dosed as mesylates, whose "active" ingredient is called dihydroergotoxine mesylate,

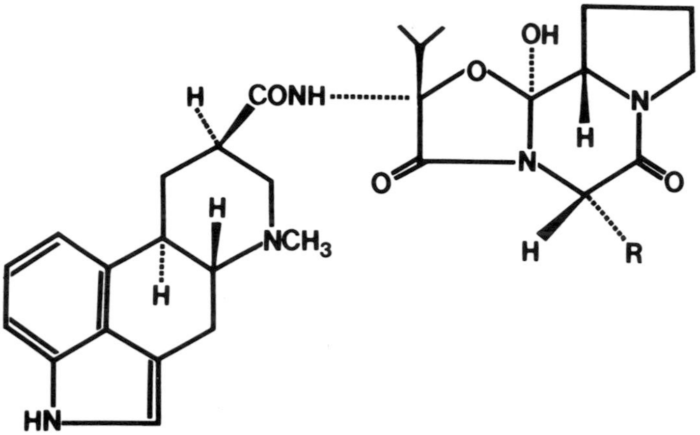

FIG. 6. R= Pri, ergocornine; = $CH_2C_6H_5$, ergocristine; = Bu-i, ergocryptine

is in a historical sense one of the primary lead compounds in this area. This material has almost certainly received the greatest amount of study due largely to the efforts of the Sandoz group, whose work will be discussed further by Dr. Meier-Ruge, et al. *(this volume).* Reviews of this material have been mixed and, on balance, reaction to it has been lukewarm. The *Medical Letter* has consistently given it negative reviews. In the latest issue (33), which was concerned with the Mead Johnson version—namely, Deapril-ST—they report as follows:

> A recent advertisement for Deapril-ST asserts that mood depression, unsociability, confusion, dizziness and lack of self-care in the elderly may be related to 'Idiopathic Cerebral Dysfunction (ICD).' The ad cites postmortem studies in patients with severe dementia which indicate that such symptoms may not be caused by arteriosclerotic changes or impaired cerebral blood flow ((147)). The manufacturer claims that the 'target' symptoms of 'ICD' may be associated with altered metabolic activity in the brain and that the dihydrogenated ergot alkaloids in Deapril-ST have been shown experimentally (in cat brain tissue) to improve cerebral metabolism.
>
> *Clinical Trials.* The advertisements for Deapril-ST cite three double-blind studies as demonstrating the effectiveness of the ergot alkaloids (12,103,121). All three compared Hydergine with papaverine (many manufacturers) which is offered for relief of cerebral and peripheral ischemia but has never been proven effective for any indication. None of these reports indicate which symptoms were present before treatment, or the degree of severity of individual symptoms. None of the studies lasted longer than three months or included a 'crossover' control. Among the individual symptoms, 'lack of self-care' was not present in enough patients to evaluate in one study; it was reported to improve significantly more with Hydergine than with papaverine in the second, but not in the third study. Improvement in dizziness was reported significantly greater with ergot alkaloids in one of the three studies. Improvement in both depression and anxiety was said to occur more often with the ergot alkaloids than with papaverine. In all of the studies, improvement in symptoms was

> graded from 1 to 7 but the published data are insufficient to evaluate the clinical importance of numerical gains in the ratings.
>
> Since Deapril-ST has the same active ingredients as Hydergine, any evidence that Hydergine is effective would be applicable to the new product. A recent review ((71)) of 12 controlled trials of Hydergine concluded that most of the trials were inadequate, that the only crossover study showed no significant changes due to Hydergine, and that the only long-term study indicated that initial improvement with this drug was lost over a one-year period.
>
> *Conclusion.* There is no convincing evidence that dihydrogenated ergot alkaloids are effective for treatment of 'Idiopathic Cerebral Dysfunction' or any other disorder of elderly patients.

I have quoted this negative review so completely because the compound is such a key material and also because the case *for* the compound is strongly documented. However, let us first consider some other reviews. Nickerson and Collier, in Goodman's and Gilman's standard text [(106)], state: ". . . the parameters affected (in the hydergine studies) are not consistent and it has not been shown that the drug can augment the vasodilatation due to accumulated carbon dioxide in low-flow areas of the brain (93). . . ." Hollister is somewhat more optimistic (68):

> The α-adrenergic receptor blocking agent, Hydergine, has long been thought to act by decreasing vasomotor tone in the cerebral vasculature. Recent evidence suggests a far more complicated mode of action. Hydergine increased the pyruvate-lactate ratio in hypothermic or ischemic brain disease, negating the hypometabolism induced by these states. The biochemical change was accompanied by a return to normal EEG energy. The effect on the opening of the cerebral micro-circulation is now viewed as a consequence of increased brain metabolism rather than its cause. The drug inhibits phosphodiesterase which could increase metabolic activity in terms of new protein synthesis. The latter is stimulated by the drug in brain preparations inhibited by barbiturates. All this new information suggests that Hydergine may have direct metabolic, rather than purely vascular actions. This mechanism would be more compatible with our present understanding of plaque and tangle formation—the major cause of disability.

As Meier-Ruge, et al. state *(this volume),* "In contrast to the action of papaverine, DH-ergotoxine mesylate stabilizes the EEG energy without influencing the cerebral blood flow. The improvement of EEG activity correlates well with local pO_2 microelectrode measurements in the brain cortex. The ameliorative effect of DH-ergotoxine mesylate is not due to increased metabolism, but to an improved metabolic economy of the brain tissue."

Before briefly considering other studies, let us put the *Medical Letter* report into perspective. The abstract of Hughes, et al. (71) reads as follows:

> A critical review is presented of 12 clinical trials with Hydergine (a hydrogenated ergot alkaloid preparation) in the treatment of dementia. Qualitative and quantitative comparisons of improvement in symptoms showed that Hydergine consistently produced statistically significant ($p \leq 0.05$) improvement in 13 symptoms associated with dementia. However, because of the small magnitude of the improvement and the absence of indications of long-term benefits, Hydergine would seem of minor value in dementia therapy. Further research with better methodology and design might lead to a different conclusion.

It is certainly interesting, if not instructive, to compare the two statements concerning the same manuscript.

Because Meier-Ruge, et al. discuss the overall Sandoz contributions, as well as those of others in this area *(this volume),* I shall conclude with a review of a few additional papers. The first is by Gaitz, et al. (42), a Houston group, which offers a balanced critique of many clinical studies. I shall use this paper to state the case for Hydergine. Their abstract reads as follows:

> Evaluation of treatment modalities, including pharmacotherapy, for organic brain syndrome (OBS) has been difficult because of sampling and methodological problems, and comparison of research studies are all but impossible. In this study, an ergot derivative, a combination of dihydroergocornine mesylate, dihydroergocristine mesylate and dihydroergokryptine mesylate (Hydergine) was compared with placebo, using a double-blind technique in a sample of nursing home residents with evidence of OBS. An 18-category symptom rating scale [the Sandoz Assessment of Clinical Status Rating Form—Geriatric (SCAG)] was used for periodic assessment over a six-month interval. Comparisons of the two groups of subjects disclosed that the Hydergine-treated group *showed statistically significantly more improvement in most of the variables measured, especially during the last three months of treatment.* Furthermore, sophisticated analysis revealed that positive changes in cognitive function cannot be accounted for as a mere reflection, or 'halo effect' associated with improved mood and general sense of well-being.

Several other studies involve the use of dihydroergotoxin in postoperative and posttraumatic periods of neurosurgical patients (after brain surgery or severe head injuries) because of the demand for sufficient blood supply and oxygenation of cerebral tissue (88). A significant increase of cerebral venous CO_2 concentration following the administration of dihydroergotoxine in these patients was attributed to an improved O_2-utilization in the cerebral tissue.

The possible action of ergot alkaloids of this class as dopamine agonists has been recently examined in two rodent models (3); their α-adrenergic blocking properties utilized via [^{3}H]-dihydroergocryptine, which has been developed as a reagent for α-adrenergic receptor identification (145)—and both of these properties, dopamine receptor involvement as well as α-adrenergic receptor studies, have been concurrently studied (50).

Finally, a number of very recent, relevant studies have been reported (26, 52,76). First, the effect of papaverine and dihydroergotoxine mesylate (DHET) in cerebral microflow, and EEG and po_2 in oligemic hypotension (52) was examined, an effect discussed further by Meier-Ruge, et al. *(this volume).* There, DHET exerted a protective effect on the oligemically disturbed brain metabolism (stabilization of EEG activity and shift of po_2 distribution in the direction of the normotonic state). On the other hand, papaverine showed a marked vasoactive effect without preventing the breakdown of the EEG activity. In spite of increasing local cerebral blood flow, papaverine had no positive effect on the oligemia-induced decline in cerebral pO_2 values (as a consequence of a shunt perfusion).

Several other such studies involved (a) comparative effects of DHET on cerebral blood flow and metabolism changes produced by experimental cerebral edema, hypoxia, and hypertension (26); and (b) a study of the incorporation, after single and repeated application of radiolabeled DH-ergot alkaloids in different organs of the cat, with special reference to the brain (76). One hour after intravenous administration, ^{3}H-DH-ergot alkaloids showed maximal uptake in the range of 10^{-5}M in various visceral organs, and of 10^{-7}M in most parts of the CNS of the cat. Repeated administration demonstrated a higher retention in the CNS than in the other organs. The single-dose level in the CNS was reinforced and, in contrast to liver and lung, was maintained for at least 24 hr.

Vincamine and Ethyl Apovincaminate

Two materials receiving extensive current European study are the reserpine-like vincamine long used (for about 20 years) in Eastern Europe as a cerebral vasodilator, and its derivative, ethyl apovincaminate.

A recent issue of *Arzneimittel Forschung* (Drug Research) was devoted in its entirety to the vincamine work (10). The opening paper by Witzmann and Blechacz (146) laid the groundwork for the discussion:

> Stroke and mental decline as a result of a so-called 'cerebral atherosclerosis' represent the largest group of diseases in neuropsychiatry. They are characterized by one of the highest rates of mortality and morbidity among all diseases. A stroke is mainly caused by thromboembolism; in dementia we differentiate today between a 'multiinfarct' group and a primary degenerative group. In these diseases regional and global disturbances of cerebral metabolism (CMR) and cerebral blood flow (CBF) are interdependent. To this day, the evolution of these diseases cannot be prevented by any therapeutic measure; treatments are conventional and characterized by un-specific measures, such as stabilization of blood pressure, normalization of cardiac functions and circulation, treatment of edema and syndrome. Specific measures such as the prevention of an expansion of tissue necrosis are still in the experimental stage. A small group of drugs therefore become of ever increasing interest as it is assumed that they will cause secondary increases of CBF by an activation of specific metabolic steps. 14,15-Dihydro-14β-hydroxy-[3α,16α]-eburnamen-ine-14-carbonic acid methylester (vincamine) belongs to this group.

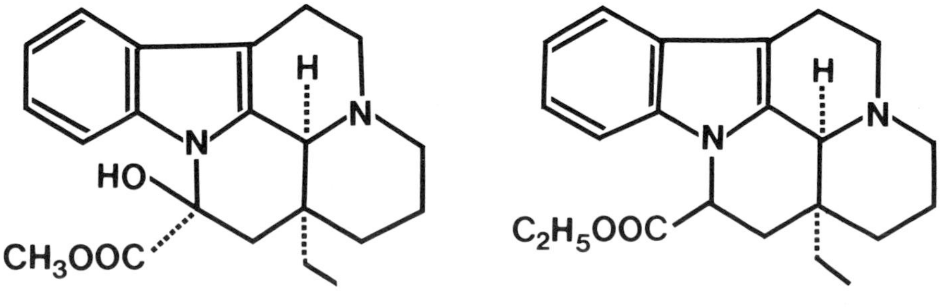

FIG. 7. Vincamine *(left),* ethyl apovincaminate *(right).*

Other studies discussed the influence of vincamine on global and regional CBF in acute and subchronic human cerebral ischemia (63). CBF measurements were made after intraarterial injection of ^{133}Xe. A single i.v. application of 30 mg vincamine in 200 ml levulose 5% was infused during 20 min, and a significant increase in CBF (6%) was noted. The regional CBF was influenced to a varying degree, the areas marked by an insufficient blood supply showing an increase of 13% and the areas marked by normal basic values showing an increase of 5%. A second, more elaborate study (82) on this same aspect concluded: "The analysis of the regional CBF values revealed that vincamine belongs to the group of drugs effecting a heterogenous reaction of the cerebral circulation, i.e., an increasing focal ((sic)) flow was observed in ischemic or relative ischemic areas (up 28% on average). On the other hand, hyperemic areas remain unchanged, and areas within the mean hemispheric flow values showed an increased focal ((sic)) flow of 16%." The effect of vincamine on CBF as a function of application rate was also reported (60).

There have been two major reports on the clinical studies. Witzmann and Blechacz (146) summarize their findings as follows:

> The analysis of 33 clinical reports, published in the international literature, revealed an improvement in 75% of all patients treated while the remainder did not show any alteration or deterioration of the clinical picture. Daily doses of 60 mg proved favorable in oral long-term treatment. Vincamine was tested versus placebo in 7 clinical double blind studies in 4 countries (dose 60 mg, duration of treatment 20 to 60 days). The results were obtained with the aid of psychometric tests, rating scales and function tests (EEG, audiogram, etc.). The 7 trials all yielded significant ($p<0.05$–0.001) improvements of various symptoms, in particular, of disturbances of memory and attention. The overall evaluation of the double blind studies showed 105 improvements in 151 patients in the verum group and 32 improvements in 125 patients in the placebo group. This difference has statistical significance ($p<0.01$).

Needless to say, the earlier discussion of the Hydergine cases suggests many qualifications on these data.

An even more recent clinical study has since been reported (40), and again, the investigators' own summary seems worth recording:

> The neuropsychiatric symptoms of old patients with disturbed cerebral metabolism or blood flow mostly leads to great individual difficulties and make those patients difficult to handle; in the family as well as in hospital, such patients develop alienation, isolation and, therefore, adaptation to a social structure deteriorates with time. In the course of a test program for medicinal therapy of this syndrome, we studied the efficacy of a vincamine-containing formulation on the symptoms of a chronic brain disturbance. The study was randomized double blind. We found that under the influence of the vincamine formulation the subjective symptoms such as lack of interest, apathy, aggressiveness, raging, psychomotor retardation, lack of concentration, dysmnesia, decreased with a statistical significance ($p\leq0.05$). Also, the subjective symptoms reported by the patients such as tinnitus and vertigo decreased significantly under the treatment with the vincamine preparation. Therefore, some of the parameters impor-

> tant for resocialization and revitalization of old patients could be influenced in a favorable way. Based on the good results of our investigations, the treatment of psychiatric disturbances in old patients with vincamine-containing drugs seems to be justified.

The case of ethyl apovincaminate is quite similar to the above. For example, an entire issue of *Arzneimittel Forschung* (9) is devoted to the compound. The issue is introduced by Gy. Fekete, Scientific Director of Gideon Richter, Budapest, who presumably sponsored the presentations. His preface puts the vincamine/ethyl apovincaminate story in perspective. The 23 papers include coverage of the synthesis, pharmacokinetics, metabolism, and safety of the drug, as well as its use in a variety of clinical conditions such as ophthalmological therapy, certain hearing disorders and in neurosurgical patients. The overall issue has been recently summarized thusly (5):

> Ethyl apovincaminate is a synthetic derivative of the vincamine type. The new drug appears to be less toxic, more potent and more specific in its vasodilator actions than the naturally-occurring precursor. Studies in large numbers of patients with cerebral vascular insufficiency have shown that ethyl apovincaminate (10 mg, i.v.) markedly reduces cerebral vascular resistance without significant effects on total vascular resistance, and it also increases the cerebral fraction of cardiac output. Cerebral blood flow is increased, particularly in grey matter, but there is no appreciable effect on arterial pressure. These effects can be subsequently maintained for many months by oral administration of the drug, at doses of 5 mg three times daily. Sodium apovincaminate is also useful in ophthalmology by improving circulation in the eyeground, particularly in cases of atherosclerosis of the central retinal artery. In most cases, visual acuity has been improved during long-term therapy with the drug. It seemed to be effective in improving a variety of neurovascular conditions, including vascular impairment of hearing and Meniere's disease.

In a comparative trial in 143 patients with cerebrovascular disease, ethyl apovincaminate was significantly superior to xanthinol nicotinate in improving paresis and decreasing morbidity and mortality (140). Incidentally, in the CNS, it had neither antidepressive nor anticholinergic activities. It displayed neither sedative nor analgesic effects in rodents (112).

Nicergoline

Nicergoline [1,6-dimethyl-8β-(S-bromonicotinyloxymethyl)-10α-methoxy-ergoline]Sermion is an ergoline derivative with α-adrenergic blocking and vasodilating activities, being developed by Farmitalia for almost a decade (8). Like Hydergine, it increases cerebral microcirculation by primary protective effects on neuronal metabolism. In a study in the cat, for example, Boismare and Lorenzo (21) found that in a hypoventilated hypercapnic animal with an already dilated cerebral network, nicergoline still affords protection against the effects of the cerebral ischemia with a rapid ischemic recovery of EEG activity. In brain metabolic studies, Benzi, et al. (15) found that it increased the general metabolizing ability (demethylation and acetylation of aminopyrine and glucuro-

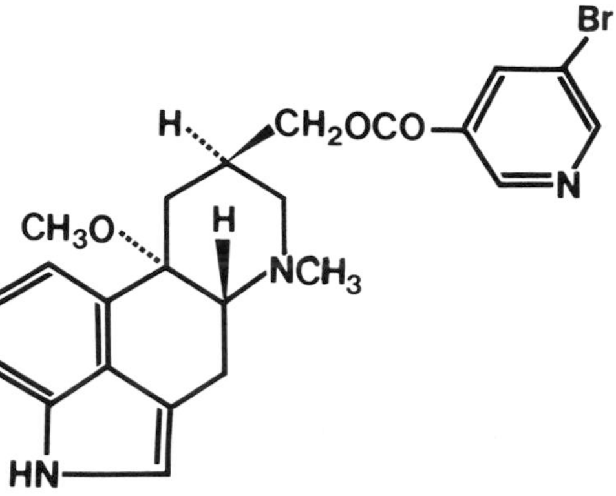

FIG. 8. Nicergoline.

noconjugation of oxazepam) of the brain, and, under conditions of experimental ischemia, it increased both neuronal glucose uptake and reduced pyruvate and lactate formation (16). Its molecular mechanism of action seems related to effects on brain ATP levels, which are more rapidly restored after experimental ischemia when pretreatment with nicergoline has been performed (15). It stimulates cAMP accumulation in rat cerebral cortex slices by its inhibitory action on phosphodiesterase and/or by a direct stimulation of adenylate cyclase (100). The influence of a molecule of this type on the biosynthesis of brain macromolecules has been reported very recently (114).

In human studies with the drug, the data are sparse and occasionally conflicting. Thus, while Pasotti, et al. stated that nicergoline increased cerebral blood flow and the symptoms associated with acute or chronic reduction of brain perfusion rates (113), Prencipe, et al. found no increase in cerebral blood flow following nicergoline treatment in patients with chronic or diffuse cerebrovascular pathology (116). A British group, Iliff, et al. (74), even more recently reported an appreciable short-term increase of cerebral blood flow (^{133}Xe clearance) in patients with established cerebrovascular disease. Surprisingly, this last study did not cite the earlier two.

It was reported that the drug increases cerebral oxygen and glucose consumption in elderly patients suffering from cerebral arteriosclerosis (91), and, from appropriate EEG changes post-drug dosing, it was concluded that arteriopathic cerebral insufficiency constitutes an indication of choice for nicergoline treatment (126).

In one open uncontrolled study, 15 elderly patients suffering from cerebral arteriosclerosis were considered to have benefited from nicergoline therapy (139). Bernini, et al. (17) report on a similar study with 25 patients suffering from chronic cerebral insufficiency. Nicergoline treatment lasted one month, and both clinical and cerebral angioseriographic results were recorded. From the improvements in symptoms and in the angiographic data in a high percentage of the patients studied, the authors concluded that the drug was useful in the treatment of chronic cerebral insufficiency.

Miscellaneous

In a recent review on the topic of cerebral vasodilators, Hauth and Richardson divided such agents into two groups according to their site of action (59). In the first group are those which have primary vascular action, dilatation usually being effected either by a direct relaxant action of the drug on the smooth muscle cells in arteriolar walls or by an inhibitory effect on the endogenous vasoconstrictive nerve fibers. A majority of these agents were developed initially as peripheral vasodilators. They called this group "vasotropic dilators," and their list included papaverine, bencyclane, cyclandelate, betahistine, and cinnarizine. The other group contains those agents which stimulate neuronal metabolism; the resultant increase in local perivascular CO_2 production consequently causes vasodilatation. They called this group "Cerebrometabolic Stimulants." It included Hydergine, nicergoline, and vincamine.

Since the two classification functions are not independent, this distinction will be simplistic to many, but it does emphasize the metabolic aspects. Our present discussion has also listed drugs in more or less the same sequence without mechanistic emphasis. To conclude this section, brief mention will be made of several other active compounds.

Cinnarizine

Cinnarizine does increase cerebral blood flow in man (13) and animals (144) via vasodilatation (141), a process that may proceed via its ability to block calcium uptake at the membranal level (48). However, its clinical picture has been ambiguous. In a typical study, Toledo et al. (137) conducted a double-blind crossover trial of cinnarizine on 30 psychiatric patients, all presenting concomitant symptoms of impaired cerebral circulation. The trial lasted 18 weeks and comprised three 6-week treatment periods alternating cinnarizine-placebo-cinnarizine or placebo-cinnarizine-placebo, at a dosage of two 25-mg tablets

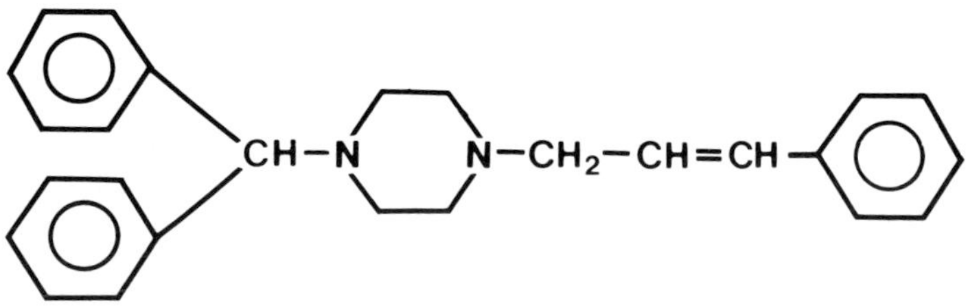

$$(C_6H_5)_2CH-N\langle\rangle N-CH_2-CH=CH-C_6H_5$$

FIG. 9. Cinnarizine.

t.i.d. Clinical evaluation of efficacy was made by one group of investigators, psychometric evaluation by another. Cinnarizine was found effective in all five types of clinical situation (schizophrenic, oligophrenic, alcoholic, involutional psychotic, psychopathic) in which it was tested, whether the symptoms of impaired cerebral circulation were primary manifestations or complications of other diseases. Despite the success of this Argentinian study and of several others, a British group (2), in an earlier general practitioner multicentered study, concluded that their trial had not demonstrated any effect from the drug in the chronic cases of cerebral arteriosclerosis evaluated. It was a double-blind, 8-week, crossover study, using the Crichton geriatric scale to evaluate the results (the dose was identical to that used by the Argentinian group).

Bencyclane

Critical reaction to Bencyclane has been mixed, and it is discussed extensively in Hauth and Richardson's account (59). Thus, in one study, it was not found to improve cerebral blood flow (CBF) in patients with insufficient flow; in fact,

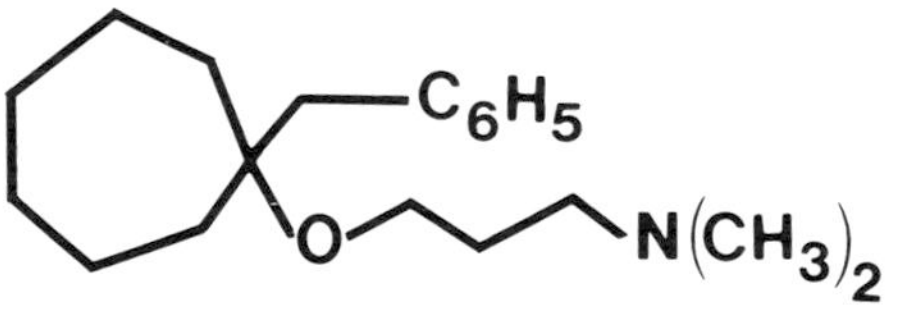

C_6H_5

O

$N(CH_3)_2$

FIG. 10. Bencyclane.

in one third of the cases, it reduced it (61), whereas in another study it did improve CBF (81). In animals (pigs and cats), the substance caused an increase in CBF with vasodilatation (43), and in rats it stimulated glucose metabolism (56). The mixed clinical picture (59) leaves the value of this compound very much in question.

General Compounds

The general compounds that follow are drugs that have not been evaluated substantially, or indeed at all, in man. The Japanese drug, a dihydropyridine compound, YC-93, (Fig. 11A) in dogs, monkeys and cats (119) produced a dose-dependent increase in cerebral and coronary blood flow. The drug had 100 to 300 times the potency of other common vasodilators, and its duration of action was also longer than any of these. It was well-absorbed from the gastrointestinal tract. In humans, its profile suggested that it might be a cerebral vasodilating drug (130). Another Japanese product, PF-244 or 3-[bis-(3,3-diphenylpropyl)amino]-propan-1-o1, (Fig. 11B) was also found to be a potent cerebral vasodilator without significant effects on myocardial hemodynamics

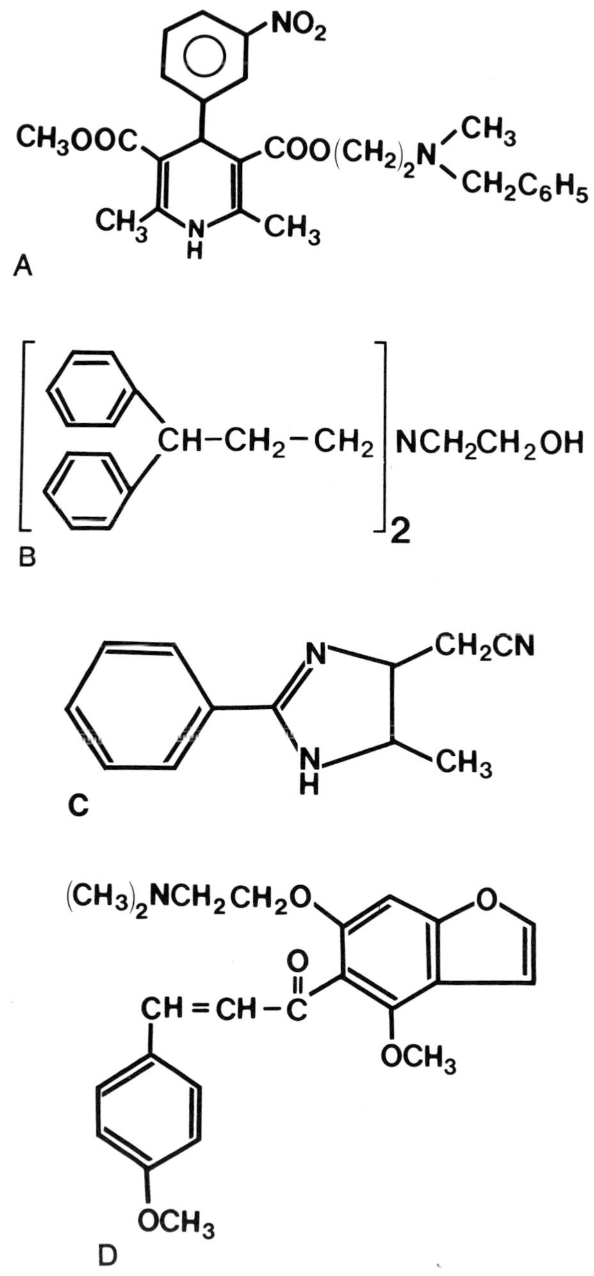

NO_2

CH_3OOC $COO(CH_2)_2N$ CH_3 $CH_2C_6H_5$

CH_3 N H CH_3

A

$[$ CH–CH_2–CH_2 $]_2$ NCH_2CH_2OH

B

N CH_2CN

N H CH_3

C

$(CH_3)_2NCH_2CH_2O$ O

O

CH = CH – C

OCH_3

OCH_3

D

FIG. 11. YC–93 **(A),** PF–244 **(B),** MCN–2378 **(C),** and mecinarone **(D).**

or femoral blood flow (77). The imidazoline (MCN-2378) (Fig. 11C) has also been reported to be a cerebral vasodilator, with a cardiovascular profile resembling that of papaverine (54).

Mecinarone (Fig. 11D) has been found to be both a peripheral and a cerebral vasodilator, its mode of action involving selective inhibition of the calcium influx in depolarized cells of isolated arteries, with a consequent reduction in the capacity of free Ca^{2+} ions to induce and maintain contraction (115). It will have been noted that three of the four compounds just mentioned contain the structural unit ($-O-C_x-NR_2-$). Some others in this class include naftidrofuryl, bencyclane, and isoxsuprine (discussed earlier), as well as moxisylate[3] (Fig. 12A) and proxazole, (Fig. 12B) the oxygen atom being endocyclic, i.e., part of a ring. A rather similar functional group $-N-C_x-NR_2-$ is found in betahistine (one N is endocyclic), in the alkaloidal moieties discussed earlier (both Ns are endocyclic), and in such compounds as butalamine (Fig. 12C) and fenoxedil (Fig. 12D) (59). A sulfur analog of that same kind of functional unit $-S-C_x-N$ is present in pyritonol, (Fig. 12E) which has recently been evaluated for its effects on blood flow and on cerebral oxidative metabolism in patients with dementia (70).

Other Major Alkanolamine Drugs

In this category, I want to discuss three compounds that have received considerable study—namely, deanol, centrophenoxine, and procaine.

Deanol

Deanol has received study in the treatment of tardive dyskinesia (134), minimal brain dysfunction (87), and a range of other involuntary movement disorders. Its reputed ability to increase brain acetylcholine levels (58) has been reviewed (49) but not confirmed (148). Based on its possible involvement in the brain cholinergic systems and because of the latter's apparent role in certain memory processes, the drug was given openly to 14 senile outpatients for 4 weeks (up to 1800 mg/day) (38). Ten patients improved globally, whereas 4 were unchanged. No improvement in memory or in cognitive functions was revealed. The behavioral changes involved reduction in anxiety, depression, and irritability, and increases in motivation-initiative.

The material has also been assessed from an entirely different angle. The lysosome hypothesis of aging (65) postulates an initial deterioration of lysosomal membranes that results in uncontrolled leakage of lysosomal enzymes. Any treatment that inhibits such deterioration should, therefore, increase longevity.

[3] References to all these cerebral vasodilators may be found in Hauth and Richardson's review (59).

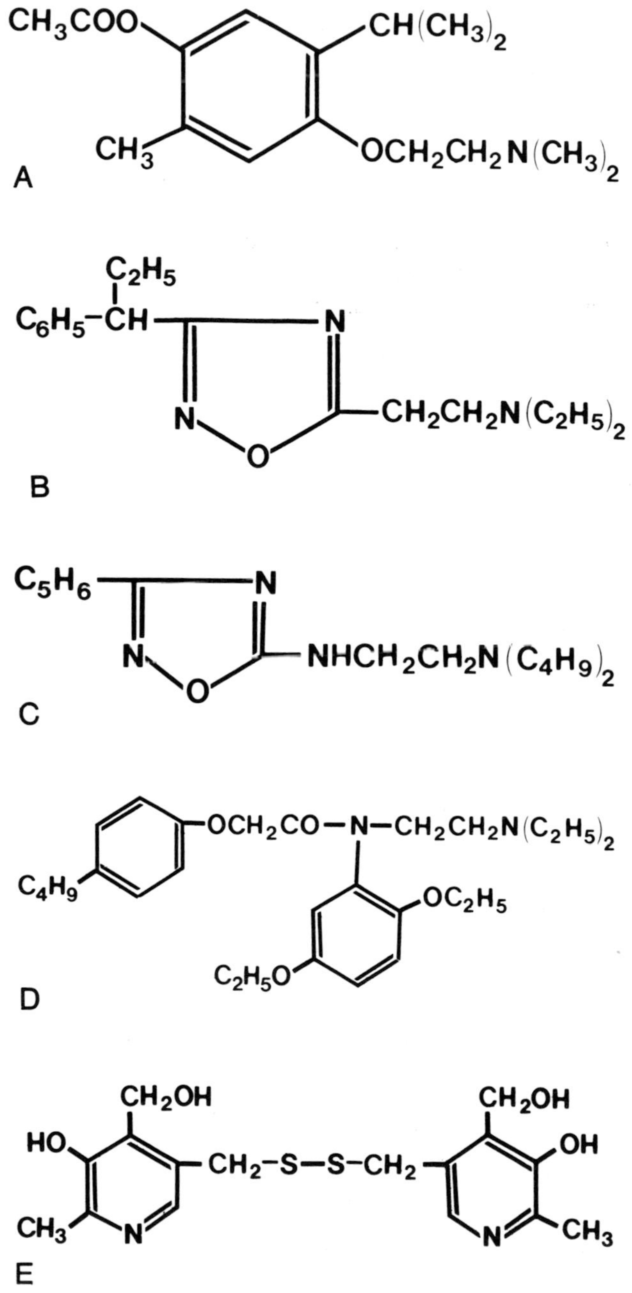

CH_3COO $CH(CH_3)_2$
CH_3 $OCH_2CH_2N(CH_3)_2$
A

C_2H_5
C_6H_5-CH N
N O $CH_2CH_2N(C_2H_5)_2$
B

C_5H_6 N
N O $NHCH_2CH_2N(C_4H_9)_2$
C

$OCH_2CO-N-CH_2CH_2N(C_2H_5)_2$
C_4H_9 OC_2H_5
C_2H_5O
D

CH_2OH CH_2OH
HO $CH_2-S-S-CH_2$ OH
CH_3 N N CH_3
E

FIG. 12. Moxisylate **(A)**, proxazole **(B)**, butalamine **(C)**, fenoxedil **(D)**, and pyritonol **(E)**.

$$(CH_3)_2NCH_2CH_2OH$$

FIG. 13. Deanol.

Stabilizers containing the dimethylaminoethanol moiety have been reported to extend the life-span of Drosophila (64) and mice (66). However, a recent study using Japanese quail (28) revealed that contrary to predictions, the deanol-treated group had a shorter life-span than controls. No drug effects were found upon activity response to light flash, sexual mounting response to a female quail, or in a classical conditioning of the heart rate study.

Centrophenoxine

Centrophenoxine has been shown to markedly reduce the accumulation of lipofuscin granules in aged guinea pigs (57,101), rats (118), and in C1300 mouse neuroblastoma cells in culture (125), most probably by phagocytosis (133). Histochemical studies with guinea pig brains demonstrated (102) that the compound reduces the activity of succinic and lactic dehydrogenases and enhances the activity of glucose-6-phosphate dehydrogenase, suggesting that one of its effects is to enhance cellular metabolism by activating the "pentose shunt" pathway. It enhances the resistance of cerebral cells in rats, mice, and rabbits to various forms of oxygen deprivation, including cyanide intoxication (123), reduced atmospheric pressure (104), and reduced oxygen tension in the inspired air (34). It increased cerebral blood flow (62) and, in contrast to deanol and procaine, caused a marked increase in glucose transfer from blood to cerebral tissues. It has also been reported to increase the life-span of mice by 30% (67). Its membrane effects have been cited earlier (64,66), and these are currently being studied (149). Its membrane properties may also be involved in its ability to potentiate the cytotoxic action of most alkylating agents (132). A double-blind study of its effects on memory performance of 76 fit, able, elderly subjects has been carried out (92). A number of performance measures designed to evaluate various aspects of memory function were employed. These revealed that centrophenoxine appears to increase the consolidation of new information into long-term memory, but does not affect other aspects of remembering. It was also found that signifi-

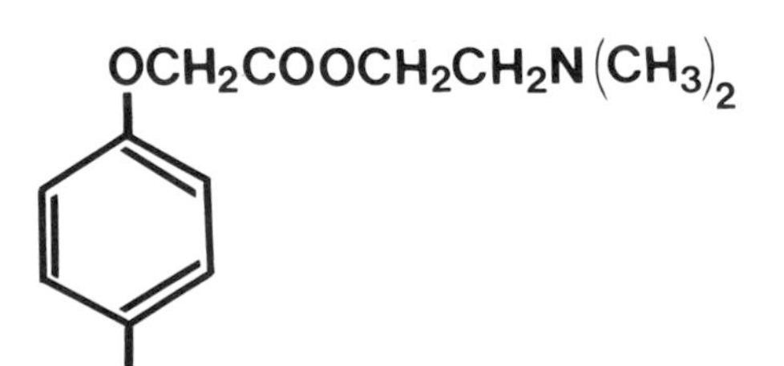

FIG. 14. Centrophenoxine.

cantly more of the subjects receiving centrophenoxine reported an increased level of mental alertness.

Procaine

Procaine, or Gerovital H3 (GH3), which is basically a 2% solution of procaine, has been a controversial drug ever since Aslan first reported on its reputed benefits in 1956 (79). Its general pharmacologic profile includes local anesthetic action, apparent stimulation of the central nervous system, quinidine-like action on the heart, and inhibition of monoamine oxidase (MAO). It is rapidly hydrolyzed by pseudocholinesterase in the blood, a fact that mitigates against very high circulating blood levels of drug. Such hydrolysis leads to paraaminobenzoic acid and diethylaminoethanol. It has been described (90) as "a weak reversible fully competitive inhibitor of MAO"; consequently, it may show antidepressant action. In 1974, the FDA approved Phase 2 clinical testing of Gerovital as an antidepressant in a geriatric population. A recent review by Ostfeld, et al. (110), based on an evaluation of the literature on the systemic use of procaine in the treatment of the aging process and the common chronic diseases of later life, included data from 285 articles and books, and described treatment in more than 100,000 patients during the past 25 years. The conclusions of that review were as follows:

> This review of the literature yields no convincing evidence that, except for a possible antidepressant effect, the systemic use of procaine (or Gerovital, of which the major component is procaine) is of value in the treatment of diseases in older patients.
>
> The literature on procaine reveals that the quality of clinical trials of new agents in the treatment of the elderly may be very poor. There is need for conferences and symposia to discuss the current status of evaluating new drugs in the elderly and the special problems of clinical trials in the aged.
>
> If procaine has an antidepressant effect, there is some likelihood that this may account for the impression among some observers that in procaine-treated patients there is a decrease of complaints attributable to the musculoskeletal, cardiovascular, endocrine, sexual, gastrointestinal and respiratory systems. Depression may play a greater role than previously suspected in these multiple discomforts of the elderly. Controlled clinical trials of standard antidepressant drugs among aged persons deserve careful consideration.

A recent paper has reexamined the MAO-inhibiting role of procaine (41). Fuller and Roush showed that inhibition by procaine hydrochloride of monoamine oxidase from either rat brain or liver was substrate-dependent. It was

$$NH_2-C_6H_4-COOCH_2CH_2N(C_2H_5)_2$$

FIG. 15. Procaine.

more effective in inhibiting serotonin oxidation than phenylethylamine oxidation, and had an intermediate effect on tryptamine oxidation. MAO activity in tissue homogenates from rats treated with procaine (150 mg/kg, i.p.) was inhibited most in liver, less in heart, and only very slightly in brain for a duration of up to 8 hr. Their data suggest that in high doses procaine may inhibit MAO weakly *in vivo,* and they conclude: "If the reported usefulness of procaine preparations in treating geriatric patients indeed depends upon MAO inhibition, more effective inhibitors would seem to be available."

During the week of October 10, 1977, the FDA announced that the Gerovital IND had been withdrawn, stating: "At present, there is no evidence that Gerovital is safe and effective for treating mental depression or manifestations of aging" (6).

Memory-Enhancing Agents

Piracetam

Because of the key role fading memory plays in the problems of the geriatric group, drugs targeted to memory improvement represent an important therapeutic class. One compound in this area, Piracetam, has not only created vigorous controversy but also appears to be the prototype of a new drug class.

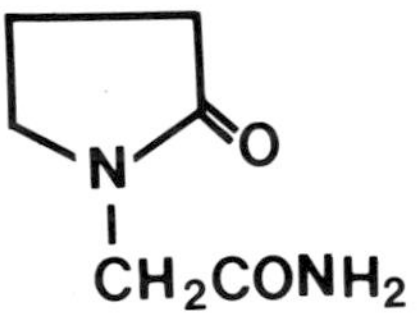

FIG. 16. Piracetam.

Piracetam, 2-pyrrolidone acetamide, is a compound believed to exert a direct selective effect on the telencephalon (45). The drug appears to exert no sedative or stimulant effects and to display no effects upon behavior; thus, Giurgea has coined the term nootropic agent (*noos,* mind; *tropein,* towards) to describe it and similar compounds (44). Nootropic compounds are presumed to influence telencephalic plasticity and the integrative action of the brain. While the action of piracetam on cortical cells has been hypothesized to stimulate the transformation of ADP into ATP (47), these results have not been confirmed by Nickolson and Wolthuis (108). In this rat study, neither piracetam nor naftidrofuryl affected the cerebral contents of adenine nucleotides and, accordingly, both substances were without effect on the adenylate energy charge. The disagreement between these results and those in the earlier literature is explained by Wolthuis by methodological differences. Piracetam was found to increase adenylate kinase activity, and it was suggested that this action is responsible for the protective effect of piracetam against cerebral hypoxia. In a recent human study, Richardson and Bereen examined the effects of 10-g daily doses of piracetam upon patients undergoing neurosurgery (117). The random nonstratified study of 100 patients showed that a significantly higher percentage of patients receiving piracetam attain or maintain a normal or near-normal level of consciousness postoperatively than do those receiving a placebo.

The effect of the material on learning and memory, typified by Wolthuis's (108) description of the drug as acquisition-enhancing, has been well documented in rats and goldfish (39). It facilitates interhemispheric transfer of visual information in rats (25), stimulates the uptake of labeled leucine by rat cerebral cortex slices (107), and inhibits the breakdown of newly formed protein therein (107). In human studies, there has not been agreement. Dimond (35) has reported that after 14 days dosing at 400 mg/day, verbal learning was significantly increased among 16 university students, a study that received considerable notice (7). Abuzzahab has recently reported on a much more extensive study (1). Fifty hospitalized geriatric patients (between 65 and 80 years old) were given piracetam (2400 mg/day) or placebo on a double-blind basis over a two-month period. Every patient submitted to a battery of psychological tests before and after the two-month trial. These tests included the Similarities, Vocabulary, Digit Symbol, and Block Design subtests of the Wechsler Adult Intelligence Scale; Graham Kendall Memory for Design; Benton Visual Retention; Hooper Visual Organization; Raven Colored Progressive Matrices A, AB, and B; Wechsler Memory Scale A and B; Porteus Maze; and Finger Tapping. In addition, at pretreatment, 4 and 8 weeks, the patient completed a Profiles of Mood States, and a Clinical Global Evaluation was performed.

There was no significant statistical difference between the two groups of patients (25 on piracetam, 25 on placebo) on all measures utilized except for the Clinical Global Evaluation, where 52% of the patients on piracetam showed minimal improvement versus 25% of the placebo group. Abuzzahab discusses a variety of reasons why their study failed to demonstrate the efficacy of piracetam. These reasons included inadequate testing procedures, inadequate dosing, too short a dosing period, and poor selection of patient population. The study considered some of the pitfalls in this area of research and may well help to shape newer studies.

Piracetam was evaluated in a double-blind study involving 196 elderly patients, compared to placebo in the "treatment of psychoorganic symptoms occurring in the course of senile involution" (135). It was administered orally, 800 mg, t.i.d., over an 8-week period. The general mental condition of the patients treated with piracetam showed a significant improvement compared with that of the placebo group (improvement classified as good in 26% of the cases and excellent in 15%).

Two other recent studies of interest include a study of the drug's effect on cortical bioelectrical activity in rabbits (85) and an autoradiographic study of the distribution of ^{14}C-piracetam in the primate brain (111) where, incidentally, it is preferably concentrated in the cortex of cerebrum and cerebellum.

FINAL COMMENTS

Because of limitations in space, I have not attempted to treat this subject comprehensively. For example, I have excluded all the work done using CNS stimulants in this area. (For coverage of some such classes that I omitted, the reviews by Bindra (19), Bender, et al. (14), Kormendy and Bender (83), and LaBella (86) should be consulted.) It was with particular regret that I had to omit the burgeoning area of peptide research in this field, perhaps well-illustrated by the Organon's group work (99).

ADDENDUM

Rao, et al. (120a) assessed the effectiveness of cyclandelate under double-blind conditions in 58 geriatric patients. The cyclandelate and placebo groups (32 and 26 patients, respectively) received either 1600 mg/day of cyclandelate in fractional doses or placebo over a period of 12 weeks. During the initial examination and every four weeks thereafter, patients were assessed for possible clinical changes. In addition, the Sandoz Clinical Assessment—Geriatric (SCAG) and the Nurses Observation Scale for Inpatient Evaluation (NOSIE) were completed, with particular attention to symptom clusters. A final global clinical assessment of the patients was also made. The results indicated that cyclandelate was a safe and effective agent for treating certain symptoms (e.g., improvement in all of the SCAG attributes except fatigue) of senility in properly selected patients, provided the therapy is carried on for at least eight weeks and, if indicated, for a longer period.

REFERENCES

1. Abuzzahab, F. S., Merwin, G. E., Zimmermann, R. L., and Sherman, M. C. (1977): A double blind investigation of Piracetam (Nootropil) vs placebo in geriatric memory. *Pharmakopsychiatr.,* 10:49–56.
2. Addlestone, G., et al. (1969): General practitioner clinical trials: Manifestations of cerebral arteriosclerosis unaffected by a vasodilator. *Practitioner,* 203:695–698.
3. Anlezark, G., Pycock, C., and Meldrum, B. (1976): Ergot Alkaloids as dopamine agonists: comparison in two rodent models. *Eur. J. Pharmacol.,* 37:295–302.
4. Anon (1971): Cerebral vasodilators. *Br. Med. J.,* 702–703.
5. Anon (1977): Ethyl apovincaminate: new and potent cerebral vasodilator. *Inpharma,* 69:13 (15th Jan.)
6. Anon (1977): *F-D-C Reports,* Trade & Government Memos-4, Oct. 17.
7. Anon (1976): Pop a pill and remember more. *New Scientist,* 383; Land, T. (1976): Drug lessens brain damage, increases intelligence. *The Medical Post,* p. 37 (June 8).
8. Arcari, G., Bernardi, L., Bosisio, G., Coda, S., Fregnan, G. B., and Glässer, A. H. (1972): 10-Methoxyergoline derivatives as α-adrenergic blocking agents. *Experientia,* 28:819–820.

9. *Arzneim.-Forsch.,* (1976): 26:1905–1989.
10. *Arzneim.-Forsch.,* (1977): 27:6A, complete issue, pp. 1238–1247.
11. *Assessment in Cerebrovascular Insufficiency,* (1971): edited by G. Stöcker, et al. Thieme, Stuttgart.
12. Bazo, A. J. (1973): An ergot alkaloid preparation (Hydergine) versus papaverine in treating common complaints of the aged: double blind study. *J. Am. Geriatr. Soc.,* 21:63–71.
13. Behrens, E. (1966): Clinical effect on the brain perfusion of Stutgeron. *Med. Welt.,* 38:2029–2031.
14. Bender, A. D., Kormendy, C. G., and Powell, R. (1970): Pharmacological control of aging. *Exp. Gerontol.,* 5:97–129.
15. Benzi, G., Manzo, L., De Bernardi, M., Ferrara, A., Sanguinetti, L., Arrigoni, E., and Berté, F. (1971): Action of lysergide, ephedrine and nicergoline on brain metabolizing activity. *J. Pharm. Sci.,* 60:1320–1324.
16. Benzi, G., De Bernardi, M., Manzo, L., Ferrara, A., Panceri, P., Arrigoni, E., and Berté, F. (1972): Effect of lysergide and nicergoline on glucose metabolism investigated on the dog brain isolated *in situ. J. Pharm. Sci.,* 61:348–352.
17. Bernini, F. P., Muras, I., Maglione, F., and Smaltino, F. (1976): The action of Nicergoline on the cerebral circulatory insufficiency of arteriosclerosis: clinical and angioseriographic evaluation. *Farmaco [Prat.],* 32:32–46.
18. Bijlsma, U. G., Funcke, A. B. H., Tersteege, H. M., Bekker, R. F., Ernsting, M. J. E., and Nauta, W. T. (1956): The pharmacology of cyclospasmol. *Arch. Int. Pharmacodyn. Ther.,* 105:145–174.
19. Bindra, J. S. (1974): Anti-aging drugs. In: *Annual Reports in Medicinal Chemistry,* edited by R. V. Heinzelman, pp. 214–221. Academic Press, New York.
20. Blaine, E. H. (1977): Renin secretion after papaverine and furosemide in conscious sheep. *Proc. Soc. Exp. Biol. Med.,* 154:232–237.
21. Boismare, F., and Lorenzo, J. (1975): Study of the protection afforded by Nicergoline against the effects of cerebral ischemia in the cat. *Arzneim.-Forsch.,* 25:410–413. See also: Suchewsky, G. K., and Pegrassi, L. (1974): Action of Nicergoline on electroencephalographic recovery after cat brain ischemia. *Naunyn-Schmiedebergs Arch. Pharmacol.,* 248:311–318.
22. Branconnier, R. J., and Cole, J. O. (1977): Effects of chronic papaverine administration on mild senile Organic Brain Syndrome. *J. Am. Geriatr. Soc.,* 25:458–462.
23. Branconnier, R. J., and Cole, J. O. (1977): A memory assessment technique for use in geriatric psychopharmacology: drug efficacy trial with naftidrofuryl. *J. Am. Geriatr. Soc.,* 25:186–188.
24. Brodie, N. H. (1977): A double blind trial of naftidrofuryl in treating confused elderly patients in general practice. *Practitioner,* 218:274–278.
25. Buresova, O., and Bures, J. (1976): Piracetam induced facilitation of interhemispheric transfer of visual information in rats. *Psychopharmacol.* 46:93–102.
26. Cahn, J., and Borzeix, M. G. (1978): Comparative effects of Dihydroergotoxine (DHET) on CBF and metabolism changes produced by experimental cerebral edema, hypoxia and hypertension. *Gerontology,* 24(Suppl. 1):34–42.
27. Caplan, L. R. (1977): Drug therapy reviews: vasodilating drugs and their use in cerebral symptomatology. *Am. J. Hosp. Pharm.,* 34:1075–1079.
28. Cherkin, A., and Eckardt, M. J. (1977): Effects of dimethylaminoethanol upon life span and behavior of aged Japanese quail. *J. Gerontol.,* 32:38–45.
29. Cooper, D. S., and Jacobs, L. S. (1977): Failure of papaverine to alter L-dopa-influenced GH and PRL secretion. *J. Clin. Endocrinol. Metabol.,* 44:585–587.
30. Corsellis, J. A. N. (1962): *Mental Illness and the Aging Brain.* Oxford University Press, London.
31. Cosnier, D., Cheucle, M., Rispat, G., and Streichenberger, G. (1977): Influence of hypercapnia on the cerebrovascular activities of some drugs used in the treatment of cerebral ischemia. *Arzneim. Forsch.,* 27:1566–1569.
32. Culebras, A. (1976): Effect of papaverine on cerebral electrogenesis. *Neurol.,* 26:673–679.
33. "Deapril-ST for Senile Dementia" (1977): *Med. Lett. Drugs Ther.,* 19:61–62.
34. Dereymacker, A., Theeuwissen-Lesuisse, F., Buu-Hoi, N. P. and Lapiere, C. (1962): Experimental cerebral anoxia: protective effect of derivatives of p-chlorophenoxyacetic acid. *Med. Exp.,* 7:239–244.
35. Dimond, S. J., and Brouwers, E. Y. M. (1976): Increase in the power of human memory in normal man through the use of drugs. *Psychopharmacol.,* 49:307–309.

36. Elliott, C. G., Brown, A. L., and Smith, T. C. G. (1973): Multicentre general practitioner trial of isoxsuprine in cerebrovascular disease: a pilot study. *Curr. Med. Opin.,* 1:554–562.
37. Ernst, A. M. (1962): Experiments with an O-methylated product of dopamine in cats. *Acta Physiol. Pharmacol. Neerl.,* 11:48–53.
38. Ferris, S. H., Sathananthan, G., Gershon, S., and Clark, C. (1977): Senile dementia: treatment with deanol. *J. Am. Geriatr. Soc.,* 25:241–244.
39. File, S. E., and Hyde, J. R. G. (1977): The effects of piracetam on acquisition and retention of habituation. *Proc. Br. Pharm. Soc.,* 475P; and Wolthuis, O. L. (1971): Experiments with UCB 6215, a drug which enhances acquisition in rats: its effects compared with those of amphetamine. *Eur. J. Pharmacol.,* 16:283–297.
40. Foltyn, P., Groh, R., Lücker, P. W., and Steinhaus, W. (1978): On the problems of demonstrating the efficacy of cerebroactive drugs in the aged patient. Results of a random double blind study of a vincamine containing preparation. *Arzneim. Forsch.,* 28:90–94.
41. Fuller, R. W., and Roush, B. W. (1977): Procaine hydrochloride as a monoamine oxidase inhibitor: implications for geriatric therapy. *J. Am. Geriatr. Soc.,* 25:90–93.
42. Gaitz, C. M., Varner, R. U., and Overall, J. E. (1977): Pharmacotherapy for Organic Brain Syndrome in late life. *Arch. Gen. Psychiatr.,* 34:839–845.
43. Gärtner, E., Enzenross, H. G., Vlahov, V., Schanzenbacher, P., Brandt, H., and Betz, E. (1975): Blood circulation, oxygen pressure and pH of the cerebral cortex under the influence of bencyclan. *Arzneim. Forsch.,* 25:887–891.
44. Giurgea, C. (1973): The nootropic approach to the pharmacology of the integrative action of the brain. *Cond. Refl.,* 8:108–115.
45. Giurgea, C. (1976): Piracetam: Nootropic pharmacology of neurointegrative activity. *Curr. Dev. Psychopharmacol.,* 3:221–276.
46. Gloning, K., and Klausberger, E. M. (1958): Investigations on the cerebral vessel function with the aid of motion pictures. *Wien. Klin. Wochenschr.,* 70:145–149.
47. Gobert, J. G. (1972): Genesis of a drug: Piracetam: metabolism and biochemical research. *J. Pharm. Belg.,* 26:281–304.
48. Godfraind, T., and Kaba, A. (1969): Blockage or reversal of the contraction induced by calcium and adrenaline in depolarized arterial smooth muscle. *Br. J. Pharmacol.,* 36:549–560.
49. Goldberg, A. L. (1977): Is deanol a precursor of acetylcholine? *Dis. Nerv. Syst.,* 38:16–20.
50. Goldstein, M., Lew, J. F., Hata, F., and Lieberman, A. (1978): Binding interactions of ergot alkaloids with monoaminergic receptors in the brain. *Gerontol.,* 24(Suppl. 1):76–85.
51. Gonzalez-Vegas, J. A. (1974): Antagonism of dopamine-mediated inhibition in the nigrostriatal pathway. *Brain Res.,* 80:219–228.
52. Gygax, P., Wiernsperger, N., Meier-Ruge, W., and Baumann, T. (1978): Effect of papaverine and dihydroergotoxine mesylate on cerebral microflow, EEG and po_2 in oligemic hypotension. *Gerontol.,* 24(Suppl. 1):14–22.
53. Hachinski, V. C., Lassen, N. A., and Marshall, J. (1974): Multiinfarct dementia: a cause of mental deterioration in the elderly. *Lancet,* 207–210.
54. Hageman, W. E., and Pruss, T. P. (1975): Cardiovascular profile of 5-methyl-2-phenyl-4-imidazole-acetonitrile (MCN-2378), a cerebral vasodilator. *Eur. J. Pharmacol.,* 30:100–106.
55. Hall, P. (1975): Cited in *Medical World News,* 52 (Nov. 3).
56. Hapke, H.-J. (1973): Experimental animal investigations for the characterization of the central nervous system: action of Bencyclane. *Arch. Int. Pharmacodyn. Ther.,* 202:231–243.
57. Hasan, M., Glees, P., and Spoerri, P. E. (1974): Dissolution and removal of neuronal lipofuscin following dimethylaminoethyl p-chlorophenoxyacetate administration in guinea pigs. *Cell Tissue Res.,* 150:369–375.
58. Haubrich, D. R., Wang, P. F. L., Clody, D. E., and Wedeking, P. W. (1975): Increase in rat brain acetylcholine induced by choline or deanol. *Life Sci.,* 17:975–980.
59. Hauth, H., and Richardson, B. P. (1977): Cerebral vasodilators. In: *Annual Reports in Medicinal Chemistry,* Vol. 12, edited by F. H. Clarke, pp. 49–59. Academic Press, New York.
60. Heiss, W.-D., Podreka, I., and Samec, P. (1977): The effect of vincamine on cerebral blood flow as a function of application rate. *Arzneim. Forsch.,* 27:1291–1293.
61. Herrschaft, H., Gleim, F., and Duus, P. (1974): Influence of bencyclan on regional cerebral blood flow in patients with insufficient cerebral blood flow. *Klin. Wochenschr.,* 52:293–295.
62. Herrschaft, H. (1975): The efficacy and course of action of vaso- and metabolic-active substances on regional cerebral blood flow in patients with cerebrovascular insufficiency. In: *Blood Flow*

and Metabolism in the Brain, Proc. Int. Symp. 7th., Aviermore, Scotland, June, 1975, edited by M. Harper, B. Jennelt, and D. Miller. 11.24–11.28. Churchill-Livingstone, London.
63. Herrschaft, H. (1977): The influence of vincaminc on global and regional cerebral blood flow in acute and subchronic human cerebral ischemia. *Arzneim. Forsch.,* 27:1278–1284.
64. Hochschild, R. (1971): Effect of membrane stabilizing drugs on mortality in *Drosophila melanogaster. Exp. Gerontol.,* 6:133–151.
65. Hochschild, R. (1971): Lysosomes, membranes and aging. *Exp. Gerontol.,* 6:153–166.
66. Hochschild, R. (1973): Effect of dimethylaminoethyl p-chlorophenoxyacetate on the life span of male Swiss Webster albino mice. *Exp. Gerontol.,* 8:177–183.
67. Hochschild, R. (1974): Action of meclofenoxate on the life span of male albino Swiss Webster mice. *Ann. Anaesthesiol. Fr.,* 15:595–599.
68. Hollister, L. E. (1975): Drugs for mental disorders of old age. *J.A.M.A.,* 234:195–198.
69. Horton, G. E., and Johnson, J. C. (1964): The application of radioisotopes to the study of cerebral blood flow, comparison of three methods. *Angiol.,* 15:70–74.
70. Hoyer, S., Oesterreich, K., and Stoll, K.-D. (1977): Effects of pyritinol-HC1 on blood flow and oxidative metabolism of the brain in patients with dementia. *Arzneim. Forsch.,* 27:671–674.
71. Hughes, J. R., Williams, J. G., and Currier, R. D. (1976): An ergot alkaloid preparation (Hydergine) in the treatment of dementia: critical review of the clinical literature. *J. Am. Geriatr. Soc.,* 24:490–497.
72. Hughes, M. J. (1977): A non-histaminic response of rabbit atria to betahistine. *J. Clin. Pharmacol.,* 17:91–92.
73. Hussain, S. M. A., Gedye, J. L., Naylor, R., and Brown, A. L. (1976): The objective measurement of mental performance in cerebrovascular disease. *Practitioner,* 216:222–228.
74. Iliff, L. D., Du Boulay, G. H., Marshall, J., Ross Russell, R. W., and Symon, L. (1977): Effect of Nicergoline on cerebral blood flow. *J. Neurol. Neurosurg. Psychiatr.,* 40:746–747.
75. Innes, I. R., and Nickerson, M. (1975): Norepinephrine, epinephrine and the sympathomimetic amines. In: *The Pharmacological Basis of Therapeutics,* edited by L. S. Goodman and A. Gilman, 5th ed., p. 508. Macmillan, New York.
76. Iwangoff, P., Enz, A., and Meier-Ruge, W. (1978): Incorporation, after single and repeated application of radioactive labeled DH-ergot alkaloids in different organs of the cat, with special reference to the brain. *Gerontol.,* 24(Suppl. 1):126–138.
77. Kadokawa, T., Fujitani, B., Kuwashima, J., Hatano, N., and Shimizu, M. (1975): Pharmacological studies of 3-[bis(3,3-diphenylpropyl)-amino]-propan-l-ol hydrochloride (PF-244), a new cerebral vasodilator. *Arzneim. Forsch.,* 25:632–638.
78. Karlsberg, P., Eliott, H. W., and Adams, J. E. (1963): Effect of various pharmacologic agents on cerebral arteries. *Neurol.,* 13:772–778; see also Wang, H. S., and Obrist, W. D. (1976): Effect of oral papaverine on cerebral blood flow in normals: evaluation by the Xenon 133 inhalation method. *Biol. Psychiatr.,* 11:217–225.
79. Kent, S. (1976): A look at Gerovital—the "youth" drug. *Geriatrics,* 31:95–102.
80. Kent, S. (1977): Classifying and treating organic brain syndromes. *Geriatrics,* 32:87–96.
81. Kohlmeyer, K. (1973): Measurements of regional cerebral blood circulation with Xe-133 before and after intravenous administration of various potentially vasoactive and cerebral metabolically active drugs in acute testing. *Verh. Dtsch. Ges. Kreislaufforsch,* 39:96–101.
82. Kohlmeyer, K. (1977): On the influence of Vincamine on human cerebral blood flow in the acute test/studies with ^{133}Xe clearance method. *Arzneim. Forsch.,* 27:1285–1290.
83. Kormendy, C. G., and Bender, A. D. (1971): Chemical interference with aging. *Gerontologia,* 17:52–64.
84. Kostenbauder, H. B. (1977): Sustained-release papaverine hydrochloride. *J. Am. Pharm. Assoc.,* 17:303–306; Maggi, G. C., Cerchiari, D., and Coppi, G. (1977): Papaverine blood levels after administration of a sustained-release preparation. *Arzneim. Forsch.,* 27:1214–1215.
85. Krug, M., Ott, T., Schulzeck, K., and Matthies, H. (1977): Effects of orotic acid and pirazetam on cortical bioelectrical activity in rabbits. *Psychopharmacol.,* 53:73–78.
86. LaBella, F. S. (1972): Pharmacolongevity: control of aging by drugs. In: *Search for New Drugs,* edited by A. A. Rubin, pp. 347–383. Marcel Dekker, Inc., New York.
87. Lewis, J. A., and Lewis, B. S. (1977): Deanol in minimal brain dysfunction. *Dis. Nerv. Syst.,* 38:21–24.
88. Liesegang, J., Back, W. J., Seibert, H., and Schumacher, W. (1976): Blood gas analytical study on cerebral circulation under the influence of dihydroergotoxin. *Arzneim.-Forsch.,* 26:1619–1622.

89. Lish, P. M., Dungan, K. W., and Peters, E. L. (1960): A survey of the effects of isoxsuprine on non-vascular smooth muscle. *J. Pharmacol. Exp. Ther.*, 129:191–199.
90. MacFarlane, M. D. (1975): Procaine HCl (Gerovital H3), a weak, reversible, fully competitive inhibitor of monoamine oxidase. *Fed. Proc.*, 34:108–110; see also MacFarlane, M. D., and Besbris, H. (1974): Procaine (Gerovital H3) therapy: mechanism of inhibition of monoamine oxidase. *J. Am. Geriatr. Soc.*, 22:365–371.
91. Maiolo, A. T., Bianchi Porro, G., Galli, C., and Sessa, M. (1972): Effect of Nicergoline on cerebral hemodynamics and metabolism in primary arterial hypertension and arteriosclerosis. *Clin. Ter.*, 62:239–252.
92. Marcer, D., and Hopkins, S. M. (1977): The differential effects of Meclofenoxate on memory loss in the elderly. *Age Ageing*, 6:123–131.
93. McHenry, L. C., Jaffe, M. E., Kawamura, J., and Goldberg, H. I. (1971): Hydergine effect on cerebral circulation in cerebrovascular disease. *J. Neurol. Sci.*, 13:475.
94. McHenry, L. C. (1972): Cerebral vasodilator therapy. *Stroke*, 3:686–691.
95. *Med. Lett. Drugs Ther.* (1970): Drugs for improvement of cerebral function in the elderly, 12:38–39.
96. Meier-Ruge, W. (1976): Experimental pathology and pharmacology in brain research and aging. *Life Sci.*, 17:1627–1635.
97. Meyer, J. S. (1975): [cited in ref. 54. See also *Medical World News* (1976): 4–5 (Oct. 25).]
98. Meynaud, A., Grand, M., and Fontaine, L. (1973): Effect of Naftidrofuryl upon energy metabolism of the brain. *Arzneim. Forsch.*, 23:1431–1436.
99. Miller, L. H., Harris, L. C., Van Riezen, H., and Kastin, A. J. (1976): Neuroheptapeptide influence on attention and memory in man. *Pharmacol. Biochem. Behav.*, 5(Supp.):17–22.
100. Montecucchi, P. (1976): Stimulation of cyclic AMP formation in rat brain by nicergoline *in vitro*. *Farmaco [Prat.]*, 31:10–17.
101. Nandy, K., and Bourne, G. H. (1966): Effect of centrophenoxine on the lipofuscin pigments in the neurons of senile guinea pigs. *Nature*, 210:313–314.
102. Nandy, K. (1968): Further studies on the effects of centrophenoxine on the lipofuscin pigments in the neurons of senile guinea pigs. *J. Gerontol.*, 23:82–92.
103. Nelson, J. J. (1975): Relieving select symptoms of the elderly. *Geriatrics*, 30:133.
104. Nickel, J., Breyer, U., Claver, B., and Quadbeck, G. (1963): The effect of aminoethanol derivatives on the central nervous system. *Arzneim. Forsch.*, 13:881–883.
105. Nickerson, M. (1975): Vasodilator drugs. In: *The Pharmacological Basis of Therapeutics*, edited by L. S. Goodman and A. Gilman, 5th ed., pp. 727–743. Macmillan, New York.
106. Nickerson, M., and Collier, B. (1975): Drugs inhibiting adrenergic nerves and structures innervated by them. In: *The Pharmacological Basis of Therapeutics*, edited by L. S. Goodman and A. Gilman, 5th ed., p. 541. Macmillan, New York.
107. Nickolson, V. J., and Wolthuis, O. L. (1976): Protein metabolism in the rat cerebral cortex *in vivo* and *in vitro* as affected by the acquisition-enhancing drug, piracetam. *Biochem. Pharmacol.*, 25:2237–2240.
108. Nickolson, V. J., and Wolthuis, O. L. (1976): Effect of the acquisition-enhancing drug Piracetam on rat cerebral energy metabolism. Comparison with Naftidrofuryl and Methamphetamine. *Biochem. Pharmacol.*, 25:2241–2244.
109. Obrist, W. D. (1972): Cerebral physiology of the aged: influence of circulatory disorders. In: *Aging and the Brain*, edited by C. M. Gaitz, pp. 117–133. Plenum Publishing Corp., New York.
110. Ostfeld, A., Smith, C. M., and Stotsky, B. A. (1977): The systemic use of procaine in the treatment of the elderly: a review. *J. Am. Geriatr. Soc.*, 25:1–19.
111. Ostrowski, J., and Keil, M. (1978): Autoradiographic studies on the distribution of ^{14}C-Piracetam in the primate brain. *Arzneim. Forsch.*, 28:29–35.
112. Pálosi, É., and Szporny, L. (1976): Effects of ethyl apovincaminate on the central nervous system. *Arzneim. Forsch.*, 26:1926–1929.
113. Pasotti, C., Liverta, C., Cacciatorio, D., and Pollini, C. (1974): Therapeutic activity of Nicergoline in the treatment of cerebral and peripheral vasculopathy. *Farmaco [Prat.]*, 29:508–519.
114. Paul, A., Mildner, B., and Chandra, P. (1978): Influence of Metergoline on the biosynthesis of brain macromolecules. *Arzneim.-Forsch.*, 28:25–29.
115. Pourrias, B., Sergant, M., Thomas, J., Gouret, C., and Raynaud, G. (1975): Pharmacological study of a new substance, Mecinarone. *Arzneim. Forsch.*, 25:782–786.

116. Prencipe, M., Cecconi, V., and Pisarri, F. (1974): Preliminary observations on the action of Nicergoline on cerebral blood flow. *Farmaco [Prat.],* 29:278–284.
117. Richardson, A. E., and Bereen, F. J. (1977): Effect of Piracetam on level of consciousness after neurosurgery. *Lancet,* 1110–1111.
118. Riga, S., and Riga, D. (1974): Effects of centrophenoxine on the lipofuscin pigments in the nervous system of old rats. *Brain Res.,* 72:265–275.
119. Roberts, P. J. (1977): YC-93: cerebral vasodilator. *Drugs of the Future,* 2:409–411.
120. Robinson, R. A. (1961): Some problems of clinical trials in elderly people. *Gerontol. Clin.,* 3:247–257.
120a. Rao, D. B., Georgiev, E. L., Paul, P. D., and Guzman, A. B. (1977): Cyclandelate in the treatment of senile mental changes: a double blind evaluation. *J. Am. Geriatr. Soc.,* 25:548–551.
121. Rosen, H. J. (1975): Mental decline in the elderly: pharmacotherapy (ergot alkaloids vs. papaverine). *J. Am. Geriatr. Soc.,* 23:169–174.
122. Roth, M., Tomlinson, B. E., and Blessed, G. (1966): Correlation between scores for dementia and counts of senile plaques in cerebral gray matter of elderly subjects. *Nature,* 209:109.
123. Rump, S., and Edelwejn, Z. (1968): Effects of centrophenoxine on electrical activity of the rabbit brain in sodium cyanide intoxication. *Int. J. Neuropharmacol.,* 7:103–113.
124. Sathananthan, G. L., and Gershon, S. (1975): Cerebral vasodilators: a review. In: *Aging,* edited by S. Gershon and A. Raskin, pp. 155–168. Raven Press, New York.
125. Schneider, F. H., and Nandy, K. (1977): Effects of centrophenoxine on lipofuscin formation in neuroblastoma cells in culture. *J. Gerontol.,* 32:132–139.
126. Sebille, A., Sanson, J., and Lhuissier, J. (1976): Results of an electroencephalographic study of a new alpha adrenolytic, nicergoline. *Arch. Med. Normandie,* 2:87–90.
127. Seipel, J. H., and Floam, J. E. (1975): Rheoencephalographic and other studies of betahistine in humans. I. The cerebral and peripheral circulatory effects of single doses in normal subjects. *J. Clin. Pharmacol.,* 15:144–154; Seipel, J. H., Fisher, R., Floam, J. E., and Bohm, M. (1975): Rheoencephalographic and other studies of betahistine in humans. II. The cerebral and peripheral microcirculatory effects of single doses in geriatric patients with "pure" arteriosclerotic dementia. *J. Clin. Pharmacol.,* 15:155–162.
128. Seipel, J. H., Fisher, R., Floam, J. E., and Bohm, M. (1977): Rheoencephalographic and other studies of betahistine in humans. III. Improved methods of diagnosis and selection in arteriosclerotic dementia. *J. Clin. Pharmacol.,* 17:63–75.
129. Seipel, J. H., Fisher, R., Blatchley, R. J., Floam, J. E., and Bohm, M. (1977): Rheoencephalographic and other studies of betahistine in humans. IV. Prolonged administration with improvement in arteriosclerotic dementia. *J. Clin. Pharmacol.,* 17:140–161.
130. Seki, T., and Takenaka, T. (1977): Pharmacological evaluation of YC-93: a new vasodilator in healthy volunteers. *Int. J. Clin. Pharmacol. Biopharm.,* 15:267–274.
131. Shader, R. I., Harmatz, J. S., and Salzman, C. (1974): A new scale for clinical assessment in geriatric populations: Sandoz Clinical Assessment—Geriatric (SCAG). *J. Am. Geriatr. Soc.,* 22:107–113.
132. Sladek, N. E. (1977): Potentiation of antitumor drug action by centrophenoxine: Specificity. *J. Pharmacol. Exp. Ther.,* 201:518–526.
133. Spoerri, P. E., and Glees, P. (1975): Mode of lipofuscin removal from hypothalamic neurons. *Exp. Gerontol.,* 10:225–228.
134. Stafford, J. R., and Farr, W. E. (1977): Deanol acetamidobenzoate (Deaner) in tardive dyskinesia. *Dis. Nerv. Syst.,* 38:3–6.
135. Stegink, A. J. (1972): The clinical use of piracetam, a new nootropic drug: The treatment of symptoms of senile involution. *Arzneim.-Forsch.,* 22:975–977.
136. Takayanagi, I., Kondo, N., and Takagi, K. (1976): Effect of a cholinergic stimulant on interaction of isoprenaline with β-adrenoceptors. *Eur. J. Pharmacol.,* 38:179–182.
137. Toledo, J. B., Pisa, H., and Marchese, M. (1972): Clinical evaluation of cinnarizine in patients with cerebral circulatory deficiency. *Arzneim.-Forsch.,* 22:448–451.
138. Tomlinson, B. E., Blessed, G., and Roth, M. (1970): Observations on the brains of demented old people. *J. Neurol. Sci.,* 11:205–242.
139. Toni, E. (1973): Therapeutic activity of nicergoline against cerebrovascular syndromes. *Minerva Med.,* 64:4466–4473.
140. Vamosi, B., Molnar, L., Demeter, J., and Tury, F. (1976): Comparative study of the effect

of ethyl apovincaminate and xanthinol nicotinate in cerebrovascular diseases: Immediate drug effects on the concentrations of carbohydrate metabolites and electrolytes in blood and CSF. *Arzneim.-Forsch.,* 26:1980–1984.
141. Van Neuten, J. M. (1969): Comparative bioassay of vasoactive drugs using isolated perfused rabbit arteries. *Eur. J. Pharmacol. (Amst.),* 6:286–293.
142. Washington correspondent (1975): *Pharmascope,* 15:1013–75/12.
143. Westreich, G., Alter, M., and Lundgren, S. (1975): Effect of Cyclandelate on Dementia. *Stroke,* 6:535–538.
144. Wilke, O. (1966): *Med. Welt.,* 17:1472–1474.
145. Williams, L. T., and Lefkowitz, R. J. (1977): Molecular pharmacology of *alpha* adrenergic receptors: Utilization of [^{3}H]-dihydroergocryptine binding in the study of pharmacological receptor alterations. *Mol. Pharmacol.,* 13:304–313.
146. Witzmann, H. K., and Blechacz, W. (1977): On the role of vincamine in the therapy of cerebrovascular diseases and impairment of cerebral function. *Arzneim. Forsch.,* 27:1238–1247.
147. Worm-Peterson, J., and Pakkenberg, H. (1968): Atherosclerosis of cerebral arteries, pathological and clinical correlations. *J. Gerontol.,* 23:445–449.
148. Zahniser, N. R., Chou, D., and Hanin, I. (1977): Is 2-dimethylaminoethanol (Deanol) indeed a precursor of brain acetylcholine? A gas chromatographic evaluation. *J. Pharmacol. Exp. Ther.,* 200:545–559.
149. Zs-Nagy, I. (1977): *(Personal communication.)* The study is being conducted at the Biological Institute, Medical University, Debrecen, Hungary.

Physiology and Cell Biology of Aging (Aging, Volume 8), edited by A. Cherkin, et al.
Raven Press, New York © 1979.

Aging and Pharmacokinetics: Impact of Altered Physiology in the Elderly

Robert E. Vestal

Department of Medicine, University of Washington School of Medicine, Seattle, Washington 98195; and Medical Service, Veterans Administration Hospital, Boise, Idaho 83702

The elderly constitute a growing segment of the American people. Almost 11% of our population is now over age 65 and by the year 2030, this group will number nearly 52 million and will represent at least 17% of the population. As a group, the elderly have more illness and inevitably take more medications than younger patients. In 1976, the aged spent almost $2.8 billion for drugs and drug sundries, or about 25% of the national total (22). If these trends continue, in several decades drug expenditures by the elderly may exceed 40% of the national total.

Several studies have demonstrated that older patients have a roughly twofold greater incidence of adverse reactions to drugs than younger patients (10,29,56). Because the incidence of adverse drug reactions increases with the number of drugs administered (62), the elderly with multiple diseases requiring multiple medications are predisposed to complications arising from drug use. In addition there are important and sometimes subtle physiological changes that occur with aging, independent of more overt diseases. These changes might be expected to alter drug responsiveness as a result of age-related differences in pharmacokinetics (the time course of absorption, distribution, biotransformation, and excretion of drugs) or in pharmacodynamics (the pharmacologic response to drugs at their sites of action). Accordingly, the clinical pharmacology of drugs in the geriatric age group is receiving considerable investigative effort.

The purpose of this paper is to briefly review some of the physiological differences that are associated with aging and to illustrate how they may lead to pharmacokinetic age differences. It should be emphasized that the available literature on drug disposition in the elderly describes the results of studies comparing individuals of different ages. These studies are cross-sectional rather than longitudinal in design and as such provide information about age *differences* rather than age *changes.* The distinction is important for the gerontologist because age differences in cross-sectional studies may not necessarily reflect the intrinsic biological effects of aging. For example, environmental changes with time may lead to cohort differences which, in the strict sense, are separate from the effects of aging per se (55).

PHYSIOLOGICAL EFFECTS OF AGING

Several alterations in gastrointestinal physiology might be expected to affect drug absorption (4). Elevated gastric pH could alter the ionization and solubility of some drugs. Reduced splanchnic blood flow could modify the rate and extent of drug absorption, as could a reduction in the number of absorbing cells, a delay in gastric emptying, and a decrease in gastrointestinal motility. Since most drugs are absorbed by passive diffusion, reduced active transport suggested for some compounds—such as galactose, 3-methyl glucose, calcium, thiamine, and iron—probably has only limited importance for drug absorption.

Body composition is an important determinant of the distribution of drugs in body fluids and tissues. Total body water is 10% to 15% less (15,71) and body fat 10% to 20% greater (19,48) in elderly than in young subjects. The effect of these age differences in total body water and body fat is a reduction in lean body mass. It may be predicted that when the dose is based on the usual estimates of body size such as total body weight or surface area, drugs whose distribution is limited to body water or fat-free body mass will produce higher blood or tissue levels in the elderly than in the young. Alterations in body fat might result in accumulation and prolongation of highly lipid-soluble drugs.

Since only the unbound or free drug is available for distribution out of blood into other body fluid and tissue compartments, alterations in the binding of drugs to plasma proteins and red blood cells may affect drug distribution and elimination. Although the total serum protein concentration remains constant, albumin is reduced in old age by as much as 20% (9,79). Despite considerable interindividual variation, our own data on healthy male volunteers who participated in pharmacokinetic studies at the Gerontology Research Center in Baltimore also show a significant decline in serum albumin with age (Fig. 1). A disturbance of the normal metabolic response to a reduced albumin pool has been demonstrated in some aged individuals (42). The reduced albumin concentration can result in higher concentrations of free drug for diffusion into body tissues where sites of action or drug metabolism and excretion may be located. It can also make the elderly more susceptible to the effects of multiple drug therapy on drug binding (74). An alteration in plasma binding of a number of drugs has been reported (Table 1). Except for meperidine (13) and chlormethiazole (46), no effects of age have been demonstrated for erythrocyte binding in the few studies available.

Drugs are eliminated from the body principally by hepatic metabolism (biotransformation) to less active and inactive metabolites and by renal excretion of parent drug and its metabolites. Biliary and pulmonary excretion play a role for some drugs. Reduced activity of hepatic microsomal enzymes (32) and alterations in microsomal enzyme induction (1) with advanced age have been demonstrated in experimental animals. Although there are no similar direct studies of the effects of age on hepatic drug metabolizing enzyme activity in

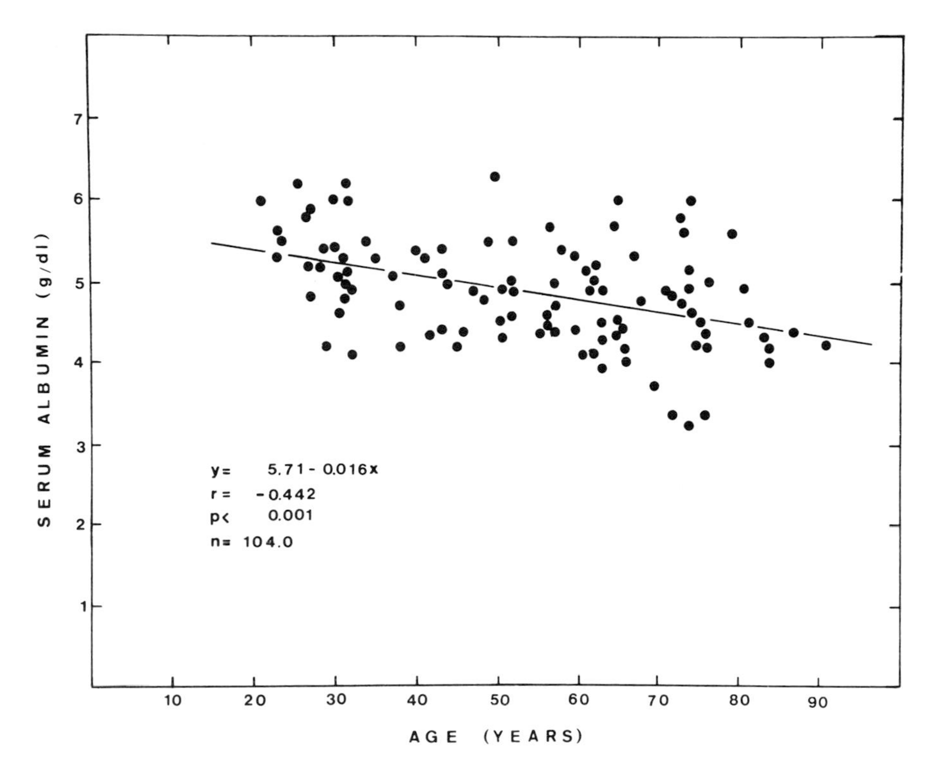

FIG. 1. Serum albumin concentrations in 104 healthy, community-dwelling male subjects ranging in age from 21 to 92 years.

man, it is known that both liver mass and liver blood flow diminish with age (3,21,51,78). Thus, the rate of removal of some drugs from the body may be lessened in the elderly, but it is not possible to generalize on the extent or clinical importance of impairment (Table 2). In some studies the half-life may be altered because of age differences in the apparent volume of distribution (Vd), whereas the total metabolic clearance (Cl), which more accurately reflects the intrinsic metabolic capacity of the liver, may not differ between young and

TABLE 1. *Effect of age on protein binding of drugs in young and old subjects*

Difference found	No difference found	
Chlormethiazole (46)	Chlordiazepoxide (77)	Phenobarbituric acid (5)
Lorazepam (77)	Desmethyldiazepam (33)	Phenytoin (5)
Meperidine (39)	Diazepam (33)	Quinidine (49)
Phenylbutazone (74)	Meperidine (13)	Salicylate (5)
Phenytoin (25,28)	Oxazepam (77)	Sulfadiazine (74)
Tolbutamide (41)	Penicillin G (5)	Warfarin (58)
Warfarin (24)		

TABLE 2. *Effect of age on the apparent hepatic elimination of drugs in young and old subjects*

Difference found		No difference found
	Biological Half-Life	
Acetaminophen (8,66)		Imipramine (47)
Acetanilid (18)		Indomethacin (68)
Aminopyrine (30)		Isoniazid (18)
Antipyrine (38,50,51,71)		Lorazepam (77)
Chlordiazepoxide (57,77)		Morphine (7)
Chlormethiazole (44,46)		Nitrazepam (12)
Desipramine (47)		Oxazepam (77)
Diazepam (33)		Phenylbutazone (50,66)
Lidocaine (45)		Phenytoin (51)
Lorazepam (35)		Tolbutamide (40)
Quinidine (49)		Warfarin (58)
	Total metabolic clearance	
Acetaminophen (8)		Acetaminophen (66)
Antipyrine (50,51,71,78)		Diazepam (33)
Chlordiazepoxide (57,77)		Ethanol (72)
Chlormethiazole (44)		Lidocaine (45)
Phenylbutazone (14)		Phenytoin (51)
Phenytoin (25)		Warfarin (58)
Propranolol (78)		
Quinidine (49)		
Tolbutamide (41)		

old. This is possible because the half-life of elimination ($t_{1/2}$) is a function of both the volume of distribution and the clearance ($t_{1/2} = 0.693 \cdot Vd/Cl$).

The effects of age on renal function are well known. Although there is considerable variation, glomerular filtration rate, as measured by inulin or creatinine clearance (Fig. 2), declines an average of 35% between ages 20 and 90 (53), and renal plasma flow declines by 40 to 45% between ages 25 and 65 (3). Urine concentrating ability (54) and renal sodium conservation (16) also decline with age. In addition the elderly may be more sensitive than the young to the stimulus of hyperosmolality (26). Because of the age-related differences in glomerular and tubular function, it is predictable that drugs excreted primarily by the kidney will have a prolonged half-life and a reduced clearance in the elderly (Table 3). A simple test of renal function, such as the creatinine clearance, can be used along with measurements of plasma drug levels to adjust the doses and dosage schedules of renally excreted drugs.

PHARMACOKINETIC EFFECTS OF AGING

The effects of aging on pharmacokinetics and drug use have been the subject of several recent reviews (14,52,67,73), and the reader is encouraged to consult these. Rather than attempting an exhaustive summary of the current literature on geriatric clinical pharmacology, it seems appropriate to illustrate how some

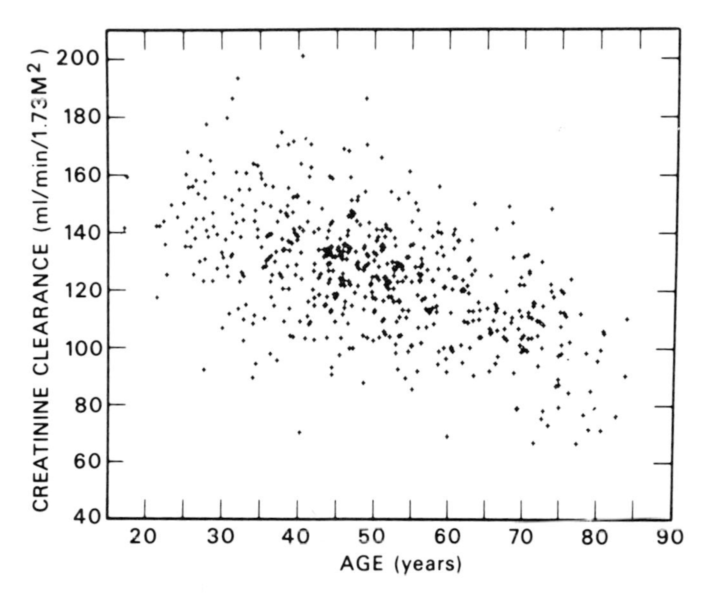

FIG. 2. Cross-sectional analysis of true creatinine clearance in 548 normal subjects ($r = -0.54$, slope $= -0.80$, and intercept at age zero $= 165.6$). These values are equivalent to values by the automated total chromagen method when multiplied by 0.80. (From ref. 53, with permission of the author and publisher.)

of the physiological alterations described above are pertinent to studies of drug disposition and elimination using examples largely drawn from the research experience of the author and his colleagues.

Several years ago at the Gerontology Research Center in Baltimore, we undertook a study of the effect of age on ethanol metabolism in man (72). Ethanol was selected for investigation because of its unique metabolism (oxidation primarily by the cytosolic hepatic enzyme alcohol dehydrogenase), its profound medical and social importance, its occasional therapeutic administration in geriatric patients, and the suggestion of age differences in ethanol metabolism and toxicity in experimental animals (75,76). The study population consisted of 50 healthy male volunteers ranging in age from 21 to 81 years. Most were partici-

TABLE 3. *Drugs eliminated by the kidney which have pharmacokinetic age differences*

Cefradine (61)	Penicillin G (37,43)
Cephazolin (61)	Phenobarbital (69)
Digoxin (17)	Practolol (11)
Dihydrostreptomycin (70)	Propicillin (59)
Doxycycline (60)	Sulphamethiazole (66)
Kanamycin (34)	Tetracycline (70)
Lithium (27,36)	

pants in the Baltimore Longitudinal Study of Aging (65). Under basal conditions (10-hr overnight fast), each subject received a 1-hr intravenous infusion of 15% (v/v) ethanol at a rate of 375 mg/m²/min (equivalent in these subjects to 0.57 g/kg body weight). The concentration of ethanol in whole blood was measured by gas-liquid chromatography, and the values were converted to concentrations in blood water, assuming that whole blood contains 83% water. The kinetics of ethanol distribution and metabolism were evaluated with the aid of a model incorporating saturable Michaelis-Menten kinetics and a central and a peripheral compartment.

The time course of blood ethanol concentration (Fig. 3) exhibited a distributive phase that persisted 45 to 60 min after termination of the infusion, followed by an apparent zero-order elimination phase. Unfortunately, limitations of other testing did not permit us to follow blood levels longer than 300 min, and precluded estimates of K_m (the *in vivo* Michaelis-Menten constant) from the terminal first-order elimination phase. However, the volumes of distribution and the V_{max} (the elimination rate during the zero-order phase) were estimated accurately.

Age had no effect on the rate of ethanol elimination (Fig. 4), but the variation was slightly greater in the older subjects than in the young. However, as can be inferred from Fig. 3, peak ethanol concentrations at the end of the infusion

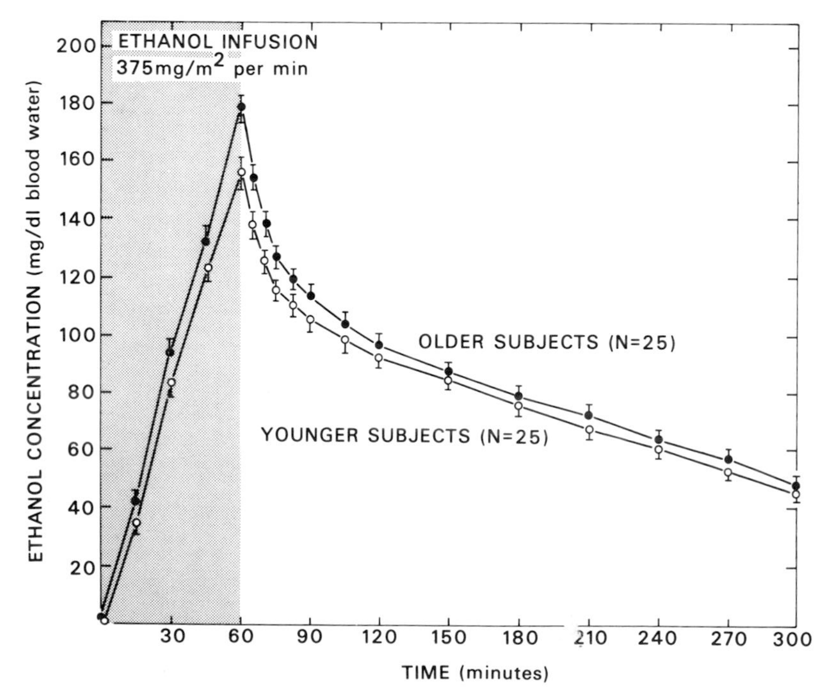

FIG. 3. Mean ethanol concentration in blood water (± SEM) for younger (ages 21 to 56) and older healthy subjects (ages 57 to 81). (From ref. 72, with permission of the publisher.)

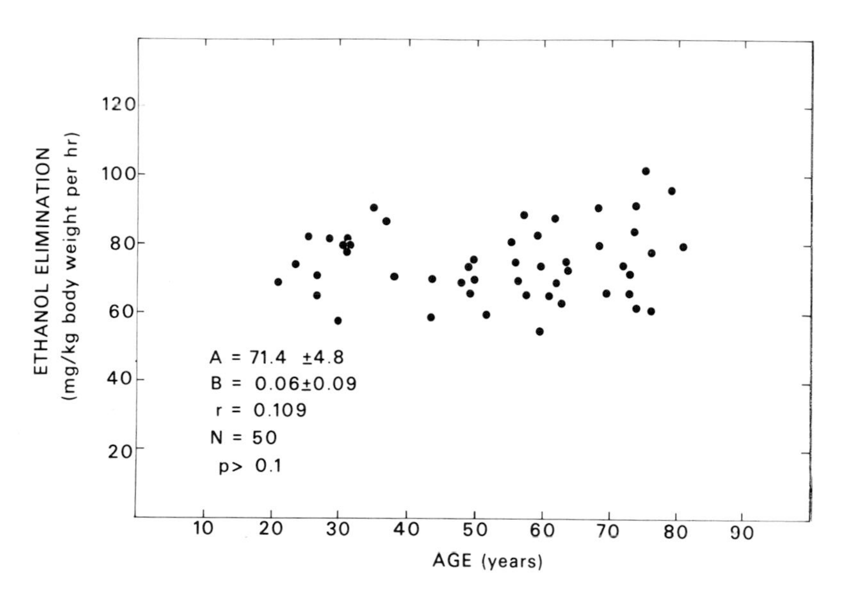

FIG. 4. Rates of ethanol elimination with respect to age. **A** indicates the intercept and **B**, the slope.

correlated with age (Fig. 5) and increased 33% over the adult age span. The explanation seemed to be related to age differences in body composition as determined by the anthropometric method of Behnke (2). The younger group of subjects was significantly leaner than the older, 23.4% ± 1.38 versus 28.4% ± 1.95 body fat (mean ± SEM, $p < 0.05$). This increased adiposity with age was also demonstrated by a negative correlation of age and lean body mass expressed as a fraction of total body weight (r = −0.312, $p < 0.05$). Although the dose of ethanol given in terms of surface area was the same for all subjects, the older subjects actually received a higher dose per lean body mass than did the young subjects (r = 0.561, $p < 0.001$). This occurred because of a negative correlation between lean body mass/m^2 surface area and age (r = −0.533, $p < 0.001$). Since ethanol is known to distribute in total body water and since both total body water and lean body mass decline with age, this effect of body composition on the distribution of ethanol is not surprising. In fact, both the initial and total volume of distribution correlated negatively with age, but there was no correlation when these were normalized to lean body mass.

The point emphasized by this study of ethanol metabolism is that body composition differences may be important in interpreting the results of pharmacokinetic studies in the elderly. The observation of Berkowitz, et al. (7), that early serum levels of morphine (2 min) following 10 mg/70 kg body weight administered intravenously correlated positively with age (r = 0.63, $p < 0.01$) may be related to age differences in body composition. The authors, however, suggested an

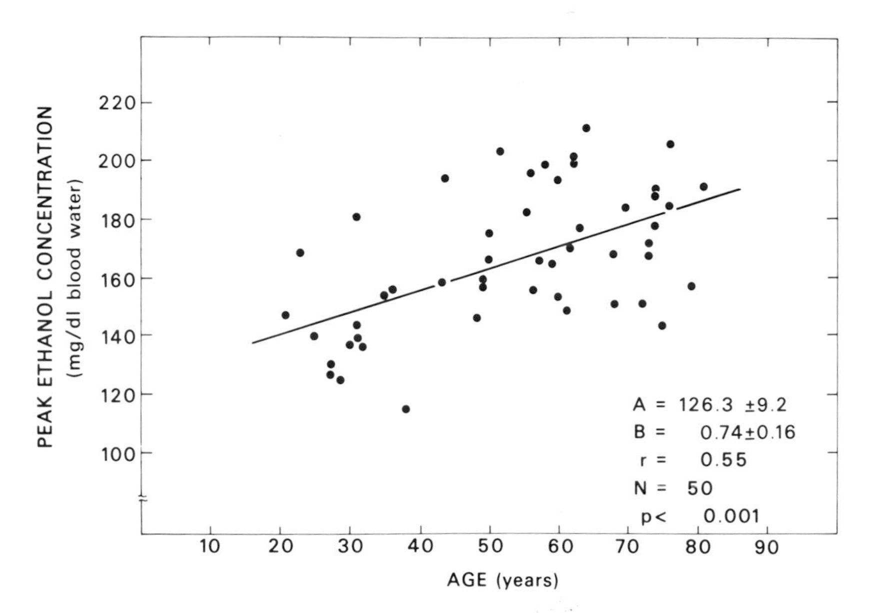

FIG. 5. Correlation of peak ethanol concentration in blood water at the end of ethanol infusion with age. **A** indicates the intercept and **B,** the slope. (From ref. 72, with permission of the publisher.)

effect on mixing or distribution of the drug by the reduced cardiac output that occurs with increasing age. There was no relationship between age and the biological half-life of morphine in that study. An age-related alteration in the distribution of diazepam has been reported by Klotz, et al. (33). A four- to fivefold increase in the terminal plasma half-life was found between age 20 and 80 years (from 20 hr to about 90 hr), but total plasma clearance was unchanged. Although there was no effect of age on either plasma protein binding or erythrocyte binding, the initial and steady-state volumes of distribution increased markedly with age and accounted for the prolongation of the half-life in the elderly subjects. The authors were unable to explain their findings, but body composition differences may have been an important factor.

Tolbutamide is an oral hypoglycemic agent that is highly bound to plasma protein and metabolized by the liver. Sotaniemi and Huhti (63) found no effect of age on the half-life in a large unselected hospital population. There was wide interindividual variation with a mean of 5.8 hr and a range of 1.7 to 19.2 hr. In general, patients with a slow elimination rate had a serious illness or several concurrent diseases, malignancy, previous treatment with noninducing drugs, or congestive heart failure. Rapid elimination was noted in patients whose cardiovascular system was intact, who had taken large amounts of alcohol, and who had a relatively mild illness. Because serious drug-induced hypoglycemic reactions have been reported in the elderly, a more detailed study of tolbutamide kinetics was undertaken in collaboration with the Clinical Pharmacokinetics Laboratory at the University of Maryland School of Pharmacy. The analysis is still incomplete and only preliminary results are available (40,41).

Healthy, nonsmoking male volunteers ranging in age from 23 to 87 received a single dose of tolbutamide (14.3 mg/kg i.v.), and plasma levels at timed intervals up to 24 hr were determined by electron-capture gas-liquid chromatography. Confirming the previous report (63), no differences in either the initial or the terminal phase plasma half-life were found between the young and old subjects. However, increases of 38% in the total plasma clearance ($p < 0.005$) and 43% in its volume of distribution ($p < 0.001$) were observed in the old group as compared with the young group. Total plasma levels of tolbutamide in the elderly were less than in the young. These findings seem to be explained by a 14% increase in the free or unbound fraction of tolbutamide ($p < 0.05$) as determined by equilibrium dialysis at a total tolbutamide concentration of 100 μg/ml. The increase in free fraction in turn seems to be the result of an 11% decrease in serum albumin in the old group ($p < 0.01$) (41). Because of reduced plasma protein binding in the elderly, more free tolbutamide is available for distribution and binding in body tissues, as well as for metabolism by the liver. Similar pharmacokinetic data have been reported for phenytoin (Fig. 6) (25).

The data obtained following acute intravenous administration of tolbutamide do not suggest that the elderly are more susceptible to hypoglycemia because of prolonged half-lives and higher plasma levels. Another explanation must be sought. However, single-dose studies may not necessarily reflect the kinetics that obtain during therapeutic conditions with repeated oral dosage to steady state. It is also important to recognize that controlled studies in healthy elderly subjects may help define effects of aging as distinct from the effects of disease, but in elderly patients the effects of chronic illness may be more important than effects of old age. For example, congestive heart failure may have marked effects on pharmacokinetics (6).

Up to this point we have been concerned with effects of age on factors that primarily affect drug distribution, such as body composition and protein binding. What about the effect of age on the intrinsic capacity of the liver to metabolize drugs?

In 1971 O'Malley, et al. (50) reported that after oral administration of antipyrine a young control group of 61 men and women ranging in age from 20 to 40 years had a mean plasma half-life of 12.0 $\pm$ 0.4 hr (SEM). In contrast, 19 elderly subjects ranging in age from 70 to 100 years had a half-life of 17.4 $\pm$ 1.6 hr. This was the first study in man to suggest that old age was associated with impaired hepatic drug metabolism. Antipyrine is a useful model compound for such a study because it is only minimally protein bound, distributes in total body water, and is extensively metabolized by the liver prior to excretion.

Prompted by the report from the group in Dundee, we analyzed data compiled in the Baltimore Longitudinal Study of Aging (65) on 307 male participants,

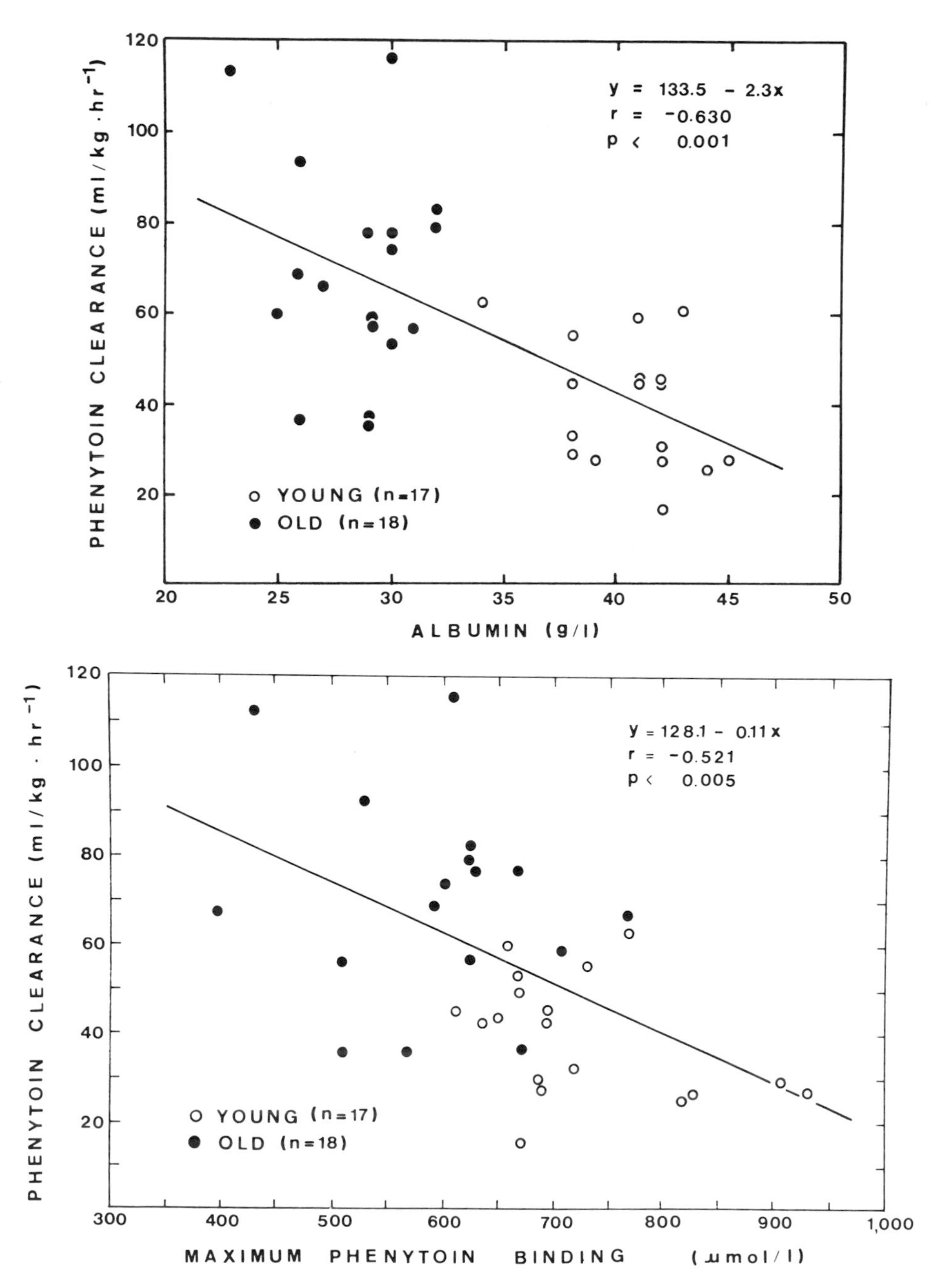

FIG. 6. Correlations of phenytoin clearance with serum albumin **(upper panel)** and maximum phenytoin binding **(lower panel).** Total plasma clearance was determined in young subjects (ages 20 to 38) and old subjects (ages 65 to 90) after 250 mg phenytoin. Maximum plasma binding was determined from measurements of free and bound drug at four different drug concentrations. (Figure based on data from ref. 25.)

aged18 to 92, who had not received medications but who had received antipyrine (1.0 g i.v.) for the measurement of total body water. The biological half-life was 12.7 ± 0.50 hr in the young group and 14.8 ± 0.65 hr in the old ($p <$ 0.025). Middle-aged subjects had intermediate results. Although the magnitude of the age differences was less than that in the previous report (50), these results (71), along with those of others (38,51,78), seemed to confirm an apparent effect of age on antipyrine metabolism. Although the correlations were weak, both the apparent volume of distribution (Fig. 7) and the metabolic clearance rate (Fig. 8) declined significantly with age.

The subjects were also classified into groups on the basis of cigarette smoking, caffeine consumption, and alcohol consumption. Most of the variables examined were significantly associated with age, with each other, and with antipyrine metabolism (Table 4). Because of the complexity of these associations, multiple regression analysis was utilized to determine the contribution of each variable to the variance explained. Both alcohol and caffeine consumption contributed less than 1% to the variance explained. However, cigarette smoking was found to contribute 12% and age 3% to the variance explained when the metabolic clearance rate of antipyrine was significantly predicted ($p < 0.001$) by the equation: MCR = 35.14 + 2.14(smoking class) − 0.13(age).

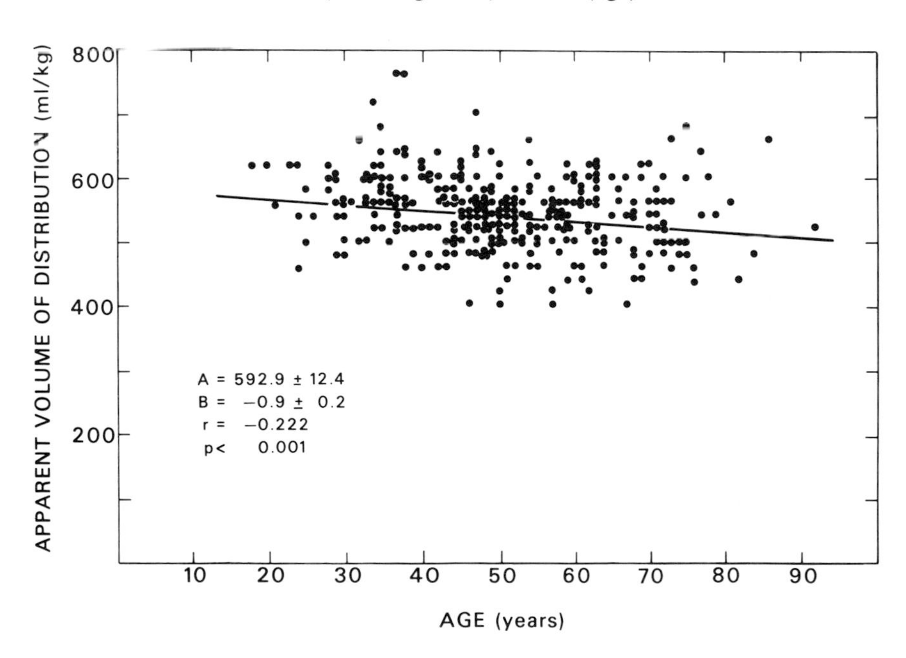

FIG. 7. Correlation of the apparent volume of distribution (Vd) of antipyrine with age for 307 healthy male subjects assuming a single compartment model. (Vd = Dose/C_0 when C_0 = the theoretical plasma concentration at zero time obtained from regression analysis of the natural log of sequential plasma concentrations in each subject with respect to time.) By this method, Vd for antipyrine is an estimate of total body water. **A** indicates the intercept and **B,** the slope.

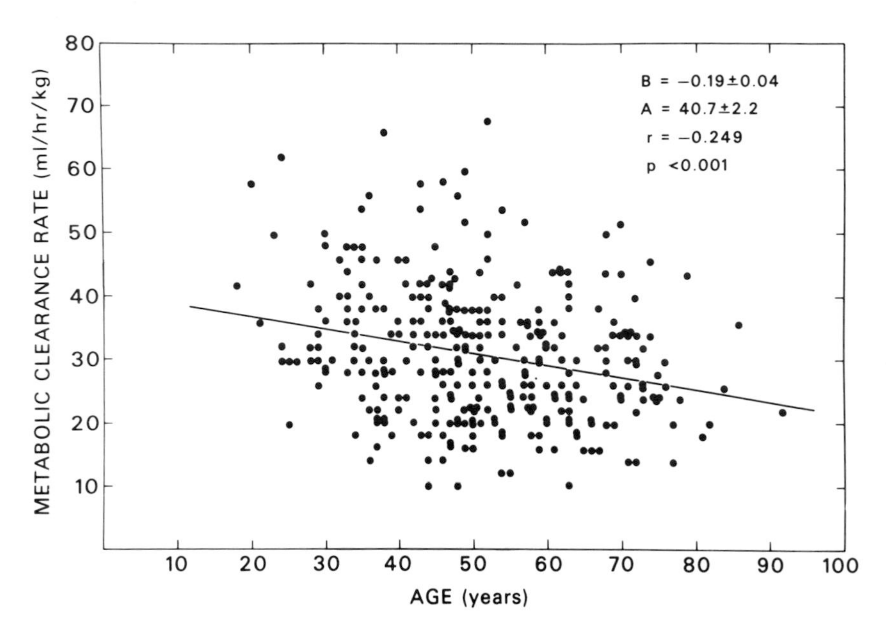

FIG. 8. Decline in metabolic clearance rate (MCR) of antipyrine with age in 307 healthy male subjects. (MCR = Vd · k_e where k_e = the overall elimination rate constant for antipyrine and Vd = apparent volume of distribution.) The correlation of MCR with age alone contributed only 6.2% ($r^2 \times 100$) of the total variance explained. (From ref. 71, with permission of the publisher.)

Clearly, however, the sixfold interindividual variation exceeds the effects of both smoking and age. This wide interindividual variation is due to a variety of genetic and environmental factors. There is good evidence for an effect of diet (20,31) and confirmatory evidence for an effect of cigarette smoking (20,23) on antipyrine metabolism. Our data actually suggest that the elderly are resistant to the apparent induction of drug metabolism by cigarette smoking (Table 5).

TABLE 4. *Summary of associations among metabolic clearance rate, age, and habits*[1]

	MCR	Age	Smoking	Caffeine	Alcohol
MCR	—				
Age	↓	—			
Smoking	↑	↓	—		
Caffeine	↑	↓	↑	—	
Alcohol	0	↓	↑	↑	—

[1] Simple associations by χ^2 analysis. The arrows give the direction of the association and all are statistically significant at $p < 0.025$; 0 indicates no significant association.

(From ref. 71, with permission of the publisher.)

TABLE 5. *Effect of cigarette smoking on half-life and metabolic clearance rate (MCR) of antipyrine*

Age group	Smoking group[a]	Number of subjects	Percent of age group	$t_{1/2}$ (hr)[b]	MCR[b] (ml/kg/hr)
Young	Nonsmokers/light	41	56.2	14.1 ± 0.61	30.5 ± 1.29
(age 18–39)	Moderate/heavy	32	43.8	10.9 ± 0.73[c]	39.7 ± 2.11[c]
Middle	Nonsmokers/light	108	72.0	14.6 ± 0.53	28.1 ± 0.86
(age 40–59)	Moderate/heavy	42	28.0	11.5 ± 0.89[c]	37.9 ± 1.89[d]
Old	Nonsmokers/light	72	85.7	14.9 ± 0.71	28.0 ± 1.05
(age 60–92)	Moderate/heavy	12	14.3	14.1 ± 1.63	29.6 ± 3.16
All ages	Nonsmokers/light	221	72.0	14.7 ± 0.37	28.5 ± 0.60
(age 18–92)	Moderate/heavy	86	28.0	16.6 ± 0.57[d]	37.4 ± 1.32[d]

[a] Smoking groups are defined as follows: nonsmokers did not smoke cigarettes (smoking class 0). Light: smoked "once in a while" or less than 10 cigarettes/day (smoking class 1 and 2). Moderate: smoked 10 to 20 cigarettes/day(smoking class 3). Heavy: smoked 21 or more cigarettes/day (smoking class 4 and 5).

[b] Mean ± SEM.

[c] Significantly different from nonsmokers/light group ($p < 0.005$).

[d] Significantly different from nonsmokers/light group ($p < 0.001$).

Data from ref. 71.

Wood, et al. (78) have recently reported the same relationship between age and cigarette smoking for the metabolism of propranolol, as well as for antipyrine. Another recent report shows an increase in the plasma clearance of antipyrine in young subjects following treatment with the hypnotic dichloralphenazone, but no significant alteration in the elderly group (64). Thus, the elderly seem to show a reduced capacity for enzyme induction. Additional studies are needed to confirm these observations and elucidate the mechanism.

CONCLUSION

There are a number of physiological factors that lead to alterations in drug disposition (Table 6). There is as yet insufficient evidence to assess the effect of age on drug absorption, but age-related alterations in body composition and protein binding have been shown to influence the distribution of drugs in the body. Higher plasma levels with a greater proportion of unbound drug may be present in the elderly than in the young. Hepatic drug metabolism of some drugs seems to be impaired with aging, but other factors that differ with age—such as smoking, nutrition, and disease states—may be more important than age alone. Interindividual variation is great and makes plasma level determinations and careful clinical assessment mandatory for drugs with a low therapeutic index. Renal excretion of drugs and their metabolites is reduced in older patients because of an age-related decline in renal function. Age differences in pharmacokinetics would seem to have important implications for the clinical testing of drugs and for the assessment of dose-response relationships in the elderly.

TABLE 6. *Factors affecting drug disposition in the elderly*

Pharmacokinetic parameter	Altered physiology	Altered pharmacology
Absorption	Elevated gastric pH Reduced GI blood flow ? Reduced number of absorbing cells ? Reduced GI motility	Limited available data shows no significant effect of age
Distribution	Body composition	
	Reduced total body water	Higher concentration of drugs distributed in body fluids
	Reduced lean body mass/kg body weight Increased body fat	? Longer duration of action of fat soluble drugs
	Protein binding	
	Reduced serum albumin	Higher free fraction of highly protein bound drugs
Elimination	Hepatic metabolism	
	? Reduced enzyme activity Reduced hepatic mass Reduced hepatic blood flow	Apparently slower biotransformation of some drugs Influenced by environmental factors and disease Large interindividual variation
	Renal excretion	
	Reduced glomerular filtration rate Reduced renal plasma flow Altered tubular function	Slower excretion of some drugs and metabolites

ACKNOWLEDGMENTS

The original work reviewed in this paper was performed while the author was a Clinical Associate at the Gerontology Research Center, National Institute on Aging, National Institutes of Health, in Baltimore, Maryland, and was made possible through the collaborative efforts of a number of investigators. The author also wishes to acknowledge the secretarial assistance of Mim Carpenter.

REFERENCES

1. Adelman, R. C. (1971): Age-dependent effects in enzyme induction—A biochemical expression of aging. *Exp. Gerontol.,* 6:75–87.
2. Behnke, A. R. (1961): Anthropometric fractionation of body weight. *J. Appl. Physiol.,* 16:949–954.
3. Bender, A. D. (1965): The effect of increasing age on the distribution of peripheral blood flow in man. *J. Am. Geriatr. Soc.,* 13:192–198.
4. Bender, A. D. (1968): Effect of age on intestinal absorption: Implications for drug absorption in the elderly. *J. Am. Geriatr. Soc.,* 16:1331–1339.

5. Bender, A. D., Post, A., Meier, J. P., Higson, J. E., and Reichard G. (1975): Plasma protein binding of drugs as a function of age in adult human subjects. *J. Pharm. Sci.,* 64:1711–1713.
6. Benowitz, N. L., and Meister, W. (1976): Pharmacokinetics in patients with cardiac failure. *Clin. Pharmacokinet.,* 1:389–405.
7. Berkowitz, B. A., Ngai, S. H., Yang, J. C., Hempstead, B. S., and Spector, S. (1975): The disposition of morphine in surgical patients. *Clin. Pharmacol. Ther.,* 17:629–635.
8. Briant, R. H., Dorrington, R. E., Cleal, J., and Williams, F. M. (1976): The rate of acetaminophen metabolism in the elderly and the young. *J. Am. Geriatr. Soc.,* 24:359–361.
9. Cammarata, R. J., Rodnan, G. P., and Fennell, R. H. (1967): Serum anti-gamma-globulin and anti-nuclear factors in the aged. *J.A.M.A.,* 199:115–118.
10. Caranasos, G. J., Stewart, R. B., and Cluff, L. E. (1974): Drug-induced illness leading to hospitalization. *J.A.M.A.,* 228:713–717.
11. Castleden, C. M., Kaye, C. M., and Parsons, R. L. (1975): The effect of age on plasma levels of propranolol and practolol in man. *Br. J. Clin. Pharmacol.,* 2:303–306.
12. Castleden, C. M., George, C. F., Marcer, D., and Hallett, C. (1977): Increased sensitivity to nitrazepam in old age. *Br. Med. J.,* 1:10–12.
13. Chan, K., Kendall, M. M., Mitchard, M., and Wells, W. D. E. (1975): The effect of aging on plasma pethidine concentration. *Br. J. Clin. Pharmacol.,* 2:297–302.
14. Crooks, J., O'Malley, K., and Stevenson, I. H. (1976): Pharmacokinetics in the elderly. *Clin. Pharmacokinet.,* 1:280–296.
15. Edelman, I. S., and Leibman, J. (1959): Anatomy of body water and electrolytes. *Am. J. Med.,* 27:256–277.
16. Epstein, M., and Hollenberg, N. K. (1976): Age as a determinant of renal sodium conservation. *J. Lab. Clin. Med.,* 87:411–417.
17. Ewy, G. A., Kapadia, G. G., Yao, L., Lullin, M., and Marcus, F. I. (1969): Digoxin metabolism in the elderly. *Circulation,* 39:449–453.
18. Farah, F., Taylor, W., Rawlins, M. D., and James, O. (1977): Hepatic drug acetylation and oxidation: Effects of aging in man. *Br. Med. J.,* 2:255–256.
19. Forbes, G. B., and Reina, J. C. (1970): Adult lean body mass declines with age: Some longitudinal observations. *Metabolism,* 19:653–663.
20. Fraser, H. S., Mucklow, J. C., Bulpitt, C. J., Khan, C., Mould, G., and Dollery, C. T. (1977): Environmental effects on antipyrine half-life in man. *Clin. Pharmacol. Ther.,* 22:799–808.
21. Geokas, M. C., and Haverback, B. J. (1969): The aging gastrointestinal tract. *Am. J. Surg.,* 117:881–892.
22. Gibson, R. M., Mueller, M. S., and Fisher, C. R. (1977): Age differences in health care spending, fiscal year 1976. *Soc. Sec. Bull.,* 40:3–14.
23. Hart, P., Farrell, G. C., Cooksley, W. G. E., and Powell, L. W. (1976): Enhanced drug metabolism in cigarette smokers. *Br. Med. J.,* 3:147–149.
24. Hayes, M. J., Langman, M. J. S., and Short, A. H. (1975a): Changes in drug metabolism with age. 1. Warfarin binding and plasma proteins. *Br. J. Clin. Pharmacol.,* 2:69–72.
25. Hayes, M. J., Langman, M. J. S., and Short, A. H. (1975b): Changes in drug metabolism with increasing age. 2. Phenytoin clearance and protein binding. *Br. J. Clin. Pharmacol.,* 2:73–79.
26. Helderman, J. H., Vestal, R. E., Rowe, J. W., Tobin, J. D., Andres, R., and Robertson, G. L. (1978): The response of arginine vasopressin to intravenous ethanol and hypertonic saline in man: The impact of aging. *J. Gerontol.,* 33:39–47.
27. Hewick, D. S., and Newbury, P. A. (1976): Age: Its influence on lithium dosage and plasma levels. *Br. J. Clin. Pharmacol.,* 3:354P.
28. Hooper, W. D., Bochner, R., Eadie, M. J., and Tyrer, J. H. (1974): Plasma protein binding of diphenylhydantoin: Effects of sex hormones, renal and hepatic disease. *Clin. Pharmacol. Ther.,* 15:276–282.
29. Hurwitz, N. (1969): Predisposing factors in adverse drug reactions to drugs. *Br. Med. J.,* 1:536–539.
30. Jori, A., Di Salle, E., and Quadri, A. (1972): Rate of aminopyrine disappearance from plasma in young and aged humans. *Pharmacology,* 8:273–279.
31. Kappas, A., Anderson, K. E., Conney, A. H., and Alvares, A. P. (1976): Influence of dietary protein and carbohydrate on antipyrine and theophylline metabolism in man. *Clin. Pharmacol. Ther.,* 20:643–653.

32. Kato, R., and Takanaka, A. (1968): Metabolism of drugs in old rats. I. Activities of NADPH-linked electron transport and drug metabolising enzyme systems in liver microsomes of old rats. *Jpn. J. Pharmacol.,* 18:381–388.
33. Klotz, U., Avant, G. R., Hoyumpa, A., Schenker, S., and Wilkinson, G. R. (1975): The effects of age and liver disease on the disposition and elimination of diazepam in adult man. *J. Clin. Invest.,* 55:347–359.
34. Kristensen, M., Molholm Hansen, J., Kampmann, J., Lumpholtz, B., and Siersbaek-Nielsen, K. (1974): Drug elimination and renal function. *J. Clin. Pharmacol.,* 14:307–308.
35. Kyriakopoulos, A. A. (1976): Bioavailability of lorazepam in humans. In: *Pharmacokinetics of Psychoactive Drugs,* edited by L. A. Gottschalk and S. Merlis, pp. 45–60. Spectrum Publications, New York.
36. Lehmann, K., and Merten, K. (1974): Die elimination von lithium in abhangigkeit vom lebensalter bei gesunden und niereninsuffizienten. *Int. J. Clin. Pharmacol.,* 10:292–298.
37. Leikola, E., and Vartia, K. O. (1957): On penicillin levels in young and geriatric subjects. *J. Gerontol.,* 12:48–52.
38. Liddell, D. E., Williams, F. M., and Briant, R. H. (1975): Phenazone (antipyrine) metabolism and distribution in young and elderly adults. *Clin. Exp. Pharmacol. Physiol.,* 2:481–487.
39. Mather, L. E., Tucker, G. T., Pflug, A. E., Lindop, M. J., and Wilkerson, C. (1975): Meperidine kinetics in man: Intravenous injection in surgical patients and volunteers. *Clin. Pharmacol. Ther.,* 17:21–30.
40. Miller, A. K., Adir, J., and Vestal, R. E. (1977): Effect of age on the pharmacokinetics of tolbutamide in man (Abst 5). *Pharmacologist,* 19:128.
41. Miller, A. K., Adir, J., and Vestal, R. E. (1978): Tolbutamide binding to plasma proteins of young and old human subjects. *J. Pharm. Sci.,* 67:1192–1193.
42. Misera, D. P., Loudon, J. M., and Staddon, G. E. (1975): Albumin metabolism in elderly patients. *J. Gerontol.,* 30:304–306.
43. Molholm Hansen, J., Kampman, J., and Laursen, H. (1970): Renal excretion of drugs in the elderly. *Lancet,* 1:1170.
44. Nation, R. L., Learoyd, B., Barber, J., and Triggs, E. J. (1976): The pharmacokinetics of chlormethiazole following intravenous administration in the aged. *Eur. J. Clin. Pharmacol.,* 10:407–415.
45. Nation, R. L., Triggs, E. J., and Selig, M. (1977a): Lignocaine kinetics in cardiac patients and aged subjects. *Br. J. Clin. Pharmacol.,* 4:439–448.
46. Nation, R. L., Vine, J., Triggs, E. J., and Learoyd, B. (1977b): Plasma levels of chlormethiazole and two metabolites after oral administration to young and aged human beings. *Eur. J. Clin. Pharmacol.,* 12:137–146.
47. Nies, A., Robinson, D. S., Friedman, M. J., Green, R., Cooper, T. B., Ravaris, C. L., and Ives, J. O. (1977): Relationship between age and tricyclic antidepressant plasma levels. *Am. J. Psychiatry,* 134:790–793.
48. Novak, L. P. (1972): Aging, total body potassium, fat-free mass, and cell mass in males and females between ages 18 and 35 years. *J. Gerontol.,* 27:438–443.
49. Ochs, H. R., Greenblatt, D. J., Woo, E., Franke, K., and Smith, T. W. (1977): Reduced clearance of quinidine in elderly humans (Abst). *Clin. Res.* 25:513A.
50. O'Malley, K., Crooks, J., Duke, E., and Stevenson, I. H. (1971): Effect of age and sex on human drug metabolism. *Br. Med. J.,* 3:607–609.
51. Rasmussen, S. N., Hansen, J. M., Kampmann, J. P., Skovstad, L., and Bach, B. (1976): Drug metabolism in relation to liver volume and age (Abstr). *Scand. J. Gastroenterol.,* 11:86.
52. Richey, D. P., and Bender, D. (1977): Pharmacokinetic consequences of aging. *Ann. Rev. Pharmacol. Toxicol.,* 17:49–65.
53. Rowe, J. W., Andres, R., Tobin, J. D., Norris, A. H., and Shock, N. W. (1976a): The effect of age on creatinine clearance in man: A cross-sectional and longitudinal study. *J. Gerontol.,* 31:155–163.
54. Rowe, J. W., Shock, N. W., and DeFronzo, R. A. (1976b): The influence of age on renal response to water deprivation in man. *Nephron,* 17:270–278.
55. Rowe, J. W. (1977): Clinical research on aging: Strategies and directions. *N. Engl. J. Med.,* 297:1332–1336.
56. Seidl, L. G., Thornton, G. R., Smith, J. W., and Cluff, L. E. (1966): Studies on the epidemiology of adverse drug reactions. III. Reactions in patients on a general medical service. *Bull. Johns Hopkins Hospital,* 119:299–315.

57. Shader, R. I., Greenblatt, D. J., Harmatz, J. S., Franke, R. I., and Koch-Weser, J. (1977): Absorption and disposition of chlordiazepoxide in young and elderly male volunteers. *J. Clin. Pharmacol.*, 17:709–718.
58. Shepherd, A. M. M., Hewick, D. S., Moreland, T. A., and Stevenson, I. H. (1977): Age as a determinant of sensitivity of warfarin. *Br. J. Clin. Pharmacol.*, 4:315–320.
59. Simon, C., Malerczyk, V., Muller, U., and Muller, G.; Zur pharmakokinetik von propicillin bei geriatrischen patienten im vergleich zu jungeren erwachsenen. *Dtsch. Med. Wochenschr.*, 97:1999–2003.
60. Simon, C., Malerczyk, V., Engleke, H., Preuss, I., Grahmann, H., and Schmidt, K. (1975): Die pharmakokinetik von doxycyclin bei niereninsuffizienz und geriatrischen patienten im vergleich zu jungeren erwachsenen. *Schweiz. Med. Wochenschr.*, 105:1615–1620.
61. Simon, C., Malerczyk, V., Tenschert, B., and Mohlenbeck, F. (1976): Die geriatrische pharmakologie von cefazolin, cefradin und sulfisomidin. *Arzneim. Forsch.*, 26:1377–1382.
62. Smith, J. W., Seidl, L. G., and Cluff, L. E. (1966): Studies on the epidemiology of adverse drug reactions. V. Clinical factors influencing susceptibility. *Ann. Int. Med.*, 65:629–640.
63. Sotaniemi, E. A., and Huhti, E. (1974): Half life of intravenous tolbutamide in the serum of patients in medical wards. *Ann. Clin. Res.*, 6:146–154.
64. Stevenson, I. H., Salem, S. A. M., and Shepherd, A. M. M. (1977): Studies on drug absorption and metabolism in the elderly (Abst). *Symposium on Drugs and the Elderly*, Ninewalls Hospital, Dundee.
65. Stone, J. L., and Norris, A. H. (1966): Activities and attitudes of participants in the Baltimore Longitudinal Study. *J. Gerontol.*, 21:575–580.
66. Triggs, E. J., Nation, R. L., Long, A., and Ashley, J. J. (1975): Pharmacokinetics in the elderly. *Eur. J. Clin. Pharmacol.*, 8:55–62.
67. Triggs, E. J., and Nation, R. L. (1975): Pharmacokinetics in the aged: A review. *J. Pharmacokinet. Biopharmaceut.*, 3:387–418.
68. Traeger, A., Kunze, M., Stein, G., and Ankerman, H. (1973): Zur pharmakokinetik von indomethazin bei alten menschen. *Z. Alternsforsch.*, 27:151–155.
69. Traeger, A., Kiesewetter, R., and Kunze, M. (1974): Zur pharmakokinetik von phenobarbital bei erwachsenen und greisen. *Dtsch. Gesundheitswes.*, 29:1040–1042.
70. Vartia, K. O., and Leikola, E. (1960): Serum levels of antibiotics in young and old subjects following administrations of dihydrostreptomycin and tetracycline. *J. Gerontol.*, 15:392–394.
71. Vestal, R. E., Norris, A. H., Tobin, J. D., Cohen, B. H., Shock, N. W., and Andres, R. (1975): Antipyrine metabolism in man: Influence of age, alcohol, caffeine, and smoking. *Clin. Pharmacol. Ther.*, 18:425–432.
72. Vestal, R. E., McGuire, E. A., Tobin, J. D., Andres, R., Norris, A. H., and Mezey, E. (1977): Aging and ethanol metabolism. *Clin. Pharmacol. Ther.*, 21:343–354.
73. Vestal, R. E. (1978): Drug use in the elderly: A review of problems and special considerations. *Drugs*, 16:358–382.
74. Wallace, S., Whiting, B., and Runcie, J. (1976): Factors affecting drug binding in plasma of elderly patients. *Br. J. Clin. Pharmacol.*, 3:327–330.
75. Wiberg, G. S., Trenholm, X., and Coldwell, B. B. (1970): Increased ethanol toxicity in old rats: Changes in LD_{50} *in vivo* and *in vitro* metabolism, and liver alcohol dehydrogenase activity. *Toxicol. Appl. Pharmacol.*, 16:718–727.
76. Wiberg, G. S., Samson, J. M., Maxwell, W. B., Coldwell, B. B., and Trenholm, H. L. (1971): Further studies on the acute toxicity of ethanol in young and old rats: Relative importance of pulmonary excretion and total body water. *Toxicol. Appl. Pharmacol.*, 20:22–29.
77. Wilkinson, G. R. (1977): The effect of aging on the disposition of benzodiazepines in man (Abst). *Symposium on Drugs and the Elderly*, Ninewalls Hospital, Dundee.
78. Wood, A. J. J., Vestal, R. E., Branch, R. A., Wilkinson, G. R., and Shand, D. G. (1978): Age related effects of smoking on elimination of propranolol, antipyrine and indocyanine green (Abst). *Clin. Res.*, 26:14A.
79. Woodford-Williams, E., Alvarez, A. A., Webster, D., Landless, B., and Dixon, M. P. (1964/65): Serum protein patterns in "normal" and pathological ageing. *Gerontologia*, 10:86–99.

Physiology and Cell Biology of Aging
(Aging, Volume 8), edited by A. Cherkin, et al.
Raven Press, New York © 1979.

Pharmacological Aspects of Ergot Alkaloids in Gerontological Brain Research

W. Meier-Ruge, P. Gygax, H. Emmenegger, and P. Iwangoff

Institute of Basic Medical Research, Sandoz, Ltd., Basel, Switzerland.

Aging is an immanent law of life. Since the aging process is insidious and progressive, its recognition by the aging individual or by that person's relatives is rather late, occurring often only after the aging individual shows evidence of an inability to meet the demands of daily life. Changes in physical and mental function during aging are irregular and variable, and depend on environmental and genetic factors specific to each individual.

With the increasing number of retired people, alterations in the social life of the elderly due to central nervous disorders have attracted more and more interest, leading finally to a demand for effective pharmacological intervention in the aging process.

This situation gave rise to a series of questions on the part of experimental pharmacology (53,63,68): Which symptoms of aging have to be treated? Do we want to increase life or only to compensate symptoms of aging? Which models are available to investigate drugs, and which have an action on symptoms of the aging process?

Even today, there are more questions than answers available. Attempts were made to develop experimental pharmacological models and concepts to permit the development of new drugs, which at least exert an effect on symptoms of brain aging.

Normally it might be thought that experimental pharmacology on aging is necessarily carried out in aging animals. This appears to be obvious, but experimental research with aging animals is presently limited to only a few selected purposes. For routine pathophysiological and pharmacological work, aging animals are too expensive, the number of test animals is restricted, and the animal species are usually limited to rat and mouse (24,53,63,68). On the other hand, in an animal model, it is obviously difficult to reproduce the aging process in its multifactorious way.

If we follow three disciplines involved in aging research (morphology, biochemistry, and physiology), no direct experimental approach is available. Only in Parkinson's disease does neurochemistry offer a pharmacological approach correlated with this particular aging disease.

From the neurophysiological point of view, some interesting concepts were developed for testing pharmacological compounds, which are active in the treatment of cerebrovascular insufficiency, transient ischemic brain lesions, and also diffuse symptoms of a mild deficit syndrome due to the normal aging process of the brain (63,68).

DO WE HAVE REASONABLE PHARMACOLOGICAL MODELS IN GERONTOLOGICAL BRAIN RESEARCH?

It is obviously difficult to reproduce in an animal model the aging process in its various forms. There are, however, symptoms of the aging process that can be simulated experimentally, if we accept that, generally speaking, the aging process is characterized by a decline of postsynaptic neuronal response, as well as a decrease of adaptation of brain function to adverse environmental stimuli.

The pathophysiology of hypoxic energy deprivation in the brain has attracted increasing attention during recent years. The correlation of neurotransmitter function and cerebral energy production (67), recovery of neuronal function after brain ischemia (42,43), metabolic changes of the brain tissue due to hypoxia (58,76,77), and correlation of cerebral blood flow and brain function (37,38,54) were some of the most important topics in this field.

It is certain that none of the animal models used today reproduces absolutely the complex aging situation of the brain. In this connection, hypoxic energy deprivation is an efficient way to disturb the adaptability of brain function to environmental stress. Hypoxic energy deprivation has some features in common with aging, in particular with cerebral vascular insufficiency, transient ischemic attack, and stroke.

The great advantage of this model are its excellently reproducible parameters, extending from reversible up to irreversible lesions (71,75). Therefore, hypoxic energy deprivation of the brain is an attractive model not only in experimental brain research, but also for testing drugs active in symptoms of pathological aging (3,10,12,15,27). In particular, the pharmacological effects of ergot alkaloids in animal models of hypoxic energy deprivation attracted interest. In the following, such effects of dihydrogenated peptide alkaloids will be demonstrated in models of temporary cerebral ischemia and hypovolemic oligemia.

THE USE OF PATHOPHYSIOLOGICAL MODELS OF HYPOXIC ENERGY DEPRIVATION IN EXPERIMENTAL PHARMACOLOGY OF GERONTOLOGICAL BRAIN RESEARCH

Temporary cerebral ischemia represents an optimal reproducible experimental model including an acute energy deprivation of the central nervous system, suitable for pharmacological studies. We used the isolated perfused cat head (28,35), as described by Meier-Ruge, et al. (66). This model excludes effects from other organs that are especially susceptible to ischemic conditions, such as pituitary and suprarenal gland, heart, and kidney. Biphasic ischemia in the cat brain causes a disturbance in electrical brain activity and increases the lactate formation by the brain. This functional disturbance was improved by DH-ergotoxine mesylate and DH-ergonine.

The quantitative EEG evaluation demonstrated that biphasic temporary ischemia affects chiefly the EEG alpha and beta frequency range (15,27,66). The decrease in lactate formation induced by DH-ergotoxine mesylate, which indicates a decreased anaerobic metabolism, correlated well with a significant increase in EEG activity (Fig. 1). Similar results obtained with DH-ergonine demonstrated that this ergot compound is about ten times more effective than DH-ergotoxine mesylate (Fig. 2).

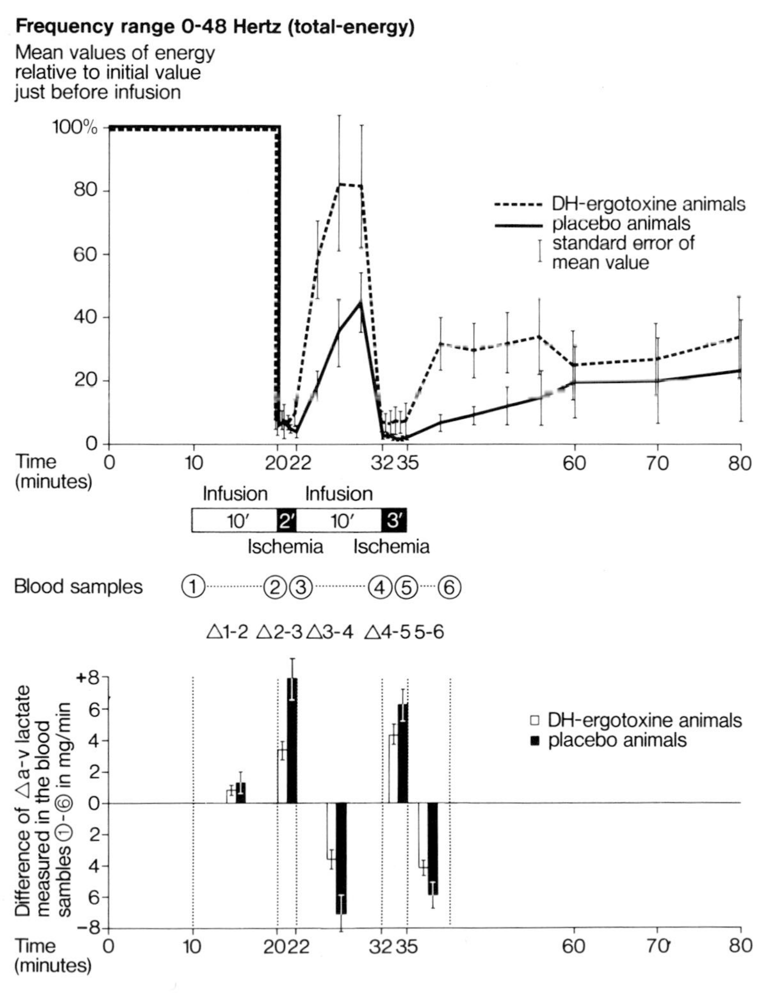

FIG. 1. Action of DH-ergotoxine mesylate on EEG activity and differences of lactate formation in the isolated perfused cat head subjected to biphasic temporary ischemia (Mean value of seven cats infused with DH-ergotoxine mesylate and eight animals infused with placebo solution) (66).

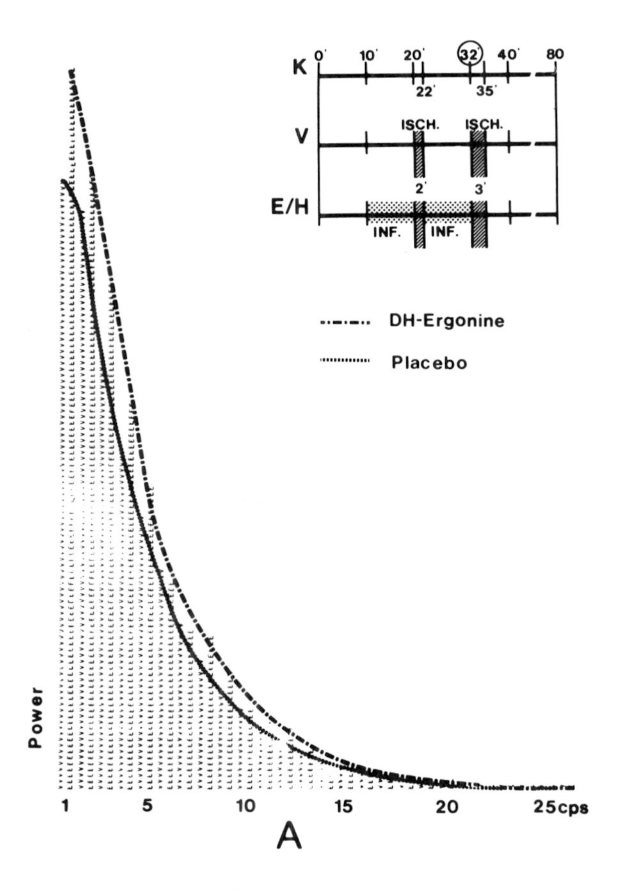

K
V
E/H
0'
10'
20'
32'
40'
80
22'
35'
ISCH.
ISCH.
2'
3'
INF.
INF.
DH-Ergonine
Placebo
Power
1
5
10
15
20
25cps
A

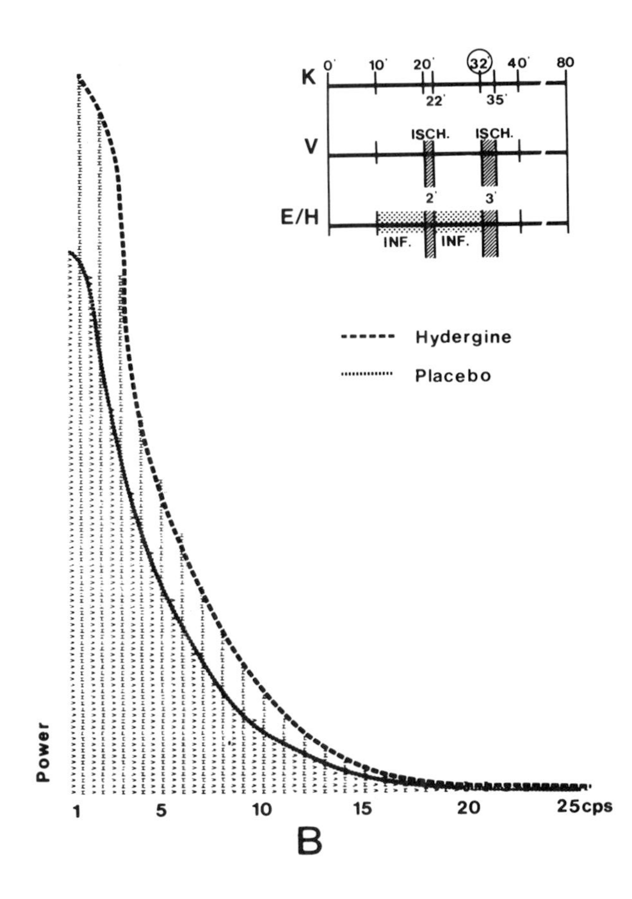

K
V
E/H
0'
10'
20'
32'
40'
80
22'
35'
ISCH.
ISCH.
2'
3'
INF.
INF.
Hydergine
Placebo
Power
1
5
10
15
20
25cps
B

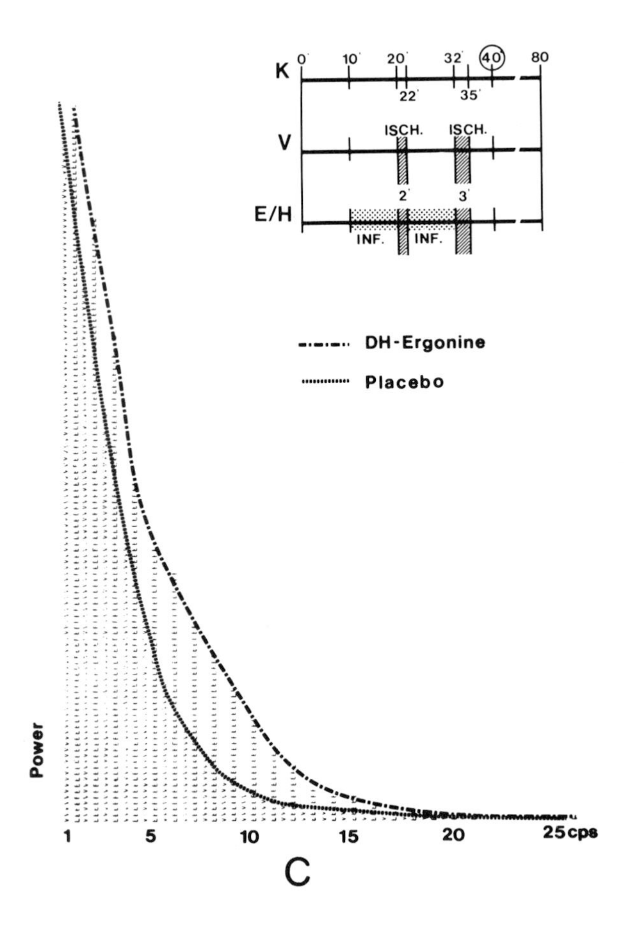

FIG. 2. EEG histogram in the isolated perfused cat head after temporary ischemia. **K:** control without ischemia and supplemental infusion ($n = 8$). **V:** Placebo group with biphasic ischemia (2 and 3 min) and infusion of 0.9% NaCl ($n = 7$). **E/H:** DH-ergotoxine mesylate (Hydergine) 8.0 μg/min for 20 min ($n = 8$) or DH-ergonine mesylate 0.6 μg/min for 20 min ($n = 9$). **CpS:** Cycles per sec. **A + B:** EEG histogram 32 min after beginning of the experiment, 20 min after drug infusion and 10 min after first (2 min) ischemia. **C:** EEG histogram 40 min after beginning of the experiment, 5 min after second (3 min) ischemia and drug infusion.

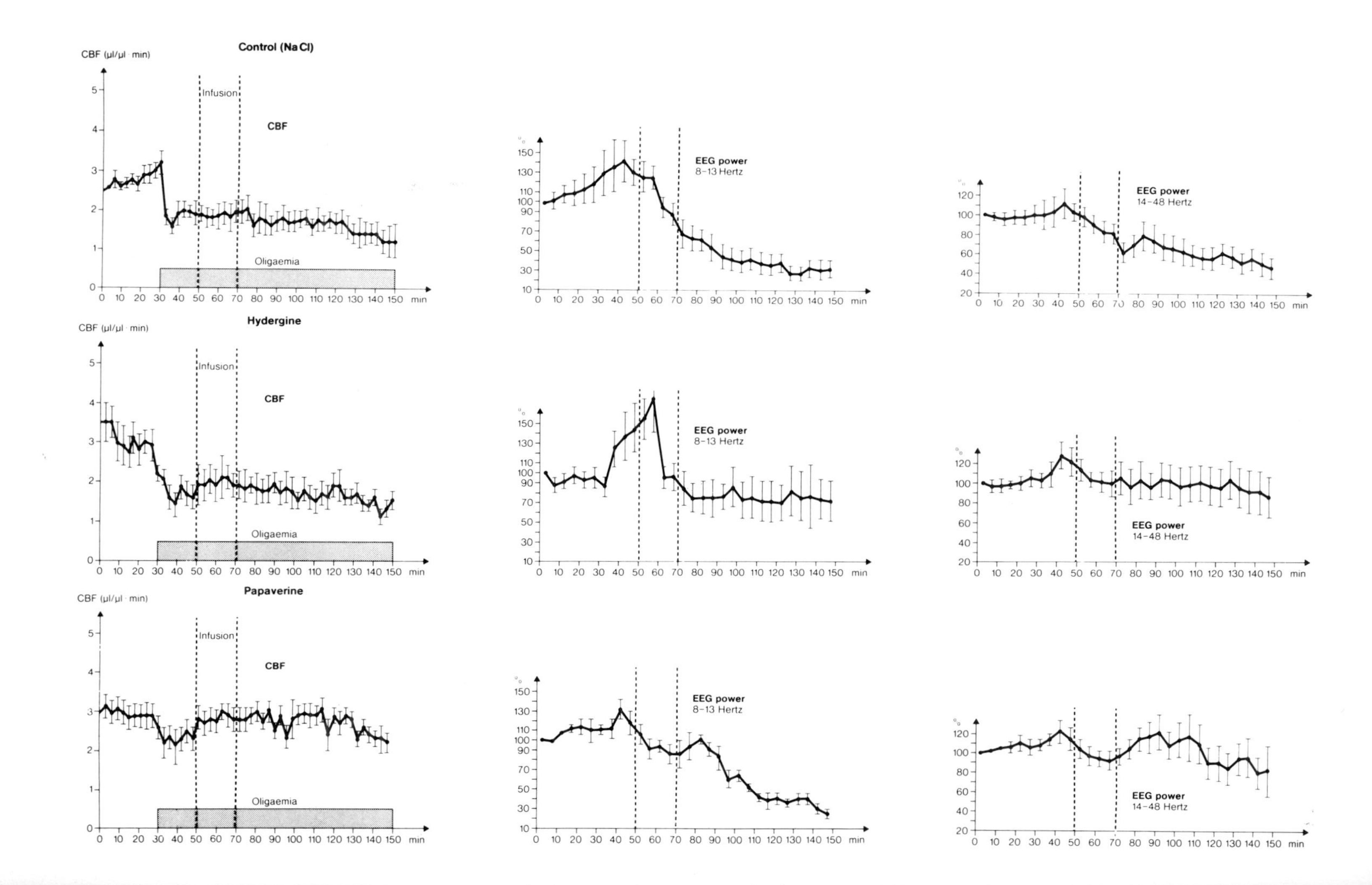

Control (NaCl)
Hydergine
Papaverine
CBF (µl/µl · min)
CBF
Infusion
Oligaemia
EEG power
8–13 Hertz
EEG power
14–48 Hertz
%
min

After these isolated brain perfusion studies, we tried, in a second series of experiments using *the model of hypovolemic oligemia,* to show whether DH-ergotoxine mesylate has analogous effects on hypoxic energy deprivation in the whole animal model. As already shown by Gygax, et al. (39,40), a 30% to 40% decrease of cerebral blood flow by bloodletting and subsequent lowering of blood pressure to 45 mmHg causes an initial EEG activation (Fig. 3). This effect is probably due to a hypoxic transmitter release (54,67), which is followed by progressive EEG depression (39,40). This symptomatology is fairly characteristic in cerebrovascular insufficiency.

This stimulated us to examine, using papaverine as a characteristic vasoactive drug, the old unproved hypothesis that a therapeutic increase of cerebral blood flow is effective in improving the symptomatology of cerebrovascular insufficiency. After intravenous injections of papaverine, the hypovolemically decreased cerebral blood flow returns to normal. Despite the normalization of the brain blood flow, a progressive decrease of EEG energy occurred (Fig. 3). This surprising effect can be explained by the fact that the increase in cerebral blood flow is a result of increased shunt circulation (Fig. 4), as a result of which the oxygen virtually bypasses the capillaries of the cerebral tissue (39,64).

In contrast to the action of papaverine, DH-ergotoxine mesylate stabilizes the EEG energy without influencing the cerebral blood flow. The improvement of EEG activity correlates well with local pO_2 microelectrode measurements in the brain cortex (Fig. 5). The histogram of pO_2 measurement shows that hypovolemic energy deprivation shifts the oxygen pressure of the brain cortex to lower values. Papaverine does not influence this situation.

Also in contrast to papaverine, DH-ergotoxine mesylate improves pO_2 distribution, returning it to preshock conditions. These pO_2 values confirm morphometric measurements of the network of brain capillaries (39,44) demonstrating pathophysiologic changes in the capillary parameters, caused by alterations of the microenvironment of the capillary bed, which are shifted by DH-ergotoxine to normal dimensions.

WHICH ARE THE NEUROCHEMICAL ACTIONS OF DIHYDROGENATED ERGOT ALKALOIDS?

The described pathophysiological studies have shown that DH-ergotoxine mesylate, as well as DH-ergonine, mitigates the deleterious effects of hypoxic

FIG. 3. Hypovolemic oligemia of the cat induced by bloodletting (down to a blood pressure of 45 mm Hg) with a 30–40% decrease of cerebral blood flow. The disturbed blood flow of the brain causes an initial EEG activation. **Control:** After an initial EEG activation a progressive decline of EEG activity, which is most pronounced in the alpha frequency range, can be observed (mean ± SEM, explanation see table 1; n = 7). **Hydergine:** The i.v. infusion of 0.08 mg/kg DH-ergotoxine mesylate (Hydergine) 20 min after the beginning of the hypovolemic oligemia causes a stabilization of the EEG energy at a 80–90% level. No influence of the drug on cerebral blood flow can be observed (mean ± SEM; n = 7). **Papaverine:** Intravenous infusion of 1 mg/kg papaverine returns cerebral blood flow to initial values. Despite this improvement of cerebral blood flow, the breakdown of the EEG energy occurs as in the controls (mean ± SEM; n = 7) (41).

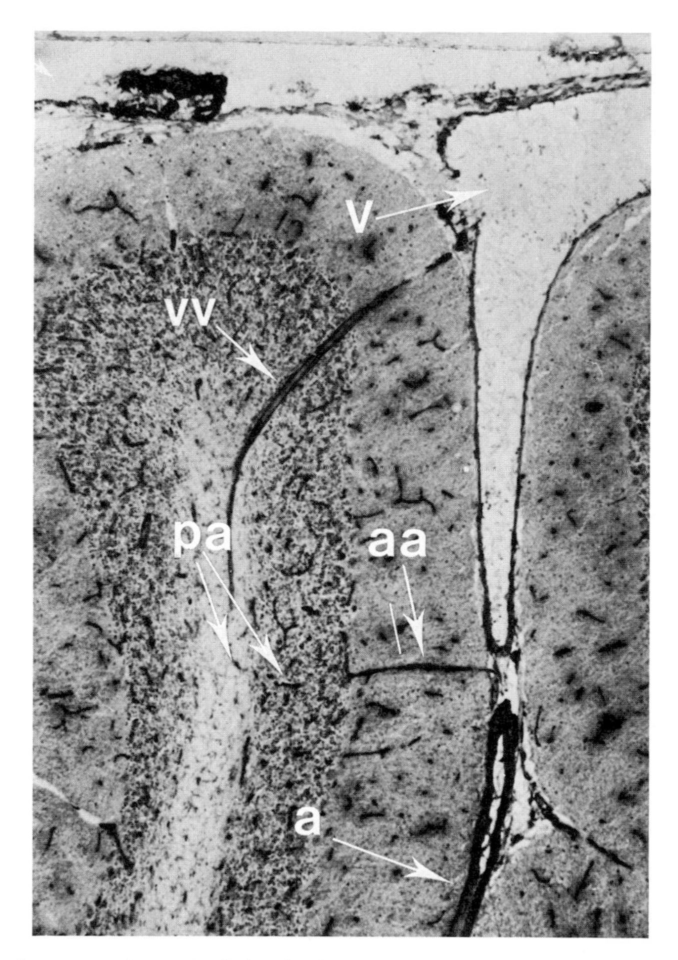

FIG. 4. Arteriovenous shunt circulation in the cerebellum caused by a vasoactive therapy. Dilatation mainly of the venous vessels. Enzyme histochemical staining of muscle containing vessels by ATPase reaction (hemalum counterstaining). a: Leptomeningeal artery; as: intracerebral artery; pa: precapillary arteries; vv: intracerebral venous vessels; v: leptomeningeal venous vessels. Mag. × 58.

energy deprivation in the brain. DH-ergot alkaloids require minimal threshold concentrations for interactions with enzymatic reactions as well as with specific receptor molecules (20,30,61). From these investigations arose the question of the tissue level of DH-ergot alkaloids—and of DH-ergotoxine mesylate, in particular—to be expected *in vivo.*

Autoradiographic and gradient centrifugation distribution studies with ^{3}H-DH-ergotoxine mesylate have demonstrated that this compound is accumulated at neuronal cells. Gradient centrifugation studies have shown that DH-ergotoxine mesylate is accumulated in the synaptosomal fraction at a rate of 60% (47,65).

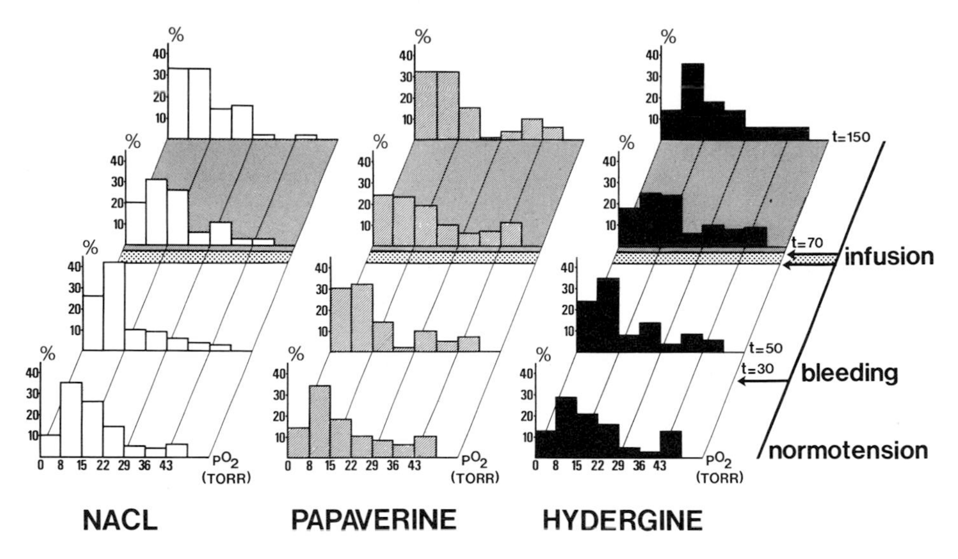

FIG. 5. Histogram of pO_2 microelectrode measurements in the cat brain cortex. The histogram shows seven classes, each of 7 mm Hg. The figure demonstrates the pO_2 distribution during the normotensive control period, 20 min after onset of hypovolemic oligemia, and 80 min after the end of drug infusion. Hypovolemic oligemia increases the incidence of low pO_2, values (0–15 mm Hg pO_2). This effect cannot be influenced by papaverine, but an increase of high pO_2 values can be seen (36–43 mm Hg pO_2) indicating an unevenly (shunt) perfused brain tissue. Administration of DH-ergotoxine mesylate (Hydergine) returns pO_2 distribution to the initial values of the normotensive state (41).

As shown by Iwangoff, et al. (48), DH-ergonine very rapidly attains a maximal brain level. The tissue level declines more slowly in the brain than in visceral organs. Repeated application of ^{3}H-DH-ergonine increases the retention of DH-ergonine in the brain (Fig. 6). These distribution studies show that a tissue level of DH-ergotoxine mesylate and of DH-ergonine is attained, which may possibly be related to the observed improvement of EEG activity.

In the following, evidence was given that the ameliorative effect of DH-ergotoxine mesylate is not due to increased metabolism, but to an improved metabolic economy of the brain tissue (30,66). Gähwiler (34a) demonstrated in nerve tissue cultures that DH-ergotoxine has a modulating effect on the spontaneous nerve cell activity with a regularization of the firing and an increase of the spike amplitudes. From studies by Markstein and Wagner (61,62) and Iwangoff, et al. (46), it can be concluded that DH-ergotoxine mesylate influences the feedback control system of cAMP level synthesis by inhibiting noradrenaline stimulation of the cAMP adenyl cyclase and also the cAMP splitting phosphodiesterase. Iwangoff and Enz (45) showed that inhibition of cAMP phosphodiesterase by DH-ergotoxine mesylate is more pronounced in brain than in visceral organs.

In order to understand more of the biological significance of cAMP phosphodi-

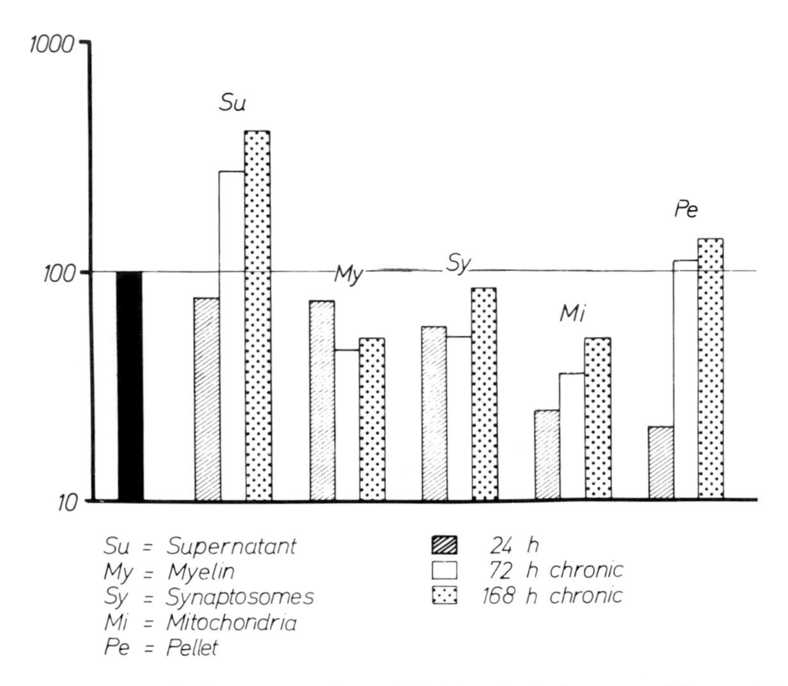

FIG. 6. Incorporation of ^{3}H-DH-ergonine (0.4 mCi/kg) into the brain cortex of the cat. Influence of repeated application (a daily injection for 3 or 7 days, once a day 2 mg/kg) on the different subfractions obtained by density gradient centrifugation. The values are indicated in relation to the 1-hr value (= 100) (48).

esterase inhibition by DH-ergotoxine mesylate, we studied the kinetics of this enzyme. Kinetic experiments on the cAMP phosphodiesterase system in the brain and other organs have clearly shown two Michaelis constants (K_m) as described by Thompson and Appleman (79). This indicates that there are two different active forms of cAMP phosphodiesterase, one active at low and the other at high cAMP levels. Accordingly, we distinguish a so-called low-K_m phosphodiesterase localization in membranes—especially in nerve endings (26)—and a high-K_m phosphodiesterase active mainly in the soluble fraction.

Iwangoff, et al. (46) demonstrated that dihydroergotoxine mesylate is a stronger inhibitor of the low-K_m than of the high-K_m phosphodiesterase (Fig. 7), which is more pronounced in the purified low-K_m phosphodiesterase than in the crude homogenate (29). This kind of phosphodiesterase inhibition may be of particular importance at normal cellular cAMP levels, but does not interfere with cAMP production by hormonal stimulation. It is relevant to the regulation of the basal cAMP level of the neuronal cell, prevents the intracellular cAMP level from dropping to zero and so improves the recovery of the neuronal cell (65).

The simultaneous inhibition of the catecholamine-stimulated cAMP adenyl

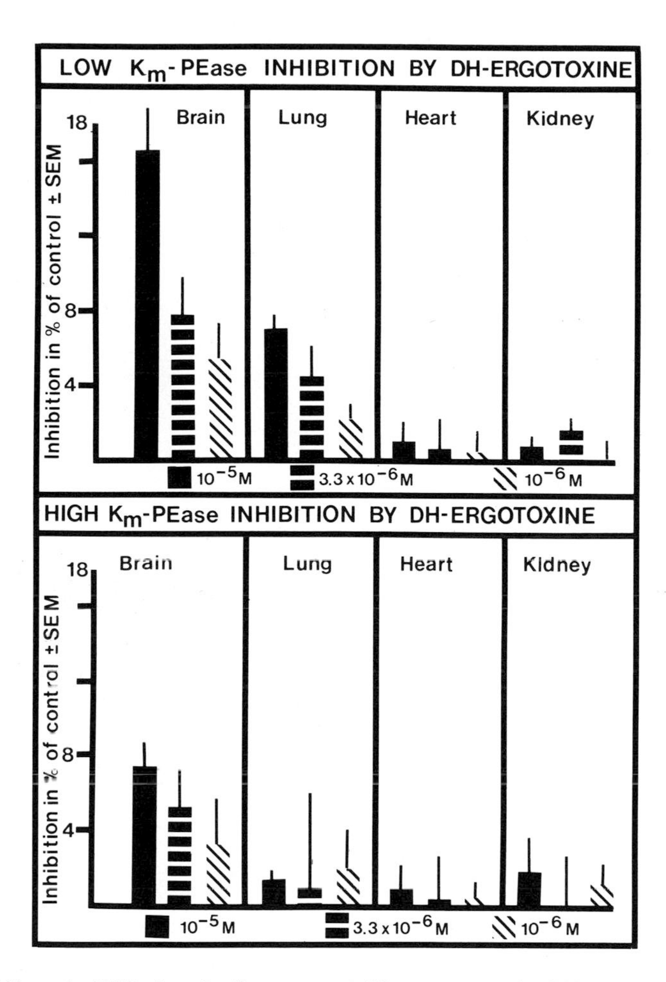

FIG. 7. Inhibition of cAMP phosphodiesterase of different organs by DH-ergotoxine mesylate. It can be seen that DH-ergotoxine is a stronger inhibitor of low-K phosphodiesterase of brain tissue (46).

cyclase and of the basal cAMP-splitting phosphodiesterase activity brings about a beneficial action of DH-ergotoxine mesylate on cAMP turnover (Fig. 8). DH-ergotoxine mesylate, however, also affects the noradrenaline-activated Na^+/K^+ ATPase (16,66). DH-ergotoxine inhibits noradrenaline effects on the $Na^+/K^\pm$ ATPase without suppressing its functional action. This helps to explain the suppressive effect of DH-ergotoxine mesylate on lactacidose caused by hypoxic energy deprivation.

In conclusion, DH-ergotoxine mesylate and DH-ergonine have introduced us to a group of compounds that are improving the metabolic economy of the central nervous system. These compounds—as well as bromocriptine, nicergoline, and lergotrile—have opened a new outlook in neuropharmacology.

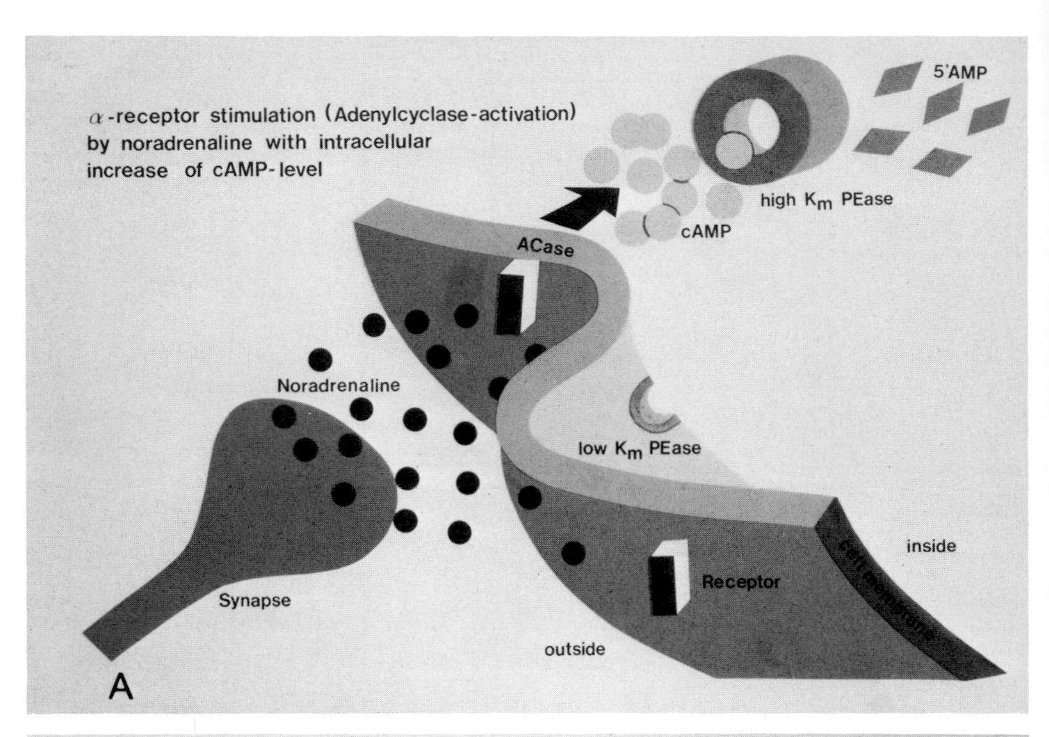

α-receptor stimulation (Adenylcyclase-activation)
by noradrenaline with intracellular
increase of cAMP-level
5'AMP
high Km PEase
cAMP
ACase
Noradrenaline
low Km PEase
inside
Receptor
cell membrane
Synapse
outside
A

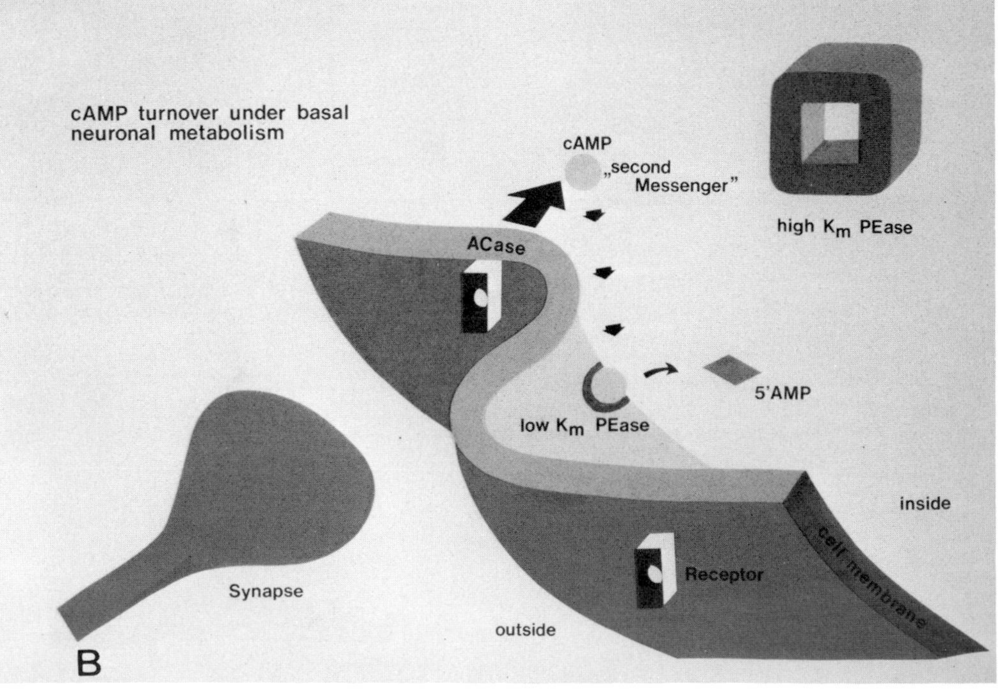

cAMP turnover under basal
neuronal metabolism
cAMP
„second
Messenger"
high Km PEase
ACase
5'AMP
low Km PEase
inside
cell membrane
Receptor
Synapse
outside
B

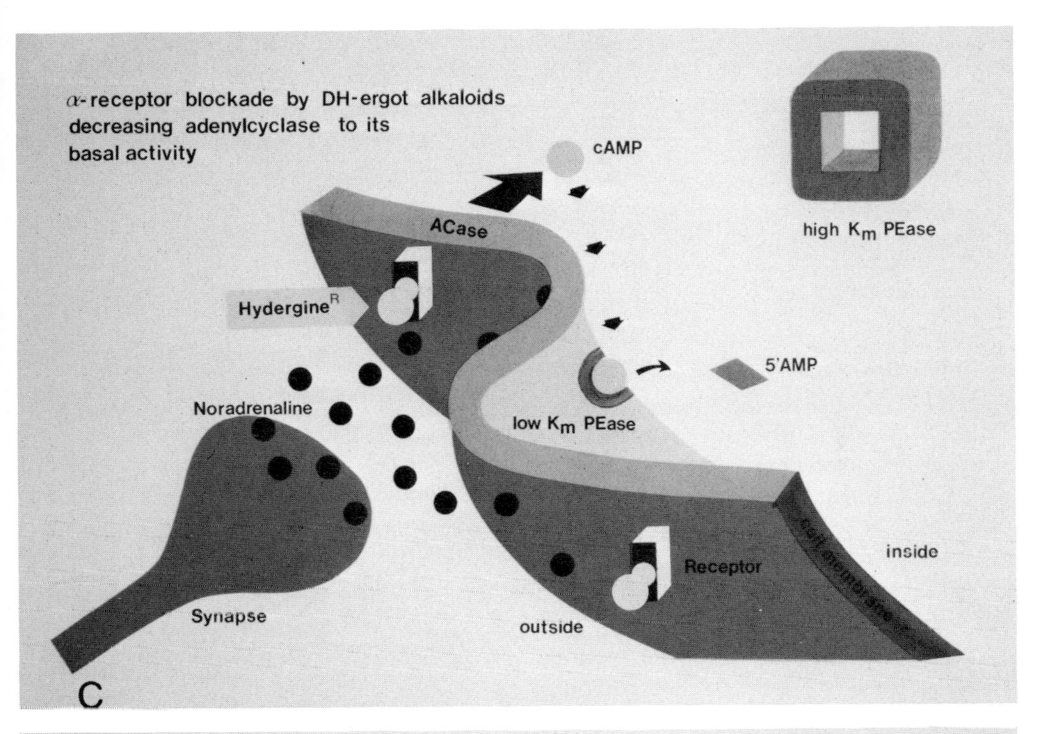

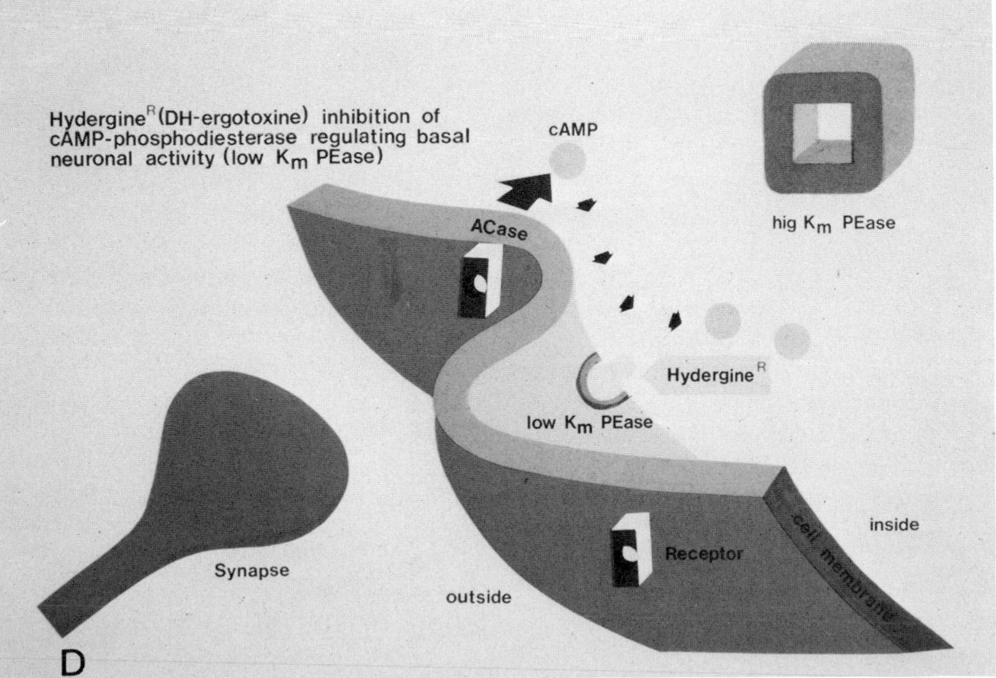

FIG. 8. Schematic representation of noradrenaline activated adenylcyclase **(A),** cAMP turnover under basal metabolism of the neuron **(B),** alpha-blocked by DH-ergotoxine with decrease of adenylcyclase to its basal activity **(C).** DH-ergotoxine inhibition of phosphodiesterase (PEase) splitting low-K_m phosphodiesterase regulating basal neuronal activity **(D)** (65).

WHICH TRENDS EXIST IN ERGOT PHARMACOLOGY OF GERONTOLOGICAL BRAIN RESEARCH?

Aging is a slow but continuous process of deviation from the situation of normal adult life (73), disposing the organism to a series of pathological or adverse reactions to environmental changes (72,74,78) as characteristic in drug-induced parkinsonism, which is the result of an antidopaminergic action. Today, well-defined animal models are available for testing dopaminergic drugs (1,2, 26,80), which may be useful in the treatment of Parkinson's disease. Anden and co-workers (1) were the first to observe the dopaminergic effect of apomorphine and its blockade by haloperidol. A series of dopaminergic ergot compounds were also studied in animals (20).

In monkeys, involuntary movements and tremor due to unilateral electrolytic-induced tegmental lesion can be effectively blocked by L-DOPA (36). More selective dopamine-like actions, different from the action of bromocriptine or levodopa, were demonstrated with a new nonpeptide ergoline derivative [a 9-10-Dihydro-6-methyl-8-beta-(2-pyridylthiomethyl)-ergoline] (49). Another effective dopaminergic ergoline derivative is lergotrile, which effectively inhibits prolactin release from the pituitary gland (18,19), an effect also exerted by bromocriptine (31,32,60). The dopaminergic effects of dihydroergotoxine mesylate (Hydergine^{-R}) were qualified as moderate and short-lasting (17,59), while bromocriptine has a long-lasting effect (59). The excellent correlation between these experimental animal observations and clinical findings (9,13,22,23,50,51,55, 56,57,82) confirmed the reliability of this model for the development of antiparkinsonian drugs.

Besides lergotrile and bromocriptine, nicergoline has recently attracted major interest in experimental pharmacology. Nicergoline, an ergoline derivative, proved to have effects on the central nervous system similar to those of dihydroergotoxine mesylate and dihydroergonine. Since 1970, nicergoline has been studied in a series of pharmacological experiments. Measurements of cortical blood flow and some basic metabolic parameters in the brain of cats and pigs were inconclusive. No correlation was found between cortical blood flow on the one hand and cortical pCO_2 and pH on the other (7,8).

In cats an improvement of the asphyxia-depressed EEG activity was shown (5). In curarized dogs with hypovolemic hypotension, nicergoline caused, during recovery, a decrease in the lactate/pyruvate ratio and an increase of the energy charge potential (4,6). In cats nicergoline caused shortening of the EEG recovery time after temporary ischemia (70) and the return of cortical evoked potentials to normal values (11).

Neurochemical studies in dogs demonstrated that nicergoline significantly increases the posthypoxic recovery of adenyl cyclase activity and of cAMP content of the brain (81). *In vivo* experiments in rat brain cortical slices showed an increase of adenyl cyclase activity (69). In the same rat experiments, inhibition of cAMP phosphodiesterase was seen (69). The prevention of an ischemic increase of free fatty acids by nicergoline (150 ug/kg/cat) tends in the same direction (70).

The increasing interest in the neuropharmacology of ergot alkaloids is mainly linked to its interaction with biogenic amines of the brain (20,33,34,59), its specific actions on the brain stem (14,25,49,52), and its beneficial effects on hypoxic energy deprivation of the brain (39,41,44). The ergot alkaloid pharmacology of the central nervous system is a new area in neuropharmacology which is of particular interest in experimental gerontology. Further pharmacological research with ergot alkaloids will provide us in the near future with a series of new results in neuropharmacology.

SUMMARY

The difficulties of neuropharmacology in experimental gerontology are outlined. It is shown that aging animals can be used today in pharmacology only for a few selected purposes, due to the limited number of aging animals, high costs, and the species limitation to rats and mice. For routine pharmacology, pathophysiological animal models are mainly used.

Hypoxic energy deprivation of the brain is a useful approach for investigating drugs active on symptoms of aging. Using the models of temporary cerebral ischemia and hypovolemic oligemia, the effects of dihydroergotoxine mesylate and dihydroergonine were studied. It could be shown that both compounds improve EEG energy and lower arteriovenous lactate difference in brains disturbed by hypoxic energy deprivation. Papaverine decreases EEG energy, although it improves cerebral blood flow, whereas dihydroergotoxine mesylate preserves EEG activity at a constant energy level. pO_2 measurements in the cerebral cortex establish this observation, showing, with dihydroergotoxine mesylate, a return of the tissue pO_2 distribution towards that of a normotonic control. Papaverine has no improving effect on the declined pO_2 values.

Neurochemical investigations demonstrated that the metabolic improvement by the investigated dihydroergotoxine may be the result of a reduction of both a catecholamine-activated ATPase, a catecholamine-activated cAMP adenyl cyclase and inhibition of cAMP low-K_m phosphodiesterase in the brain. These actions of dihydroergotoxine mesylate manifest themselves as a normalization of a norepinephrine rise in ATP turnover and as a beneficial effect on the neuronal cAMP level through regulation of the synthesizing and degrading enzymes.

In the last section, trends in ergot pharmacology of gerontological brain research are outlined. In particular, the dopaminergic compounds lergotrile and bromocriptine are mentioned for treatment of Parkinson's disease.

Finally, nicergoline is described as a drug that has in recent years attracted interest in the field of pharmacological intervention in the aging process of the brain. Nicergoline is an interesting example of the action of an ergoline derivative on brain metabolism under hypoxic deprivation.

REFERENCES

1. Anden, N. E., Dahlström, A., Fuxe, K., and Larsson, K. (1966): Functional role of the nigro-striatal dopamine neurons. *Acta Pharmacol. Toxicol. (Kbh.)*, 24:263–274.
2. Anden, N. E., Rubenson, A., Fuxe, K., and Hökfelt, T. (1967): Evidence for dopamine receptor stimulation by apomorphine. *J. Pharm. Pharmacol.*, 19:627–629.
3. Benzi, G. (1972): Effect of lysergide and nimergoline on glucose metabolism investigated on the dog brain isolated in-situ. *J. Pharm. Sci.*, 61:348–352.
4. Benzi, G. (1975): An analysis of the drugs acting on cerebral energy metabolism. *Jpn. J. Pharmacol.*, 25:251–261.
5. Benzi, G., Manzo, L., De Bernardi, M., Ferrara, A., Sanguinetti, L., Arrigoni, E., and Berté, F. (1971): Action of lysergide, ephedrine, nimergoline on brain metabolizing activity. *J. Pharm. Sci.*, 60:1320–1324.
6. Benzi, G., Arrigoni, E., Manzo, L., De Bernardi, M., Ferrara, A., Panceri, P., and Berté, F. (1973): Estimation of changes induced by drugs in cerebral energy-coupling processes in situ in the dog. *J. Pharm. Sci.*, 62:758–764.
7. Bienmüller, H., and Betz, E. (1970): Die Regulation der lokalen cortikalen Gehirndurchblutung bei Injektion von Noradrenalin, Pentobarbital und Adrenolitiko. *Aerztliche Forschung, Jg.*, 24:97–111.
8. Bienmüller, H., and Betz, E. (1972): Wirkung von Nicergoline auf die Hirndurchblutung. *Arzneim.-Forsch.*, 22:1367–1372.
9. Birkmayer, W., and Hornykiewicz, O. (1961): Der L-3, 4-dihydroxyphenylalanin(=DOPA)-Effekt bei der Parkinson-Akinese. *Wien. Klin. Wschr.*, 73:787–788.
10. Boismare, F., Boquet, J., and Courtin, J. (1975): Aspects rheographiques et pharmacologiques du syndrome subjectif chez le traumatisme cranio-cervical. *Agressologie*, 16:23–26.
11. Boismare, F., and Lorenzo, J. (1975): Study of the protection afforded by nicergoline against the effects of cerebral ischemia in the cat. *Arzneim.-Forsch.*, 25:410–413.
12. Boismare, F., and Streichenberger, G. (1974): The action of ergot alkaloids (ergotamine and dihydroergotoxine) on the functional effects of cerebral ischemia in the cat. *Pharmacology*, 12:152–159.
13. Calne, D. B., Teychenne, P. F., Claveria, L. E., Eastman, R., Greenacre, J. K., and Petrie, A. (1974): Bromocriptine in Parkinsonism. *Br. Med. J.*, IV:442–444.
14. Cerletti, A. (1959): Comparison of abnormal behavioural states by psychotropic drugs in animal and man. *Neuropsychopharmacol.*, 1:117–123.
15. Cerletti, A., Emmenegger, H., Enz, A., Iwangoff, P., Meier-Ruge, W., and Musil, J. (1974): Effects of ergot DH-alkaloids on the metabolism and function of the brain. An approach based on studies with DH-ergonine. In: *Central Nervous System—Studies on Metabolic Regulation and Function*, ed. E. Genazzani and H. Herken, pp. 201–212. Springer-Verlag, Berlin-Heidelberg-New York.
16. Chappuis, A., Enz, A., and Iwangoff, P. (1975): The influence of adrenergic effectors on the cationic pump of brain cell membrane. *Triangle*, 14:93–98.
17. Clemens, J. A., and Fuller, R. W. (1978): Chemical manipulation of some aspects of aging. In: *Pharmacological Intervention in the Aging Process*, edited by J. Roberts, R. C. Adelman, and V. J. Cristofalo, pp. 787–206. Plenum Press, New York–London.
18. Clemens, J. A., Smalstig, E. B., and Shaar, C. J. (1975): Inhibition of prolactin secretion by lergotrile mesylate mechanism of action. *Acta Endocrinol.*, 79:230–237.
19. Clemens, J. A., Smalstig, E. B., and Sawyer, B. D. (1976): Studies on the role of the preoptic area in the control of reproductive function in the rat. *Endocrinol.*, 99:728–735.
20. Corrodi, H., Fuxe, K., Hökfelt, T., Lindbrink, P., and Ungerstedt, U. (1973): Effect of ergot drugs on central catecholamine neurons: Evidence for a stimulation of central dopamine neurons. *J. Pharm. Pharmacol.*, 25:409–412.
21. Costall, B., and Naylor, R. J. (1975): A comparison of circling models for the detection of anti-Parkinson activity. *Psychopharmacol.*, 41:57–64.
22. Cotzias, G. C., Miller, S. T., Nicholson, A. R., Jr., Maston, W. H., and Tang, L. C. (1974): Prolongation of the life-span in mice adapted to large amounts of L-dopa. *Proc. Natl. Acad. Sci. USA*, 71:2466–2469.
23. Debono, A. G., Donaldson, I., Marsden, C. D., and Parkes, J. D. (1975): Bromocriptine in Parkinsonism. *Lancet*, II:987–988.

24. Denckla, W. D. (1975): A time to die. *Life Sci.,* 16:31–44.
25. Depoortere, H., and Loew, D. M. (1972): Alterations in the sleep/wakefulness cycle in rats after administration of (−)-LSD or BOL 48: a comparison with (+)-LSD. *Br. J. Pharmacol.,* 44:354P–355P.
26. De Robertis, H., Rodriguez de Lores Arnaiz, G., Butcher, R. W., and Sutherland, E. W. (1967): Subcellular distribution of adenylcyclase and cyclic phosphodiesterase in rat brain cortex. *J. Biol. Chem.,* 242:3487–3493.
27. Emmenegger, H., and Meier-Ruge, W. (1968): The actions of Hydergine on the brain. *Pharmacol.,* 1:65–78.
28. Emmenegger, H., Taeschler, M., and Cerletti, A. (1963): Neue Möglichkeit der isolierten Hirndurchströmung. *Helv. Physiol. Pharmacol. Acta,* 21:239–244.
29. Enz, A., Iwangoff, P., and Chappuis, A. (1978): The influence of dihydroergotoxine mesylate on the low-K_m phosphodiesterase of cat and rat brain in vitro. *Gerontol.,* 24(Suppl. 1):115–125.
30. Enz, A., Iwangoff, P., Markstein, R., and Wagner, H. (1975): The effect of Hydergine on the enzyme involved in cAMP turnover in the brain. *Triangle,* 14:90–92.
31. Flückiger, E. (1975): Pharmacology of prolactin secretion. *Acta Endocrinol.,* 78(Suppl. 193): 164–165.
32. Flückiger, E., Döpfner, W., Marko, M., and Niederer, W. (1976): Effects of ergot alkaloids on the hypothalamic-pituitary axis. *Postgrad. Med.* J., 52(Suppl. 1):57–61.
33. Freedman, D. X. (1961): Effects of LSD-25 on brain serotonin. *J. Pharmacol. Exp. Ther.,* 134:160–166.
34. Fuxe, K., Hökfelt, T., and Ungerstedt, U. (1970): Morphological and functional aspects of control monoamine neurons. In: *International Review of Neurobiology,* Vol. 13, ed. C. C. Pfeiffer and J. R. Smythies, pp. 93–126. Academic Press, New York.
34a. Gäwhiler, B. H. (1978): Dihydroergotoxine-induced modulation of spontaneous activity of cultured rat Purkinje cells. *Gerontology* (Suppl. 1), 24:71–75.
35. Geiger, A. (1958): Correlation of brain metabolism and function by the use of a brain perfusion method in situ. *Physiol. Rev.,* 38:1–20.
36. Goldstein, M., Battista, A., Ohmoto, T., Anagnoste, B., and Fuxe, F. (1973): Tremor and involuntary movements in monkeys: Effect of L-dopa and of a dopamine receptor stimulating agent. *Science,* 179:816–817.
37. Gygax, P., Emmenegger, H., and Stosseck, K. (1975a): Quantitative determination of cortical microflow and EEG in graded hypercapnia. In: *Cerebral Circulation and Metabolism,* ed. F. W. Langfitt, L. C. McHenry, Jr., M. Reivich, and H. Wollman, pp. 371–374. Springer-Verlag, New York.
38. Gygax, P., Stosseck, K., Emmenegger, H., and Schweizer, A. (1975b): Influence of anesthetics on cortical microflow and EEG in arterial hypotension. Comparison between pentobarbital and N_2O/O_2-anesthesia. In: *Blood Flow and Metabolism in the Brain,* ed. M. Harper, B. Jennet, D. Miller, and J. Rowan, pp. 11.14–11.15. Churchill-Livingstone, Edinburgh–London–New York.
39. Gygax, P., Hunziker, O., Schulz, U., and Schweizer, A. (1975c): Experimental studies on the action of metabolic and vasoactive substances in the brain. *Triangle,* 14:80–89. (1975).
40. Gygax, P., and Schweizer, A. (1975d): The influence of anesthesia on cortical microflow and EEG in arterial hypotension. *Arzneim.-Forsch.,* 25:1678.
41. Gygax, P., Wiernsperger, N., Meier-Ruge, W., and Baumann, Th. (1978): Effect of Papaverine and dihydroergotoxine mesylate on cerebral microflow, EEG, and po_2 in oligemic hypotension. *Gerontol.,* 24:14–22.
42. Hossmann, K. A. (1971): Cortical steady potential impedance and excitability changes during and after total ischemia of cat brain. *Exp. Neurol.,* 32:163–175.
43. Hossmann, V., and Hossmann, K. A. (1973): Return of neuronal functions after prolonged cardiac arrest. *Brain Res.,* 60:423–438.
44. Hunziker, O., Emmenegger, H., Frey, H., Schulz, U., and Meier-Ruge, W. (1974): Morphometric characterization of the capillary network in the cat's brain cortex: A comparison of the physiological state and hypovolemic conditions. *Acta Neuropathol. (Berl.),* 29:57–63.
45. Iwangoff, P., and Enz, A. (1973): The brain specific inhibition of the cAMP-phosphodiesterase (PEase) of the cat by dihydroergot alkaloids in vitro. *IRCS(Intl. Res. Comm. System Med. Sci.)* (73–4):3–10–9.

46. Iwangoff, P., Enz, A., and Chappuis, A. (1975): Inhibition of cAMP-phosphodiesterase of different cat organs by DH-ergotoxine in the micromolar substrate range. *IRCS:* 3:403.
47. Iwangoff, P., Meier-Ruge, W., Schieweck, Chr., and Enz, A. (1976): The uptake of DH-ergotoxine by different parts of the cat brain. *Pharmacol.,* 14:27–38.
48. Iwangoff, P., Enz, A., and Meier-Ruge, W. (1978): Incorporation after single and repeated application of radioactive labelled DH-ergot alkaloids in different organs of the cat, with special reference to the brain. *Gerontol.,* 24:126–138.
49. Jaton, A., Loew, D. M., and Vigouret, J. M. (1976): CF-25-397 (9,10-Didehydro-6-methyl-8-beta-(2-pyridylthiomethyl)-ergoline): A new central dopamine receptor agonist. *Br. J. Pharmacol.,* 56:371P.
50. Kartzinel, R. (1976): Bromocriptine and levodopa (with or without carbidopa) in Parkinsonism. *Lancet,* II:272–275.
51. Kartzinel, R., Perlow, M., Carter, A. C., Chase, T. N., and Calne, D. B. (1976): Metabolic studies with bromocriptine in patients with idiopathic Parkinsonism and Huntington chorea. *Arch. Neurol.,* 33:384.
52. Konzett, H., and Rothlin, E. (1953): Investigations on the hypotensive effect of the hydrogenated ergot alkaloids. *Br. J. Pharmacol. Chemother.,* 8:201–207.
53. Kormendy, Ch.G. (1974): Ageing: Can research do something about it? Pharmaceutical implications. In: *Altern,* ed. D. Platt, pp. 115–126. Edited by Schattauer-Verlag, Stuttgart and New York.
54. Kovach, A. G. B., and Sandor, P. (1976): Cerebral blood flow and brain function during hypotension and shock. *Ann. Rev. Physiol.,* 39:571–596.
55. Lieberman, A. N., Miyamoto, T., and Battista, A. (1975): Studies on the anti-Parkinsonian efficacy of lergotrile. *Neurol. (Minneap.)* 25:459–462.
56. Lieberman, A. N., Kupersmith, M., Vogel, B., Goodgold, A., and Goldstein, M. (1976a): Treatment of Parkinson's disease with bromocriptine. *Clin. Res.,* 24:511A.
57. Lieberman, A. N., Kupersmith, M., Estey, E., and Goldstein, M. (1976b): Lergotrile in Parkinson's disease. Lancet, II:515–516.
58. Ljunggren, B., Norberg, K., and Siesjö, B. K. (1974): Influence of tissue acidosis upon restitution of brain energy metabolism following total ischemia. *Brain Res.,* 173–186.
59. Loew, D. M., Vigouret, J. M., and Jaton, A. L. (1976): Neuropharmacological investigations with two ergot alkaloids, Hydergine and bromocriptine. *CNS Effects of Ergot Alkaloid Postgrad. Med. J.,* 52(Suppl. 1):40–46.
60. Lutterbeck, P. M., Pryor, J. S., Varga, L., and Wenner, R. (1971): Treatment of non-puerperal galactorrhoea with an ergot alkaloid. *Br. Med. J.,* III:228–229.
61. Markstein, R., and Wagner, H. (1975): The effect of dihydroergotoxine, phentolamine and pindolol on catecholamine-stimulated adenylcyclase in rat cerebral cortex. *FEBS Lett.,* 55:275–277.
62. Markstein, R., and Wagner, H. (1978): Effect of dihydroergotoxine on cyclic AMP generating systems in rat cerebral cortex slices. Gerontol., 21(Suppl. 1):94–105.
63. Meier-Ruge, W. (1975): Experimental pathology and pharmacology in brain research and aging. *Life Sci.,* 17:1627–1636.
64. Meier-Ruge, W. (1976): Stoffwechsel des Altersgehirns. *Aerztliche Praxis,* 28:648–653.
65. Meier-Ruge, W., and Iwangoff, P. (1976): Biochemical effects of ergot alkaloids with special reference to the brain. *Postgrad. Med. J.,* 52*(Suppl.* 1):47–54.
66. Meier-Ruge, W., Enz, A., Gygax, P., Hunziker, O., Iwangoff, P., and Reichlmeier, K. (1975): Experimental pathology in basic research of the aging brain. In: *Genesis and Treatment of Psychologic Disorders in the Elderly, Aging, Volume 2,* ed.: S. Gershon and A. Raskin, pp. 55–126. Raven Press, New York.
67. Meyer, J. S., Welch, K. M. A., Titus, J. L., Suzuki, M., Kim, H. S., Perez, F. I., Mathew, N. T., Gedye, J. L., Hrastnik, F., Miyakawa, Y., Achar, V. S., and Dodson, R. F. (1976): Neurotransmitter failure in cerebral infarction and dementia. In: *Neurobiology of Aging, Aging, Volume 3,* ed. R. D. Terry and S. Gershon, pp. 121–138. Raven Press, New York.
68. Mitruka, B. M., Rawnsley, H. M., and Vadehra, D. V. (1976): Animals for medical research. In: *Animal Models in Gerontology,* pp. 425–462. Wiley and Sons, New York–London–Sidney.
69. Montecucchi, P. (1976): Stimolazione della formazione di AMP ciclico nel cervello di ratto in vitro da parte della nicergolina. *Farmaco. Ed. Prat.,* 31:10–17.
70. Moretti, A., Pegrassi, L., and Suchowsky, G. K. (1974): Effect of nicergoline on some ischemia-induced metabolic changes in the brain of cat. In: *Central Nervous System: Studies on Metabolic*

Regulation and Function, ed. E. Genazzani and H. Herken, pp. 213–216. Springer-Verlag, Heidelberg–New York.
71. Myers, R. E. (1975): Four patterns of perinatal brain damage and their conditions of occurrence in primates. *Adv. Neurol.,* 10:223–234.
72. Ordy, J. M., and Brizzee, K. R. (1976): Cell loss, neurotransmitters and hormones as regulators of aging. *Age,* 3: Abstr. 25, Amer. Aging Assoc., Inc.
73. Peng, M.-T., Peng, Y., and Chen, F.-N. (1977): Age-dependent changes in the oxygen consumption of the cerebral cortex, hypothalamus, hippocampus, and amygdaloid in rats. *J. Gerontol.,* 32:517–522.
74. Samorajski, T. (1976): How the human brain responds to aging. *J. Amer. Geriatr. Soc.,* 24: 4–14.
75. Schneider, H., Ballowitz, L., Schachinger, H., Hanefeld, F., and Dröszus, J.-U. (1975): Anoxic encephalopathy with predominant involvement of basal ganglia, brain stem and spinal cord in the perinatal period. *Acta Neuropathol. (Berl.),* 32:287–298.
76. Siesjö, B. K., and Nilsson, L. (1971): The influence of arterial hypoxemia upon labile phosphates and upon extracellular and intracellular lactate pyruvate concentrations in rat brain. *Scand. J. Clin. Lab. Invest.,* 27:83–96.
77. Siesjö, B. K., and Plum, F. (1973): Pathophysiology of anoxic brain damage. In: *Biology of Brain Disfunction,* ed. G. Gaull, pp. 219–372. Plenum Press, New York.
78. Stein, D. G., and Firl, A. C. (1976): Brain damage and reorganization of function in old age. *Exp. Neurol.,* 52:157–167.
79. Thompson, W. J., and Appleman, M. M. (1971): Multiple cyclic nucleotide phosphodiesterase activities from rat brain. *Biochem.,* 10:311–316.
80. Ungerstedt, U. (1971): Postsynaptic supersensitivity after 6-hydroxy-dopamine induced degeneration of the nigro-neostriatal dopamine system. *Acta Physiol. Scand.,* Suppl. 367:69–93.
81. Villa, R. F. (1975): Effect of hypoxia on the cerebral energy state. *Farmaco. Ed. Sci.,* 30:561–567.
82. Yahr, M. D., Duvoisin, R. C., Schear, M. J., and Barret, R. E. (1969): Treatment of Parkinsonism with levo-DOPA. *Arch. Neurol.(Chi.),* 21:343–354.

Physiology and Cell Biology of Aging
(Aging, Volume 8), edited by A. Cherkin, et al.
Raven Press, New York © 1979.

Proper Use of Laboratory Rats and Mice in Gerontological Research

Carel F. Hollander

Institute for Experimental Gerontology TNO, Lange Kleiweg 151, Rijswijk, The Netherlands

Next to studies in man, animal models are required to understand the physiology and pathophysiology of aging in organs and organ systems. With regard to the question of extrapolation of data from animals to man, we must realize that the fundamental aspects of the aging process are the same in all mammals. Models are used to establish facts. Once these facts are established, they may lead us either to a better understanding of what happens in man or to further experimentation (10).

Animal models allow us to conduct studies in such a manner that most variables that can influence the aging process can be controlled. Mammals with relatively short life-spans are needed. The animals should be large enough to permit the collection of adequate samples of body fluids and tissues and small enough to make sure that the cost of production and of maintenance of the large numbers needed for study remain relatively low. In selecting the appropriate animals for gerontological research, the following should be kept in mind: they should be of good quality, and information should be available on their expected life-spans and their age-associated disease patterns. The laboratory rat and mouse are animals that are used extensively in biomedical research and that fulfill most of the requirements mentioned above. Both species are becoming more widely used in gerontological research, and a considerable body of knowledge has been collected on them (2,3,5–8,11–13,15,16,18).

It is mainly the type of study to be conducted that defines the required health status of the animal. The various terms used internationally in biomedical research to classify the quality or health status have been described by Wostmann (19). For use in aging research, the animals should at least be free of infectious disease and, as mentioned above, be of good quality. Good quality is often difficult to define and may mean different things to different investigators. For instance, Specified Pathogen Free (SPF) rats and mice are widely used for aging research. However, the microbiological status of SPF animals can differ tremendously among institutes (11), and this can complicate comparison of data from different laboratories. Furthermore, the costs of producing and maintaining SPF animals add an extra financial burden to the already expensive longevity studies in these animals.

In order to reduce the costs of longevity studies for aging research (as well as to lower the barrier to such an extent that the investigator can frequently visit his animals without compromising the good quality of the animals), the following scheme for rearing and maintaining rats for aging research at the Institute for Experimental Gerontology TNO has been employed in the past years and is still used with minor modifications (for detailed information, see refs. 2,11,17). Aging studies were done with cohorts of inbred rats that were born and reared in a closed colony system under strict SPF conditions. In a "closed colony" no animals from outside are permitted to enter without undergoing a rigid quarantine procedure. Animals used for aging research were either male and female virgins or retired breeders. Both virgins (at an age of approximately 12 weeks) and retired breeders (at an age of approximately 8 months) were transferred from SPF quarters to conventional animal rooms in which the longevity studies were conducted. Since the conditions in the animal rooms are also strictly controlled, we have used the term "clean conventional" for these animals. The term "clean conventional rodent" has also been used by Brennan (1) and Flynn (9). Unfortunately, neither of these authors further define this term in their publications.

In the past, mice bred and maintained under conventional conditions were employed for aging research at the Institute for Experimental Gerontology TNO. Since it has been shown that it is feasible to conduct aging research under the conditions described above for rats, an identical health regimen will be employed for longevity studies in mice. This regimen also reduces the costs of these studies while avoiding a number of problems related to a strict SPF operation in longevity studies.

It is beyond the scope of this paper to provide detailed information on the results of the longevity studies in rats conducted at the Institute for Experimental Gerontology TNO. For detailed information, the reader is referred to a monograph prepared by Burek (2) on this subject. However, some points of interest for those who are embarking on longevity studies on these animals will be given below.

Recent results of a longevity study of two different inbred strains, viz., the WAG/Rij and the BN/Bi and their (WAG × BN)F1 hybrid have shown that male rats do not always have shorter life-spans than females. The observed 50% survival age for the three strains was approximately 30 months, and the maximum survival age observed was in the order of approximately 42 months. Only in the case of the male WAG/Rij rat were both ages lower. Since no difference in survival between the F1 hybrid and its parental strains (with the exception of the male WAG/Rij rat) was observed, one should be cautious about employing F1 hybrids for aging studies only on the assumption that their "vigor" is greater than that of the parental strains and that the hybrid may therefore live longer. In addition, we found that the F1 hybrid showed no greater resistance to an infectious disease than did either of the parental strains, as shown by the 50% survival ages to be discussed below (Table 1). Furthermore, one should be very careful in interpreting data from aging studies in which an old rat is defined as such when it is 24 months of age and no further information on the survival of that strain is provided. The same observations on the survival of males and females have been made in a selected number of inbred strains of mice (12).

It is my firm belief, therefore, that, in a number of aging studies on rats and mice conducted in the past, data were obtained and interpretations were made which were not relevant to aging. This is partly due to the fact that the animals were not free of infectious diseases and partly because of the generally held opinion that a rat or a mouse of 24 months of age is an "old" animal. Recent data obtained from our life-span studies on long-lived strains of inbred rats and mice free of infectious disease showed the occurrence of multiple disease patterns in old animals, including both neoplastic and nonneoplastic diseases. This is especially relevant since a multiplicity of diseases is one of the characteristics of aging in man, and this observation increases the value of the rodent model for gerontological research. Detailed information on the age-associated disease patterns in the three strains of rats mentioned above are provided by Burek (2). See ref. (1a) for our data from the long-lived strains of inbred mice.

Finally, an example of how a change in the quality of animals used can dramatically change the survival characteristics will be presented. At the end of 1974 and early in 1975, an acute Sendai virus infection occurred simultaneously in the rodent breeding colony and in the colony of aging rats and mice (for detailed information see 4,20). Table 1 shows the effect of this infectious disease on the 50% survival ages of the three strains of rats used for aging research. The 50% survival age of animals born in early 1973 was not changed compared to data obtained from animals born prior to 1973. However, the 50% survival age of animals born in late 1975 was dramatically affected. This was caused by the gradual development of chronic respiratory disease (CRD) with bronchiectasis, lung abscess formation, and severe peribronchial lymphoid cuffing, all of which emerged after the Sendai virus infection had entered the rat colony. It is thought that the observed CRD is caused by a mycoplasma infection, probably

TABLE 1. *The 50% survival ages as a function of the birth period of 2 strains of rats and their F_1 hybrid*

	50% survival age (months) of aging rats born early 1973		50% survival age (months) of aging rats born late 1975	
	♀	♂	♀	♂
WAG/Rij	29.3	23.3	21.5	22.0
BN/Bi	30.3	31.4	15.2	11.4
(WAG × BN) F_1	27.5	28.8	15.1	15.9

triggered off by the preceding Sendai infection. However, proof for this assumption is still lacking at present.

A crucial question is how a Sendai virus infection could occur in these closed colonies of rats and mice. The origin of this infection in the colony is still unknown. A possible explanation is that is was introduced by contaminated tumor material from mice imported from a persistently Sendai virus-infected mouse colony from outside the Institute. The tumor-bearing mice were kept in quarantine but were not examined for the presence of Sendai virus infection. In fact, when these tumor-bearing mice were imported, it was unknown to us that the colony of origin of these mice was infected with Sendai-virus. Subsequently, the mice were killed and the tumors were transplanted into mice from our colony. Recently, it was reported elsewhere that Sendai virus has been found in tumor tissue of mice (14). This stresses the importance of strict quality control measures for animals used for aging research as well as for those brought into the area from outside. In fact, the occurrence of the Sendai virus infection forced us to phase out the existing colonies of rats and mice and to build up these colonies again from newly originated SPF colonies.

It is hoped that the issues raised above will make gerontologists more aware that both the quality of the animals used in their experiments, as well as the knowledge of their survival characteristics and age-associated disease patterns, are essential for the proper use of rodents in aging research.

REFERENCES

1a. Blankwater, M. J. (1978): Survival data and age-related pathology of CBA, C57Bl/Ka and NZB mice. In: Ageing and the Humoral Response in Mice, pp. 61–73. Institute for Experimental Gerontology TNO, Rijswijk, The Netherlands.

1. Brennan, P. C. (1972): Standards and procedures for the long term maintenance of microbial stability of the laboratory rodent. In: *Development of the Rodent as a Model System of Aging,* ed. Don C. Gibson, pp. 19–22. DHEW Publication No. (NIH) 72–121, Bethesda, Maryland.
2. Burek, J. D. (1978): Pathology of aging rats. A morphological and experimental study of the age-associated lesions in aging BN/Bi, WAG/Rij and (WAG × BN)F_1 rats. Thesis, Utrecht, The Netherlands.
3. Burek, J. D., and Hollander, C. F. (1978): Use in aging research. In: *The Laboratory Rat,* ed. H. J. Baker, J. R. Lindsey, and S. Weisbroth, Chapter 24. Academic Press, New York *(in press).*
4. Burek, J. D., Zurcher, C., Van Nunen, M. C. J., and Hollander, C. F. (1977): A naturally occurring epizootic caused by Sendai virus in breeding and aging rodent colonies. II. Infection in the rat. *Lab. Anim. Sci.,* 27:963–971.
5. Cohen, B. J. (1968): Effects of environment on longevity in rats and mice. In: *The Laboratory Animal in Gerontological Research,* ed. T. W. Harris, pp. 21–29. Publication 1591, National Academy of Sciences, Washington, D.C.
6. Cohen, B. J., and Anver, M. R. (1976): Pathological changes during aging in the rat. In: *Special Review of Experimental Aging. Progress in Biology,* ed. M. F. Elias, B. E. Eleftheriou, and P. K. Elias, pp. 379–403. EAR, Inc., Bar Harbor, Maine.
7. Coleman, G. L., Barthold, S. W., Osbaldiston, G. W., Foster, S. J., and Jonas, A. M. (1977): Pathological changes during aging in barrier-reared Fischer 344 male rats. *J. Gerontol.,* 32:258–278.
8. Festing, M. F., and Blackmore, D. K. (1971): Life span of specified-pathogen-free (MRC Category 4) mice and rats. *Lab. Anim.,* 5:179–192.

9. Flynn, R. J. (1972): Development and maintenance of laboratory rodents to meet the needs of aging research. In: *Development of the Rodent as a Model System of Aging,* ed. Don C. Gibson, pp. 13–17. DHEW publication no. (NIH) 72–121, Bethesda, Maryland.
10. Heston, W. E. (1972): The basis for selection and study of rodent model systems. In: *Development of the Rodent as a Model System of Aging,* ed. Don C. Gibson, pp. 7–10. DHEW Publication no. (NIH) 72–121, Bethesda, Maryland.
11. Hollander, C. F. (1976): Current experience using the laboratory rat in aging studies. *Lab. Anim. Sci.,* 26:320–328.
12. Hollander, C. F., and Burek, J. D. (1978): Animal models in gerontology. In: *Lectures on Gerontology,* ed. A. Viidik, Volume I, Chapter 14. Academic Press, London *(in press).*
13. Kunstýř, I., and Leuenberger, H. G. (1975): Gerontological data of C57BL/6J mice. I. Sex differences in survival curves. *J. Gerontol.,* 30:157–162.
14. Parker, J. C. (1978): *personal communication.*
15. Rowlatt, C., Chesterman, F. C., and Sheriff, M. U. (1976): Lifespan, age changes and tumour incidence in an ageing C57BL mouse colony. *Lab. Anim.,* 10:419–442.
16. Smith, G. S., Walford, R. L., and Mickey, M. R. (1973): Lifespan and incidence of cancer and other diseases in selected long-lived inbred mice and their F1 hybrids. *J. Natl. Cancer Inst.,* 50:1195–1213.
17. Solleveld, H. A. (1978): Types and quality of animals in cancer research. *Acta Zool. Pathol. Antverp.,* 72:5–18.
18. Storer, J. B. (1966): Longevity and gross pathology at death in 22 inbred mouse strains. *J. Gerontol.,* 21:404–409.
19. Wostmann, B. S. (1970): Gnotobiotes: Standards and guidelines for the breeding, care and management of laboratory animals. National Research Council, National Academy of Sciences, Washington, D.C.
20. Zurcher, C., Burek, J. D., Van Nunen, M. C. J., and Meihuizen, S. P. (1977): A naturally occurring epizootic caused by Sendai virus infection in breeding and aging rodent colonies. I. Infection in the mouse. *Lab. Anim. Sci.,* 27:955–962.

Subject Index